PERGAMON INTERNATIONAL LIBRARY
of Science, Technology, Engineering and Social Studies

The 1000-volume original paperback library in aid of education,
industrial training and the enjoyment of leisure

Publisher: Robert Maxwell, M.C.

INTERNATIONAL SERIES IN

MATERIALS SCIENCE AND TECHNOLOGY

VOLUME 14 – EDITOR: D. W. HOPKINS, M.Sc.

PRINCIPLES AND APPLICATIONS OF TRIBOLOGY

PRINCIPLES AND APPLICATIONS OF TRIBOLOGY

DESMOND F. MOORE

B.E., M.S., Ph.D., C.Eng.,
F.I.E.I., M.I.Mech.E., M.A.S.M.E.

Lecturer in Mechanical Engineering,
University College, Dublin, Ireland

Director,
International Mechanical Consultants Ltd.

PERGAMON PRESS

Oxford · New York · Toronto · Sydney · Paris · Braunschweig

U.K.	Pergamon Press Ltd., Headington Hill Hall, Oxford OX3 0BW, England
U.S.A.	Pergamon Press Inc., Maxwell House, Fairview Park, Elmsford, New York 10523, U.S.A.
CANADA	Pergamon of Canada, Ltd., 207 Queen's Quay West, Toronto 1, Canada
AUSTRALIA	Pergamon Press (Aust.) Pty. Ltd., 19a Boundary Street, Rushcutters Bay, N.S.W. 2011, Australia
FRANCE	Pergamon Press SARL, 24 rue des Ecoles, 75240 Paris, Cedex 05, France
WEST GERMANY	Pergamon Press GMbH D-3300 Braunschweig, Postfach 2923, Burgplatz 1, West Germany

First edition 1975

Library of Congress Cataloging in Publication Data

Moore, Desmond F. Principles and applications of tribology.

(International series in materials science and technology, v. 14)
Includes bibliographical references.
1. Tribology. I. Title. TJ1075. M59 1975   621.8′9   74–4261
ISBN 0-08-017902-9
ISBN 0-08-019007-3 (Flexi)

470113200

Printed in Great Britain by A. Wheaton & Co., Exeter.

CONTENTS

PART I PRINCIPLES

PREFACE

TRIBOLOGY is defined as the science and practice of friction, lubrication and wear applied to engineering surfaces in relative motion. In recent years it has become widely recognized as a truly diverse and interdisciplinary field of study. Physics, chemistry, metallurgy, materials science, rheology, lubrication, elasticity, viscoelasticity, elastohydrodynamics, thermodynamics and heat transfer play complex and interactive roles in determining the general condition of surface friction. It has been estimated that approximately one third to one-half of the world's energy resources in present use appears ultimately as friction in one form or another. The importance of friction and wear in our modern world cannot therefore be over-emphasized.

This book attempts to deal with the whole field of tribology in a single volume. While some may consider this an impossible task, there are those, including the author, who feel that today's engineering student deserves a relatively simple and unified treatment of the subject area. In pursuing this objective, the contents of the book have been divided into two sections dealing broadly with principles and applications respectively. The text is written from the point of view of the mechanical engineer, and the section on principles emphasizes a fundamental understanding of the subject matter before proceeding to a diversity of practical applications.

Although the complete text is presented in systematic order, the chapters are to a large extent independent and self-sufficient and may therefore be selected in a different order suitable to a particular course of instruction. Chapter 1 deals with the immense scope of tribology and range of applications in the modern world of technology. Chapter 2 is devoted entirely to the evaluation and measurement of surface texture. Chapters 3, 4 and 5 present the fundamental concepts underlying the friction of metals, elastomers and other materials respectively. The principles of hydrodynamic lubrication are dealt with briefly in Chapter 6, and the mechanisms of boundary and elastohydrodynamic lubrication receive a more comprehensive treatment in Chapters 7 and 8 respectively. Chapter 9 is a generalized treatise on wear and abrasion phenomena in metals and elastomers. Whereas surface interactions are dealt with exclusively throughout Chapters 2 to 8, internal friction in solids, liquids and gases is identified and explained in Chapter 10. Chapter 11 is an abbreviated yet thorough treatment of experimental methods used in tribological studies. The remaining five chapters in the book are devoted to specific applications, including manufacturing processes, automotive applications, transportation, locomotion, bearing design and miscellaneous.

Because of the immense scope of the subject matter, omissions must inevitably occur in a

limited text on tribology, but it is hoped that the number of these has been kept to a minimum. Above all, it is hoped to impress upon the student the relevance of tribology to
his entire mechanized society, and to stimulate his interest in this important new field
of engineering science. In attempting to achieve these objectives, a basic understanding
of tribological principles is combined with as many practical examples as possible in a text
of reasonable size.

The author expresses his sincere appreciation to Professor W. E. Meyer from the Pennsylvania State University, Professor H. C. A. van Eldik Thieme from the Technische Hogeschool in Delft and Professor Dr.-Ing. A. W. Hussmann from the Technische Universität
in Munich who stimulated and promoted in different ways the author's continuing interest
in this subject. Particular acknowledgement is due my colleague and close friend Dr.-Ing.
W. Geyer, formerly from the Technische Universität in Munich, whose enthusiasm and
support have proved invaluable during the past three years. Mr. D. W. Hopkins from the
University College of Swansea as series editor for the book deserves special thanks for
reviewing the manuscript and offering helpful suggestions. Finally, a word of appreciation
to Miss Geraldine Warren for her excellent typing skills, and to my wife Miriam for her
patience and understanding during the preparation of the manuscript.

Dublin   D. F. MOORE

PART I

PRINCIPLES

CHAPTER 1

INTRODUCTION

1.1 Definition and Scope of Tribology

Tribology is defined as the science and technology of interacting surfaces in relative motion, having its origin in the Greek word *tribos* meaning rubbing. It is a study of the friction, lubrication, and wear of engineering surfaces with a view to understanding surface interactions in detail and then prescribing improvements in given applications. The scope of tribology is, in fact, much broader than this definition implies. According to Dr. Salomon, former editor of the international journal *Wear*, "Tribology means a state of mind and an art: the intellectual approach to a flexible cooperation between people of widely differing background. It is the art of applying operational analysis to problems of great economic significance, namely: reliability, maintenance and wear of technical equipment, ranging from spacecraft to household appliances." The work of the tribologist is truly interdisciplinary, embodying physics, chemistry, mechanics, thermodynamics, and materials science, and encompassing a large, complex, and intertwined area of machine design, reliability, and performance where relative motion between surfaces is involved.

It is estimated that approximately one-third of the world's energy resources in present use appear as friction in one form or another. This represents a staggering loss of potential power for today's mechanized society. The purpose of research in tribology is understandably the minimization and elimination of unnecessary waste at all levels of technology where the rubbing of surfaces is involved. Indeed, sliding and rolling surfaces represent the key to much of the effectiveness of our technological society.

We may ask: What do rubbing surfaces cost the economy of our Western world? No one has yet offered a precise figure, but an estimate may be obtained from British figures. According to a report by the Committee on Tribology in Great Britain in 1965, approximately £500 million is lost annually in worn parts within the United Kingdom alone because of the failure of industry to understand what happens to surfaces which must move across one another. The problem is of such gigantic proportions that tribological programmes have been established by industry and government in the United Kingdom, the Soviet Union, and several western European countries, and by individual corporations in the United States.

Since World War II, the rapid rate of technological advancement has required great expansion in research on what to do about surfaces that rub. The obvious approach has

3

focused on learning how to oil, grease, or otherwise lubricate them. Thus conventional lubrication has been concerned with:

(a) evaluating lubricants in the light of standard specifications;

(b) compounding lubricants to meet new conditions, or

(c) determining how lubricants respond to cold, heat, nuclear radiation, and other environments and to the materials to be lubricated.

Much of the technology associated with rubbing surfaces still involves conventional lubrication following the above procedure. Indeed, before the 1950s, most mechanical devices were lubricated primarily by mineral oils and soap-thickened mineral-oil greases, and the selection of lubricants for given uses was largely empirical. Now, however, synthetic materials have found increasing usage in lubrication, and extensive programmes of research and development make it possible to design lubricants to meet unusual or increasingly severe requirements. Complex systems such as nuclear submarines, supersonic aircraft or Apollo spacecraft demand the solution of critical lubrication problems. Thus advancing technology has brought us to the point where every aspect of the phenomena associated with rubbing surfaces must be investigated, and the new discipline—tribology—plays an increasing role in our mechanized environment.

One of the important objectives in tribology is the regulation of the magnitude of frictional forces according to whether we require a minimum (as in machinery) or a maximum (as in the case of anti-skid surfaces). It must be emphasized, however, that this objective can be realized only after a fundamental understanding of the frictional process is obtained for all conditions of temperature, sliding velocity, lubrication, surface finish, and material properties. This book is written to explain the fundamentals of tribology in the simplest terms and to illustrate the basic concepts with a variety of everyday applications.

1.2 Macroscopic and Microscopic Viewpoints

The friction force generated between two bodies having relative tangential motion can be considered as a macroscopic or microscopic mechanism, depending on the interest and orientation of the reader. The microscopic or molecular mechanism can also be described as causative, since the molecular interaction of surface molecules of the sliding pair is treated in some detail and for a variety of experimental conditions to establish the true cause of the friction mechanism. This approach has obviously more appeal for the physicist or physical chemist. On the other hand, the macroscopic mechanism is often referred to as resultant, being based on a relatively crude simulation of frictional events and usually following a simple yet adequate model representation. Here, the simplistic treatment and emphasis on applications rather than underlying causes are of more interest to the engineer.

It is not the purpose of this text to extol the virtues of either approach at the expense of the other. Indeed, both mechanisms should be treated in unison, each complementing the other where appropriate, so that a full comprehension of tribological principles emerges. This text is written with such an objective in mind. It is interesting to note that the most commonly accepted welding–shearing–ploughing theory of metallic friction[1] can be classified as macroscopic since there is no mention of surface molecular activity. On the other hand, most of the current theories of elastomeric friction[2] describe adhesion as a thermally activated molecular–kinetic exchange mechanism, so that the microscopic viewpoint is adopted.

The essential difference between the macroscopic and microscopic approaches, apart from the scale of events, is the distinction between a continuum and reality. In the vicinity of a frictional interface, each surface molecule pervades a volume greater than atomic dimensions and is continually vibrating and twisting with thermal energy. The interchange of these molecules as one surface captures and in turn loses some of its surface molecules to the mating surface is, of course, the reality which gives rise to adhesional friction. For a finite surface area, however, the number of degrees of freedom associated with such molecular activity is prohibitively large even for the modern computer to handle, and we therefore use either of two approaches:

(a) we may apply statistical techniques to evaluate mean or averaged values of molecular motion and forces, or
(b) we may neglect molecular activity entirely and treat both sliding materials as a continuum having the same general properties as those observed from experiment.

Thus the microscopic approach, while relatively precise in its representation of surface interactions, is severely limited in terms of applications, whereas the macroscopic approach has the opposite characteristics. The use of mechanical models (embodying springs and dashpots in different combinations) is widespread in the simulation of continuum behaviour, and it must be realized that they are merely a tool for predicting dynamic performance in viscoelastic bodies. We might generally conclude that a microscopic understanding of the frictional process is of inestimable value in establishing the validity of a macroscopic model for practical applications.

1.3 Internal and External Friction

It is desirable at this stage to distinguish between internal and external friction in bodies. Surface interactions in general can be classified under the general term external friction for obvious reasons, whereas molecular–kinetic events and bulk energy dissipation occurring within the body of a material are the cause and result respectively of internal friction. The physicist will define a surface as a boundary having zero thickness, and of course, such do not exist in practice. We must therefore modify this definition and permit a surface to occupy an infinitesimal thickness layer perhaps measurable in Ångstrom units. The origin of "surface" frictional forces is molecular–kinetic interchange as described previously, and this occurs within such layers which can be measured in both sliding members. This composite activity layer approaches zero dimensions in practical terms.

Internal friction in solids is a direct consequence of the forced motion of bulk molecules which under equilibrium are closely spaced and exhibit a strong mutual attraction or repulsion for each other (see Chapter 10). Such motion causes internal shearing in the material of the body and results in internal heat generation. It is common to express internal friction as *damping capacity, loss tangent,* or *tangent modulus.* In the case of liquids and gases, internal friction is expressed as absolute or kinematic viscosity. Perhaps a pneumatic tyre offers the clearest example of both internal and external friction. During normal rolling, the body or carcass of the tyre increases in temperature as a result of continual flexing and recovery of tread elements entering and leaving the contact patch. This is the internal friction generation mechanism. At the same time, the localized "squirming" of tread rubber over road asperities during rolling, braking, driving, and cornering gives rise to the external friction component which largely controls and steers the vehicle. The relative magnitude of the internal and external frictional mechanisms in this example depends to a large extent on the mode of operation of the vehicle and experimental conditions.

1.4 Dry and Lubricated Surfaces

The most important criterion from a design viewpoint in a given application is whether dry or lubricated conditions are to prevail at the sliding interface. In many applications such as machinery, it is known that only one condition shall prevail (usually lubrication), although several régimes of lubrication may exist. There are a few cases, however, where it cannot be known in advance whether the interface is dry or wet, and it is obviously more difficult to proceed with any design. The commonest example of this phenomenon is again the pneumatic tyre. Under dry conditions it is desirable to maximize the adhesion[†] component of friction by ensuring a maximum contact area between tyre and road—and this is achieved by having a smooth tread and a smooth road surface. Such a combination, however, would produce a disastrously low coefficient of friction under wet conditions. In the latter case, an adequate tread pattern and a suitably textured road surface offer the best conditions, although this combination gives a lower coefficient of friction in dry weather.

The several lubrication régimes which exist may be classified as *hydrodynamic, boundary,* and *elastohydrodynamic.* The different types of bearing used today (journal, slider, thrust, and foil bearings) are the best examples of fully hydrodynamic behaviour, where the sliding surfaces are completely separated by an interfacial lubricant film. Boundary or mixed lubrication is a combination of hydrodynamic and solid contact between moving surfaces, and this régime is normally assumed to prevail when hydrodynamic lubrication fails in a given product design. For example, a journal bearing is designed to operate at a given load and speed in the fully hydrodynamic region, but a fall in speed or an increase in load may cause part solid and part hydrodynamic lubrication conditions to occur between the journal and bearing surfaces. This boundary lubrication condition is unstable, and normally recovers to the fully hydrodynamic behaviour or degenerates into complete seizure of the surfaces. The pressures developed in thin lubricant films may reach proportions capable of

---

† Friction comprises two principal components, adhesion and deformation, as shown in Chapter 3.

elastically deforming the boundary surfaces of the lubricant, and conditions at the sliding interface are then classified as elastohydrodynamic. It is now generally accepted that elasto-hydrodynamic contact conditions exist in a variety of applications hitherto considered loosely as belonging to the hydrodynamic or boundary lubrication régimes; for example, the contact of mating gear teeth, or that of ball-bearings in races, or lip seals on machined rotating shafts, etc. Solid lubricants exhibit a compromise between dry and lubricated conditions in the sense that although the contact interface is normally dry, the solid lubricant material behaves as though initially wetted. This is a consequence of a physico-chemical interaction occurring at the surface of a solid lubricant lining under particular loading and sliding conditions, and these produce the equivalent of a lubricating effect.

1.5 The Range of Applications

One useful method of classifying tribology applications is to distinguish between rigid–rigid, rigid–flexible, and flexible–flexible surface pairings, as indicated in Table 1.1. The rigid–rigid pairing appears to be the most common type, the usual application being the friction of metal-on-metal. Table 1.1 illustrates the application of tribological principles to manufacturing technology, the central column indicating the particular manufacturing

TABLE 1.1. TRIBOLOGICAL PROCESSES IN MANUFACTURING

Tribological classification	Manufacturing processes		Related industry
Metal-on-metal (rigid–rigid) surface pairing[a]	Forging Stamping Grinding Milling Lapping Spinning	Reaming Guillotining Drawing Extrusion Forming Swaging	Wire manufacture Iron and steel manufacturing Metal processing Tool design Mechanical components Machine design
Plastic[b]-on-metal (flexible–rigid) surface pairing[a]	Injection and blow moulding Cold working Extrusion Thermo forming Drawing Vacuum forming Coating Laminating		Tyre manufacture Plastics industry Building and construction Electrical insulation Shoe manufacture Flooring materials Solid lubricants
Fibre-on-fibre (flexible–flexible) pairing	Spinning Carding	Weaving, etc.	Textile industry Plastics Hosiery and knitwear Cable manufacture

[a] Includes vertical or tangential motion.
[b] Includes elastomers, solid lubricants, and rubbers.
Note: All operations may or may not have a lubricant.

2*

processes considered, with the related industries listed in the right-hand column. The rigid–flexible category usually embodies the friction of elastomers or plastics on a rigid base surface, whereas the friction of fibre-on-fibre comprises the flexible–flexible surface pairing classification. The general conclusion from Table 1.1 is that the range of possible applications in manufacturing technology and related fields is surprisingly large, and the number of major industries directly involved is considerable.

Typical applications of metallic friction in industry as a whole are as follows:

(a) The lubrication of reciprocating machinery.
(b) Tool design in manufacturing operations.
(c) Wire drawing, die lubrication, and extrusion.
(d) Contact stresses in gear teeth and roller bearings.
(e) Mechanical processing of iron and steel.

The most common types of elastomer–rigid surface pairing are:

(a) Skidding and slipping behaviour of automobile tyres.
(b) Dynamic leakage of elastomeric seals and O-rings in machinery.
(c) Slip–grip action of flexible belts on rough pulleys.
(d) Performance of windscreen wipers and automotive disc clutches.
(e) Use of solid lubricants in machining applications.
(f) Damping of layered structural beams in the construction industry.

Fibre-on-fibre surface pairings find particular application in the textile industry and the manufacture of cable sheathing, as indicated in Table 1.1. In general, the optimization of manufacturing operations can be said to depend to a very large extent on the understanding and application of tribological principles.

Other significant applications of tribology are to be found in the automotive, transportation, and space industries, as described in detail in Part II. The design of internal-combustion engines, air-cushion vehicles, anti-skid braking systems, pneumatic and steel-rimmed tyres, and the ultimate choice of hydrodynamic bearings, are common examples. Space exploration has created a demand for particular frictional (or lubricating) performance of engineering materials under near-vacuum conditions and at cryogenic temperatures. In such a rarefied environment, desorption or evaporation of surface coatings and lubricants tends to occur spontaneously, and cold welding of exposed surfaces which should normally be lubricated is a real danger with catastrophic consequences. The requirements for adequate lubrication in space are extremely demanding, with the further complication of ionizing radiation that varies in duration and intensity. The lubrication of the moving parts of undersea craft and other submersible mechanical devices (such as bearings and propellor shafts) is another area where extreme environmental conditions persist. The specific problems in this instance are high-pressure, corrosion, and the dilution of lubricants, and the requirement of effective sealing against leakage is especially demanding.

One of the more interesting and humane applications of tribology is in the field of biomechanics, particularly with regard to the lubrication of human joints. The crippling disease arthritis is understood to begin when the synovial discharge at human joints is no longer capable of adequately lubricating the moving surfaces. Normally, the synovial fluid is sufficiently thickened between the load-bearing surfaces (cartilage) in joints to permit free relative motion, and the arthritic patient suffers from an inability to provide this thickening action. A promising approach for the tribologist would appear to be the restoration of the thickening ability of synovial fluid by injecting suitable polymers at the troublesome joints.

1.6 The Challenge of Tribology

Technology has pressed far beyond the era where the almost casual application of lubricants is sufficient to handle the problems of moving surfaces in today's complex world. Many of these problems will be resolved satisfactorily only by bringing to bear all knowledge relevant to a system of interacting surfaces on the troublesome component of a multicomponent system. The fundamentals of tribology are to be found within the pages of this text, but many problems in nature remain unsolved. We wonder why dolphins can move through water faster than theory predicts. Is this because the dolphin establishes fully laminar flow at the surface of its skin, and if so how is turbulence avoided? The use of certain polymers and additives has been shown to reduce drag in pipelines and hydraulic systems, but again theory has failed to predict why. Such additives have significant applications in fire fighting (by permitting greater discharge pressure at the hose nozzle and thereby providing a longer jet of water and a greater volume of flow for a given size of hose) and torpedo design. What can we do to reduce skidding of automobile tyres on wet roads, or eliminating the slippage of drive wheels as our high-speed trains accelerate? How can we restore function to human arm or leg joints that are stiff or arthritic, and how can we design optimum prosthetic devices? What can we contribute as tribologists to eliminating unnecessary waste in manufacturing processes? Can we as engineers design lubricants and bearings to withstand the rigours of space? Have we found either satisfactory alternative uses or methods of disposal for waste industrial oils, solvents, cutting fluids, or coolants, so that further contamination of our water resources is precluded?

The challenge to tribologists is indeed immense, and the virtually endless list of applications demonstrates the importance of this inter-disciplinary and relatively new field of engineering science. The principles of lubrication have, of course, been broadly understood since the pioneering work of Osborne Reynolds in 1886, but who would have predicted the universal significance of friction, lubrication, and wear unified as a science and applied to our modern mechanized world? Indeed, one of the most pressing problems of our times —environmental pollution—can be shown to require (at least in part) for its solution the application of tribological principles and the elimination of waste at all levels of industry and technology.

CHAPTER 2

SURFACE TOPOGRAPHY

ENGINEERING surfaces are far from smooth when viewed under a microscope. We observe that they consist of a multitude of apparently random peaks and valleys. It is not surprising to find that, apart from knowing whether dry or lubricated conditions prevail in a given application, the most important parameter which determines frictional behaviour is surface texture. Before considering actual surface interactions, it is therefore advisable to examine in some detail the texture characteristics of surfaces which are later assumed to participate in relative sliding. Surface texture and friction can be considered inseparable in the sense that they represent cause and effect respectively in particular applications. For purposes of clarity and simplicity, the word asperity will be used in this text to identify individual texture elements, whereas the term macro-roughness will apply to a combination of such elements. Details of the geometry of a typical asperity (notably near its peak) will be given the name micro-roughness, and when a finer order of magnitude is required the term molecular roughness is used. In most engineering applications, macro-roughness may be conveniently measured in millimetres and micro-roughness in microns, whereas molecular roughness can best be represented in Ångström[†] units. Figure 2.1 shows a typical machined sur-

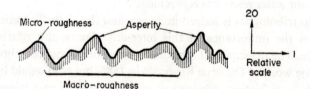

FIG. 2.1. Terminology in surface texture.

face with the vertical amplification greatly exaggerated for convenient representation. As will be evident later, the peaks of this profile participate in the generation of frictional resistance when another surface is brought into contact with it, and the voids between asperities under lubricated conditions act as reservoirs for the particular lubricant used.

[†] $1 \text{ Å} = 10^{-8}$ cm.

2.1 Texture Measurement

A distinction must be made at this stage between methods of evaluating the macroscopic and microscopic features of surface texture, this being largely dependent on the application in question. For most engineering and manufacturing surfaces, macroscopic methods suffice, and they are almost entirely mechanical in nature. Table 2.1 lists eight general methods based on the macroscopic approach. On the other hand, physicists and physical chemists require fine-scale details of surfaces and often details of molecular roughness. These details are usually provided using optical methods,[3, 4] the more usual being inter-

TABLE 2.1. MACROSCOPIC TEXTURE MEASUREMENTS[2]

Method of measurement	Parameter measured	Details
1. Stylus motion	Profile obtained directly	Vertical motion of stylus amplified and recorded as stylus is dragged at uniform speed over surface
2. Cross-sectional cuts	Profile obtained indirectly	Perpendicular, taper, and parallel sections cut through surface, taking care to avoid edge crumbling
3. Cartography	Profile obtained indirectly	Map of surface obtained by superimposing series of parallel sections
4. Void filling	Mean texture depth	Known volume of sand or grease is spread over surface and levelled: area covered gives gross measure of texture depth
5. Surface prints	Mean void spacing	Asperity-density prints obtained by uniformly pressing a sheet of paper on to surface precoated with dye, and counting number of points appearing
6. Hydraulic flow meter	Mean hydraulic radius of voids	Leakage flow between a bottomless container with peripheral rubber ring and texture, calibrated to indicate mean hydraulic radius of latter
7. Texture meter	Length of profile	Inextensible, flexible cord pressed on to macro-texture by closely spaced probes which follow surface contour
8. Photogrammetry	Three-dimensional image	Two photographs of macro-texture taken vertically from close-spaced positions gives sufficient information for stereo-photographs, which are measured with comparator

ference fringe patterns, low-energy electron diffraction, molecular-beam methods, and field-emission and field-ion microscopy.

The macroscopic methods can be broadly classified in two groups, depending on whether the entire profile of a surface or part of it is required in a specific example. The two chief methods of representing and measuring the entire profile are profilometry and cartography, as indicated in the next two sections, whereas when part of the complete profile only is required, it is common to refer either to the mean void width between asperities or details of asperity sharpness.

2.2 Profilometry

Perhaps the simplest and most convenient method of recording the complete features of surface texture is by means of a profile measuring device, as illustrated in Fig. 2.2. The instrument consists of a platform which moves horizontally on teflon runners relative to a base frame which is supported on the given texture by three levelling screws. The platform may be driven in either direction at a uniform speed of 5 mm/s by a reversing motor through

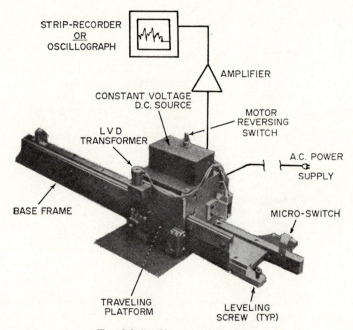

FIG. 2.2. Profile measuring device.

a rack-and-pinion mechanism. An aluminium block is mounted on two parallel leaf springs whose other extremities form part of the travelling platform, so that frictionless vertical motion of the block relative to the platform is permitted. The block holds an inclined needle-point at its lower end which follows the contour of the given texture as the platform moves, whereas a thin flexible extension rod is mounted vertically at its upper end. The latter is connected to the core of a linear–variable–differential transformer. A reversing switch for the motor and a constant voltage source for the transformer are grouped as shown in a single unit.

The output from the transformer caused by fluctuations in the position of the core serves as an input to an oscillograph. Almost any vertical sensitivity or amplification of the texture may be achieved, but for the higher sensitivities a filter network must be installed to eliminate low-frequency "drift" and to keep the oscillograph trace on the chart. Levelling of the base frame is attained by adjusting the levelling screws to centre the circular bubble mounted on the platform. The optimum angle of tilt and sharpness of the needle must be selected by trial and error to obtain a compromise between the conflicting requirements of not sticking and true reproduction of the surface profile. For very sharp textures, a hollow needle may be used with a small leakage flow through the bore to give partial support to the tip and eliminate sticking (this is not shown in Fig. 2.2).

The accuracy obtained with this profile measuring device is sufficient for most engineering applications, and examples of profiles obtained on various surfaces will appear later in this chapter. The limitation to profilometry is that the profile presented generally represents only one pass in a linear direction across a random three-dimensional surface. The cartography method described in the next section overcomes this difficulty. Care must be exercised in using the profile measuring device on brittle surfaces to avoid crumbling of the edges of asperities during passage of the needle.

2.3 Cartography

Consider once again the random profile of Fig. 2.1. Imagine a series of parallel cuts which intersect the profile and define contoured areas of the texture, each planar cut being located at a particular distance below a reference plane drawn to just touch the peak of the tallest asperity, as shown in Fig. 2.3.

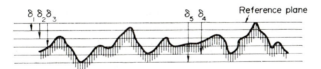

FIG. 2.3. Parallel cross-sectional cuts in texture.

The cross-sectional areas of texture intersected by each planar cut are then plotted in the form of a contour map, as shown in Fig. 2.4. We observe that it is not at all necessary to develop a two-dimensional profile of the texture (as shown in Fig. 2.3) in order to obtain the contour plot. Indeed, a profile representation in Fig. 2.3 is given only to illustrate the

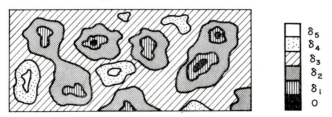

FIG. 2.4. Contour plot of cross-sectional areas.

method used in contouring rather than a prerequisite. The contour plot has the distinct advantage of representing a three-dimensional image in two dimensions, and this cannot, of course, be achieved with the profiles shown in Figs. 2.1 and 2.3.

In practice, the contour plot can be obtained by first making a three-dimensional negative of the surface in question, usually by pressing a mastic compound such as Araldite[4] on to it. When the compound has set and is removed from the original surface, a second different compound is pressed on to the first and allowed to set in this position. Subsequent machining of the combination produces the cross-sectional cuts shown in Fig. 2.3. The result of many depths of cut eventually produces the contour plot in Fig. 2.4.

The method of parallel cuts is exceedingly tedious, and one oblique cut is sometimes used as a substitute (Fig. 2.5). The length of cut is necessarily much greater to produce the same resolution, and interpretation of the resulting contour is obviously more difficult. These considerations must be balanced against the savings in preparation and machining time.

FIG. 2.5. Oblique cross-sectional cut in texture.

2.4 Photogrammetry

A third method of reproducing the complete features of surface texture is a photographic one. Two images or photographs of the texture are taken a small distance apart[5] from two points in a plane parallel with the surface. Generally, the camera points vertically downward—this is termed vertical photography, other methods usually falling into a category known as oblique photography.

The method is based upon relating the change in parallax between the top and bottom images of an object to its vertical height. If we consider a pencil photographed directly from above, the result is a circle. Now imagine a second photograph taken from a position displaced horizontally a distance b to one side of the first. The apparent movement of the top of the pencil in the lateral direction exceeds that of its lower end, and we can now observe the flank of the pencil in addition to its top end. Let h denote the vertical distance between the photographic lens and the top of the pencil and Δh the vertical height of the pencil. If p is the parallax or apparent lateral movement of the pencil top which occurs when we proceed from the first to the second photograph, it can be simply shown that

$$p/b = f/h,$$

where f is the distance between lens and photographic plate. Considering the lower end of the pencil, let the parallax be $p - \Delta p$ thus:

$$\frac{(p - \Delta p)}{b} = \frac{f}{(h + \Delta h)}.$$

By subtracting these two equations, we obtain the relationship

$$\Delta h = (h^2/bf)\, \Delta p, \tag{2.1}$$

which shows clearly that the height of the pencil can be expressed in terms of the change in parallax between its extremities. Equation (2.1) forms the basis for the method of photogrammetry.

The stereo pairs of photographs may be taken by a specially designed camera using a single lens which is caused to move the prescribed distance b laterally between each pair of photographs. When the developed photographs taken in pairs are mounted in a comparator,[5] an image of the surface is seen in relief. To measure the change in parallax Δp between any two points on the surface, a parallax bar is used, consisting of a guide with two sliding glass graticules each corresponding to one photographic image and containing an etched reference dot. The comparator permits relative motion of the photographic pair relative to the pair of graticules, so that any particular point on the left-hand photograph of the surface can be positioned under the left-hand dot. The horizontal distance between the graticules is then adjusted with a micrometer screw until the two dots appear fused and lying on a point in the surface when observed in relief. By moving the photographs so that a neighbouring point of the surface falls beneath the left-hand dot, the change in micrometer reading required to "fuse" the dots in the three-dimensional view is a measure of Δp. This in turn gives a measure of the relative height Δh of the two points in the surface following eqn. (2.1). The technique of photogrammetry may be tedious if large numbers of stereo pairs of photographs are to be examined, but it is now possible to transfer the parallax readings to punched tape and plot the results by computer methods.[5]

2.5 Texture Depth

The profilometric, cartographic, and photogrammetric measurement techniques described in the previous sections deal with a complete representation of surface roughness, usually with a view to amplifying the result for later use. The logical question then arises: Having reproduced the roughness of the surface, what remains to be done with the profile, contour plot, or photographic image? In tribology studies, a knowledge of the roughness geometry of a surface is required only as it determines subsequent frictional, lubricating, or wear behaviour. The engineer interested in purely frictional behaviour will therefore concern himself with details of asperity shape and sharpness, while the hydraulic engineer is primarily concerned with void spacing between asperities, such voids acting as local reservoirs in a lubricating environment. We must distinguish between two distinct approaches in the measurement of surface texture:

(a) In the first case, the complete profile or geometry of the surface is reproduced usually in amplified form, and a selection of features relevant to a particular application is subsequently made from the reproduction.

(b) A simpler and perhaps less precise method is to measure directly on the given texture only those geometric features which pertain to a particular operation.

Following the second approach, perhaps the three most significant individual character-istics of surface texture are given by *texture depth, outflow meter*, and *surface print* methods. The texture depth method gives an average measure of void volume according to the rela-tionship:

$$\text{Texture depth, } \varepsilon = \text{Total void volume/Area of surface covered.} \quad (2.2)$$

The usual method of estimating ε is by spreading a known volume of fine sand[6] or grease[7] on the surface in question until just the tips of the major asperities are still visible, and then measuring the area covered. It is convenient, of course, to cover a circular or rectangular area of surface. The method is used widely in many countries to estimate the texture depth of roads and runways, a typical average value of ε being 1 mm.

Although the texture depth method is simple to use, its accuracy decreases enormously on finer surfaces. Thus on machined surfaces the order of fineness precludes the method entirely, and even during normal use on roads, errors arise from humidity, compactibility, and operator "feel" for the method as a whole. The following section deals with a much more precise method of determining effective texture depth.

2.6 Outflow Meter

The outflow meter, as the name implies, is a hydraulic method of accurately estimating the effective void volume between asperities. In its simplest form, the meter[8] consists of a transparent open-ended cylinder having a thin neoprene or rubber ring of square cross-section cemented to one end. The unit is mounted vertically on the surface to be tested, as shown in Fig. 2.6. The ring is compressed on to the surface texture by a specified load ap-plied circumferentially with the aid of steel rings. The vessel is filled with water, and the time of efflux of a certain quantity of water between the rubber ring and the surface contour is obtained by observing (with a stopwatch) the time taken for the level in the vessel to drop from the upper to the lower graduation mark. According to the theory underlying the devel-opment of this instrument,[8] the important geometrical parameter of the surface texture which controls the rate of efflux is the mean hydraulic radius (MHR) of the channels pro-vided between the ring and the troughs in the texture (i.e. the surface voids). The MHR is defined as the ratio of flow area to wetted perimeter for a typical mean void in the surface, and the following equation is applicable:

$$\text{MHR} = K_1(\mu/t\sqrt{N'})^{1/4}, \quad (2.3)$$

where μ is the absolute viscosity of the water, t the time of efflux (or the time for the water level to drop through the distance $(L_i - L_f)$ in the vessel), N' the areal density or number of asperities per unit area of surface texture,[†] and K_1 an instrument constant. A relationship

† Section 2.7 illustrates how N' can be estimated in a given example.

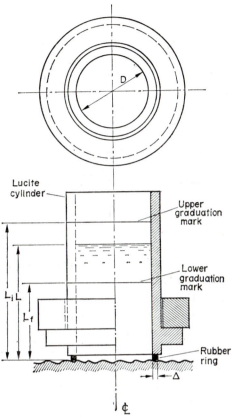

FIG. 2.6. Two views of the outflow meter.

between texture depth obtained with sand or grease and MHR using a liquid is given[9] as follows:

$$\varepsilon/\mathrm{MHR} \doteq \sqrt{(N')}\,P \tag{2.4}$$

where P is the average perimeter of a channel or void in the texture and $\sqrt{N'}$ is the linear density or number of asperities per unit length of profile. The right-hand side of eqn. (2.4) is greater than unity, and ε is therefore somewhat larger than MHR. This is also evident from Fig. 2.7 which compares the void spacing measured with the sand and hydraulic meth-

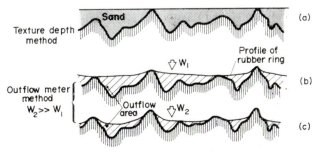

FIG. 2.7. Comparison of outflow areas or void spacing using sand and hydraulic methods.

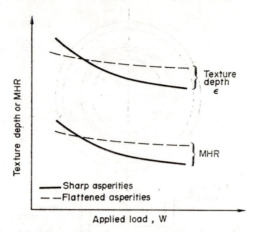

FIG. 2.8. Effects of draping on texture depth and MHR.

ods. We see that the void volume indicated by the texture depth method is considerably larger than that given by outflow meter measurements, the reason being the draping of the rubber ring about individual surface asperities in the latter case. The draping effect increases with increasing load, as we see in Fig. 2.7, and a most important application of this property is that it serves to distinguish between textures having sharp and flattened asperities. Figure 2.8 shows clearly that the rate of decrease of MHR (or texture depth) with increasing applied load is substantially lower for textures exhibiting flattened asperities compared with the sharp type illustrated in Fig. 2.7 because of the reduced change in draping pattern.

The simple outflow meter described in Fig. 2.6 is ideal for surfaces having a relatively large roughness pattern, perhaps with ε in the range 0.2–1 mm. Finer scale roughness, such as that obtained on machined metal surfaces, may be measured with an outflow meter which uses air rather than water as the working medium. A schematic of the apparatus is given in Fig. 2.9. A compressed-air supply is connected with appropriate piping, valves, and pressure gauges to an outflow meter similar in form to the hydraulic type shown earlier in Fig. 2.6. The upper end of the outflow meter is closed, and the relative hardness of the ring beneath the vessel is increased according as both the scale of texture and viscosity of

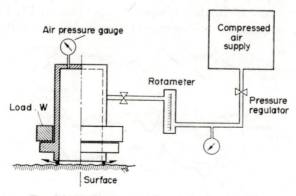

FIG. 2.9. Outflow meter using air for finer surfaces.

the working medium are reduced. Operation of the device consists simply in applying a known circumferential weight to the ring in the normal manner and observing both the flow rate from the rotameter and the resulting air pressure within the outflow meter itself. The following equation is used to find the MHR of the texture:

$$\text{MHR} = K_2 (Q/N \, \Delta p)^{1/4}, \tag{2.5}$$

where Q is the rotameter flow rate, Δp the pressure difference between the gauge reading in the outflow meter and atmospheric, N the number of voids or asperities appearing beneath the ring (measured visually), and K_2 an instrument constant.

The outflow meter, whether used as a hydraulic or pneumatic device on coarse or fine textures respectively, gives accurate and repeatable readings, and is comparatively simple to use. A knowledge of the spacing density of asperities N' or the number of asperities appearing beneath the ring is a prerequisite for calculating MHR, and this is best obtained by the method of surface prints described in the next section.

2.7 Surface Prints

The method of surface prints consists of coating the macro-roughness of the surface under consideration with Prussian blue dye and then pressing a sheet of white paper uniformly on to the asperity tips. The resulting prints for sandpaper surfaces are shown in

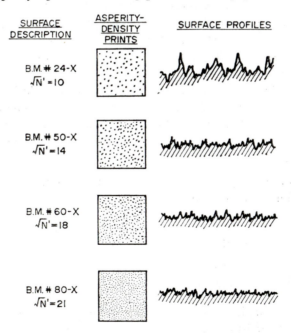

FIG. 2.10. Surface prints and profiles for sandpaper surfaces.

Fig. 2.10 in conjunction with profiles obtained on the same surfaces using the profile measuring device shown in Fig. 2.2. A square border with one-inch sides has been superimposed on the prints, and the areal density N' is read by passing a slit over the prints and counting approximately the number of points appearing. This method is most satisfactory for estimating the approximate spacing of asperities.

A second method of obtaining asperity density prints consists in placing a flat piece of aluminium foil on the surface and a small rubber disc on top of the foil. A controlled impact is then given to the disc by allowing a known weight to drop from a specified height. As a result of the impact, the major asperities of the texture pierce the foil, and the number of piercings per unit area may be readily counted.

This method has the advantage that the same pressure is used to produce the piercings in all cases, whereas the blue-dye method would seem to introduce human error in attempting to apply a uniform pressure by hand to the paper. However, operator error is relatively small particularly after continued use, and the dye method is perhaps simpler and more convenient for field application. The surface must, of course, exhibit reasonably pointed asperities for maximum effectiveness.

2.8 Statistical Features

We have dealt in the last six sections with the measurement of surface texture, and we have distinguished between complete and partial representation of texture geometry. If we consider again the complete texture (whether profile, cartograph, or stereo-photograph), we may wish to express the surface features in more convenient form. One method of achieving this goal is by the use of statistics.

The profiles of machined metal surfaces are random processes which are also both stationary and ergodic. In the language of statistics, a random process is said to be stationary when it is invariant with respect to time, and it is ergodic when the time average for a particular profile is the same for all profile samples. For single profiles, the following statistical properties are of significance:

(a) mean (or centre-line average) and root-mean-square values;
(b) autocorrelation functions; and
(c) power spectral density functions;

whereas for pairs of profiles, the corresponding properties are:

(d) cross-correlation functions; and
(e) cross-spectral density functions.

All of these properties are itemized and defined in Table 2.2. The centre-line average parameter Z_1 is also referred to as the arithmetic average or roughness height rating. We note that the root-mean-square (r.m.s.) value Z_2 is identical with the more familiar standard deviation parameter σ_x in statistical theory. The autocorrelation function Z_3 for a single profile is obtained by delaying the profile by some fixed interval Δ, then multiplying the

TABLE 2.2. STATISTICAL PROPERTIES OF SURFACES

Profile type	Statistical property	Mathematical definition
Single profile, $y = f(x)$	Mean or centre-line average value	$Z_1 = (1/L) \int_0^L |f(x)| \, dx$
	Root-mean-square value	$Z_2 = \left[(1/L) \int_0^L f^2(x) \, dx \right]^{1/2}$
	Autocorrelation function	$Z_3 = (1/L) \int_0^L f(x) f(x+\Delta) \, dx$
	Power spectral density function	$S(\omega) = (1/2\pi) \int_{-\infty}^{\infty} Z_3 e^{-i\omega t} \, dt$
Pairs of profiles, $y_1 = f_1(x)$ $y_2 = f_2(x)$	Cross-correlation function	$Z_4 = (1/L) \int_0^L f_1(x) f_2(x+\Delta) \, dx$
	Cross-spectral density function	$T(\omega) = (1/2\pi) \int_{-\infty}^{\infty} Z_4 e^{-i\omega t} \, dt$

Notation: L = representative length of profile; ω = frequency; Δ = delay interval; $S(\omega)$ and $T(\omega)$ are Fourier transforms respectively of Z_3 and Z_4.

original profile by the delayed version, and averaging the product. When the delay interval $\Delta = 0$, the autocorrelation function is identical with the square of the r.m.s. value.

It has been observed that the mean and autocorrelation functions completely characterize the profile provided that $f(x)$ is stationary and has a normal distribution. The distribution of asperity heights in a given profile is considered to be normal when it acquires the familiar bell shape shown in Fig. 2.11, and the assumption of normality is valid for most engineering

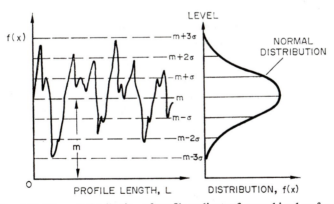

FIG. 2.11. Normal distribution of profile ordinates for machined surface.

surfaces. Figure 2.12 shows a typical autocorrelation function for a ground machine surface[10] plotted against the delay interval, Δ.

The power spectral density function for a single stationary record represents the rate of change of mean square value with frequency, where the mean square value is taken in a narrow frequency band at various centre frequencies. The total area under the power spectral density function from $-\infty$ to $+\infty$ will be the total mean square value in this record, whereas the partial area under the same function from ω_1 to ω_2 represents the mean square value in the profile corresponding to that frequency range. The power spectral density function $S(\omega)$ is usually obtained (using a digital computer) by taking the Fourier transform of the autocorrelation function Z_3, as shown in Table 2.2. Since an enormous volume of digital information must be obtained in practice from the surface profile to calculate the autocorrelation function, two substitute parameters are often used for Z_3. These are the

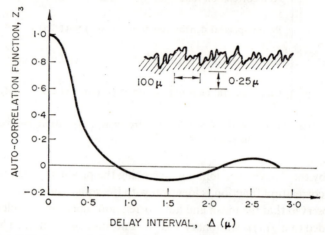

Fig. 2.12. Autocorrelation function vs. delay interval for ground surface.

standard deviation of the slope and the mean thickness of the profile itself. The cross-spectral density function $T(\omega)$ is obtained by taking the Fourier transform of the cross-correlation function Z_4, as seen in Table 2.2. We also note from the last column in this table that the autocorrelation and cross-correlation functions are identical when f_2 is replaced by f_1 or f.

Finally, it may well be that single records apply to elastomer–rigid surface contact problems, since the elastomer conforms to the roughness profile of the base. On the other hand, pairs of profiles are more appropriate for the more familiar metal-on-metal applications.

2.9 Mathematical Representation

It may be readily disputed whether or not a statistical treatment of surface profile by any of the methods listed in the previous section constitutes a mathematical representation. However, it is certain that statistical methods outline gross features of a profilometric record in a representative length, whereas a strictly mathematical treatment will identify individual features (provided the latter are typical of the record as a whole). It will therefore be presupposed in this text that mathematical and statistical methods are distinct.

A simple mathematical representation of macro-roughness for a single profile is proposed as follows. Figure 2.13 shows a typical random profile of a machined surface with a reference plane drawn through the tip of the highest asperity (or asperities) and parallel to the mean plane of the surface. A series of fictitious parallel planes may now be drawn successively below the reference plane (at distances δ_1, δ_2, δ_3, etc.) so that, depending on their location,

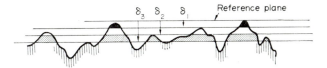

FIG. 2.13. Mathematical description of surface texture.

they intersect different numbers of asperities N_i. Then the following equations may be written

$$N_i = C_0\delta^m \tag{2.6}$$

$$A_i = C_1 + C_2\delta^n, \tag{2.7}$$

where N_i is the number of asperities intersected by the ith plane located at a distance δ below the reference plane, A_i the total contoured area of asperities intersected or touched by this plane, C_1 the plateau area (also called the approach area, or the area coincident with the reference plane), C_0 and m are constants specifying the height spacing or statistical distribution of the asperities, and C_2 and n are constants specifying the mean shape of asperities.

A close examination of eqns. (2.6) and (2.7) shows that this definition is both simple and sufficiently accurate for most purposes. In fact, the five parameters C_0, m, C_1, C_2, and n contained in these expressions yield sufficient and complete information on surface texture. Table 2.3 shows how these constants take different values for cubed, hemispherical, and

TABLE 2.3. VARIABLE ASPERITY SHAPE, CONSTANT HEIGHT

Model profiles (length L)	C_0	m	C_1	C_2	n
Cubes	$L^2/4\varepsilon^2$	0	$L^2/4$	0	—
Hemispheres	$L^2/4R^2$	0	0	$\pi L^2/2R$	1
Sq. Pyramids	L^2/ε^2	0	0	L^2/ε	1

square-pyramidal surface profiles, assuming for simplicity a constant height of asperity in each case. When the latter is allowed to vary, the constants take the values listed in Table 2.4 in this case for hemispherical asperities. It is interesting to note that for a random road

3*

TABLE 2.4. CONSTANT ASPERITY (HEMISPHERE), VARIABLE HEIGHT DISTRIBUTION

Height distribution	Profile (length L)	C_0	m	C_1	C_2	n
Linear	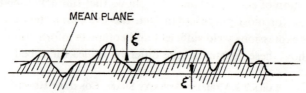	$L^2/36R^2\delta^2$	2	0	$\pi L^2/4R$	1
		$L/9R\delta^2$	2	0	$\pi L^2/8R$	1
Nonlinear		$L^a/40R\delta^4$	4^a	0	$7\pi L^2/36R$	1

a Approximate value.

surface $m = 2$ and $n = 3$. The profiles considered in both tables are, of course, three-dimensional, having the same features at right angles to the plane of the paper. We note for simplicity that the asperities selected are regular in shape rather than random. This permits a convenient tabulation of the constants, but step variations in height distribution which are thereby introduced give only average values for C_0 and C_2. In a typical random surface where the height distribution is continuous and Gaussian, these constants are obviously more meaningful.

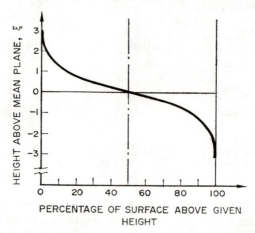

FIG. 2.14. Mean and parallel planes for random surface texture.

FIG. 2.15. Typical bearing area curve for random surface profile.

The method of selecting fictitious planes so that N_i and A_i may be defined as above is virtually identical with the cartography section method described earlier, except that the intersecting planes are imaginary in the first instance and real in the second. Another method is to select a mean plane through the texture as a reference in place of one passing through the tallest asperity, as shown in Fig. 2.14. Parallel planes are identified by their height ξ from this new reference plane, and the statistical features of the texture are presented in the form of a bearing–area curve,[11] as shown in Fig. 2.15. This is an alternative approach to the method of expressing geometric features of the texture in mathematical form.

2.10 Parameter Selection

We have at this stage stressed the importance of surface texture in tribology studies, and we have indicated both how the complete features of the texture may be reproduced (usually in amplified form), or how statistical and mathematical techniques may be used to characterize certain aspects of texture. The important question now arises: If we do not wish to use profilometric or stereophotographic methods to reproduce the surface, what parameters and how many are necessary to characterize texture?

Early investigators believed that one parameter would suffice to uniquely specify texture, but there was no consensus of opinion on which would be the most suitable. Table 2.5 lists

TABLE 2.5. SINGLE-PARAMETER MODELING OF TEXTURE

No.	Parameter	Symbol	Date
1.	Roughness factor (pipe flow)	$2d/R$	1933
2.	R.M.S. value[a]	Z_2	—
3.	Mean, or c.l.a. value[a]	Z_1	—
4.	Predominant peak roughness	d	1957
5.	$\dfrac{\text{Predominant peak roughness}}{\text{R.M.S. value}}$	$d/\text{r.m.s.}$	1958
6.	R.M.S. of first derivative[b]	Z_5	1962
7.	R.M.S. of second derivative[b]	Z_6	1962
8.	Directional parameter[b]	Z_7	1962
9.	Mean void spacing	λ	1962
10.	Texture depth	ε	1963
11.	Length of profile	S/L	1964
12.	Autocorrelation function[a]	Z_3	1965

[a] See Table 2.2. [b] See this section.
R = radius of pipe. L = representative length.

no less than twelve such parameters, including the date when they were first proposed. The earliest known parameter was proposed by Nikuradse in 1933 to explain the effects on turbulence of rough sand grains cemented to the inner bore of a pipe.

Two new parameters Z_5 and Z_6 involving root-mean-square values were defined as follows:

$$Z_5 = \left[(1/L) \int_0^L (\mathrm{d}y/\mathrm{d}x)^2 \, \mathrm{d}x \right]^{1/2} \tag{2.8}$$

and

$$Z_6 = \left[(1/L) \int_0^L (\mathrm{d}^2y/\mathrm{d}x^2)^2 \, \mathrm{d}x \right]^{1/2}. \tag{2.9}$$

The parameter Z_5 is useful in highlighting the slopes of asperities, which play an important role in scattering light rays, and Z_6 is a sensitive indicator of the degree of sharpness at asperity peaks. The directional parameter Z_7, defined by

$$Z_7 = \frac{\Sigma(\Delta x_i)_{\text{pos.}} - \Sigma(\Delta x_i)_{\text{neg.}}}{L}, \tag{2.10}$$

represents the excess of the percentage of distance along the profile where the slopes of the asperities are positive over that where the slopes are negative. Although most descriptions of texture ignore the directional effect, it has important wear implications in cases where the relative velocity of sliding varies systematically. A typical application is pavement texture at the approach to a traffic signal, where the predominantly braking mode of vehicle behaviour introduces a positive Z_7 wear asymmetry into the pavement in the direction of travel.

Predominant peak surface roughness is defined as one-half of the peak-to-trough dimension of a random profile, as shown in Fig. 2.16. Table 2.6 shows clearly that the ratio of

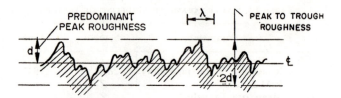

FIG. 2.16. Nomenclature for single-parameter modeling.

TABLE 2.6. ROUGHNESS MEASUREMENTS IN MACHINING OPERATIONS

Finishing operation	Grind	Hyperlay	Sandpaper	Superfinish	Lap with loose abrasives
Peak-to-trough roughness / R.M.S. roughness	4.5	6.5	7	7	10

peak-to-trough roughness to r.m.s. roughness is a useful parameter in classifying machining operations.

The mean void spacing and texture depth parameters can be obtained as described in Sections 2.5 and 2.6. The length-of-profile parameter S/L is defined as the actual length of the profile measured along the contour of asperities divided by the nominal (or horizontal) length of profile.

It can be shown simply that single-parameter representation of texture is far from unique, and we therefore consider the possibility of utilizing two parameters. Statistical theory suggests that Z_1 and Z_2 (or perhaps Z_1 and Z_3) be used to completely characterize normal distributions of asperity height, but experimenters have preferred to use three parameters rather than two. Table 2.7 lists four possible combinations each comprising three para-

TABLE 2.7. THREE-PARAMETER MODELING OF TEXTURE

No.	Selected parameters	Date
1	Histograms of profile, of first derivative (slope) and of second derivative (curvature)	1946
2	Height, shape, and spacing of asperities	1961
3	Size, spacing, and shape factors	1963
4	Surface density, height distribution, and mean radius of asperities	1968

meters. The first combination includes histograms of the profile of a random texture and of its slope and curvature, as depicted in Fig. 2.17. The histogram f_0 represents the frequency distribution of depths (or depth histogram), while the frequency distributions of first and

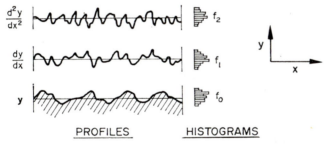

PROFILES   HISTOGRAMS

FIG. 2.17. Histograms of profile, slope, and curvature for random surface.

second derivatives of the profile are given by f_1 and f_2 respectively. The second and third combinations of parameters in Table 2.7 identify a predominant and characteristic type of asperity in the texture, and then relate this to its neighbours by some measure of spacing. The final grouping substitutes a height distribution parameter for the shape factor used in the earlier selections.

The most recent thinking on parameter representation[2] indicates that in the broadest sense at least five distinct parameters are necessary to represent the significant features of texture. The recommended parameters are:

(a) Size
(b) Spacing } factors, for typical average asperities (macro-roughness).
(c) Shape

(d) Micro-roughness at asperity peaks.

(e) Height distribution of asperities.

Indeed, any of the three-parameter models in Table 2.7 can be shown to lack uniqueness. It is readily conceded, however, that the very procedure of selecting an application reduces the practical necessity of five distinct parameters. Thus for metallic surfaces it is not absolutely essential to specify the fourth and fifth parameters above. This is because in the first place the height distribution of metallic asperities is in most cases normal or Gaussian, and, secondly, it seems irrelevant to specify the micro-roughness of asperities which later deform plastically under load. The stipulation of five distinct parameters to uniquely characterize texture is for the general case.

2.11 Model and Ideal Surfaces

Because of the difficulty of mathematically specifying the geometry of a random surface, it has been common practice to consider ideal asperity shapes taken singly or together as a group. The three basic shapes of importance are cubes, cones, and spheres, and a random

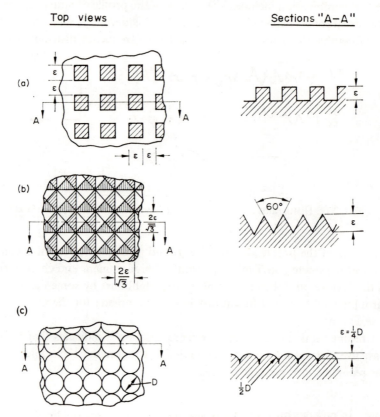

FIG. 2.18. Three idealized model surfaces exhibiting (a) cubed, (b) square-pyramidal, and (c) hemispherical asperities.

surface may be approximated by a selection in different sizes from these shapes. When the spherical projections of elastic bodies contact under the action of a normal load, the contact stresses may be predicted by Hertzian theory. Other solutions exist in the literature for the case of a cone pressed into an elastic medium. Sawtooth and sinusoidal asperity models have also been used because of their mathematical simplicity. Some sophisticated models include assemblies of spheres with different height distributions to simulate roughness on a random surface.

Figure 2.18 shows three idealized surfaces where the projections are (a) cubed, (b) square-pyramidal, and (c) hemispherical asperities. The cubed and square-pyramidal surfaces were machined carefully from aluminium stock, and the asperities are assumed to have equal height. The third surface was conveniently constructed with ball-bearings of uniform diameter cemented to a plane backplate. These surfaces are useful because they can be defined in simple mathematical terms, and they can also be used as calibration surfaces. Square-pyramidal or conical asperities give high adhesion at the expense of rapid wear, as a result of which the tips of the asperities become rounded and smooth, and the latter effect is represented by hemispheres. Furthermore, random profiles often exhibit "flats" between asperities of the sharp or rounded type, and cubic asperities symbolize this condition.

We may observe that idealized model surfaces of the type specified in Fig. 2.18 bear little resemblance to actual surfaces, even though they indicate extreme conditions of sharpness or roundedness. One solution to this problem is a representation of all three basic shapes of asperity (cube, cone, or square pyramid, and hemisphere) in one idealized model surface, while still preserving a uniform height for all asperities. In the author's opinion, this approach complicates the mathematical definition of the surface projections while becoming more unrealistic in the sense that there now exists no typical shape of asperity for the surface as a whole. A second solution is to preserve one basic form of idealized asperity while varying the height distribution, as indicated in Table 2.4 earlier. A third solution is to introduce

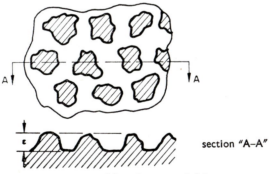

FIG. 2.19. Random asperities of constant height.

randomness of shape and spacing while preserving a constant height, as shown in Fig. 2.19. This configuration is particularly suited to a study of viscous flow between random asperities, with the objective of gradually introducing randomness of texture.

Sandpaper surfaces can be considered ideal in the sense that although the asperities are random in spacing, shape, and height distribution, their sharpness produces optimum friction

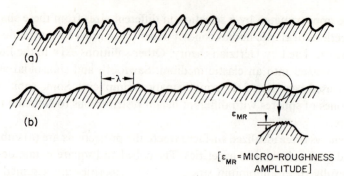

FIG. 2.20. Comparison of (a) idealized sandpaper and (b) typical surface profiles.

properties. Practical surfaces, on the other hand, have typically rounded asperities in place of the peaks which typify sandpaper textures. Figure 2.20 compares sandpaper and actual surface profiles. We note that a micro-roughness (see Fig. 2.1) must be provided at the rounded tips of actual asperities to compensate for their original lack of sharpness. This is to ensure adequate adhesion under lubricated conditions, as discussed in Chapter 8.

We conclude that simple model shapes are most useful for representing not only the physical features of actual surfaces but also their subsequent frictional performance.

2.12 Requirements for Randomness

Having discussed at length the features of random surfaces, we may be tempted to ask the question: What exactly constitutes randomness in a practical engineering surface? We may be surprised to find that contrary to what the nature of randomness may imply, certain limitations are imposed as follows:

(a) Although the randomness of surfaces may preclude periodic return to the initial features, there should be a general repeatability from one location to the next. Thus although the representative length of profile L is sufficiently large to permit reasonable statistical representation within any given location, this should also be typical of all locations within the texture.

(b) The profile of the texture proceeding from any point in the mean plane of the surface and in a specific direction should be generally the same for all directions through that point (isotropic requirement).

(c) The size, spacing, and shape of any asperity, should, where possible, not be too different from those of neighbouring asperities. There should therefore exist what might be described as a typical asperity for that surface.

(d) There should, if possible, be an absence of directional lay in the texture. This parameter is characterized by Z_7 in the text.

We note that the fulfilment of the third condition establishes at once a correlation between the mean void spacing λ and texture depth ε parameters appearing in Table 2.5. In fact, if one confines one's attention to one particular application (such as the machine finish on shafts from grinding and milling operations), this condition is automatically fulfilled in an approximate sense. The representation of surface features by one parameter is more acceptable in such limited applications. To check the validity of the first condition above, it would be desirable to obtain two or more profiles in one general location and compare them with profiles obtained in other locations. We note that at least two profiles obtained in any one general location are required to check that L is a representative length of profile.

Other criteria for randomness may be specified for particular applications. Thus the drainage aspects of surfaces require in addition that the escape channels are not too elongated but reasonably "squatty." This condition should be fulfilled for optimum results with the outflow meter discussed in Section 2.6.

2.13 Actual Contact Between Surfaces

Single profiles of surfaces have been dealt with almost exclusively in this chapter. We will now consider briefly how pairs of surfaces interact when brought into contact under load. A distinction must be made between solids of comparable hardness when brought into mutual contact, and solids of widely differing hardness. Metal-on-metal contact is a common example of the former classification, and elastomer-on-rigid-base typifies the latter, as shown in Fig. 2.21.

In the case of metals, contact is made at three or more points of zero total area. From the relationship $p = W/\Sigma A_i$, the pressure p rises rapidly to the plastic flow value p^*, whereupon the contact points deform plastically in such a manner that the total contact area ΣA_i is now finite, as shown in Fig. 2.21(a). During the occurrence of plastic flow at asperity tips,

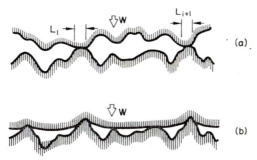

FIG. 2.21. Comparison of actual contact areas for (a) metal-on-metal, and (b) elastomer-on-rigid-base.

it is obvious that the two profiles approach each other, thus causing other contact spots to appear. It has been observed[12] that during this approach the average size of a contact spot remains constant. Thus the increasing area of individual spots at existing contact points is counterbalanced by the continual appearance of new contact points of small or zero area. Due to the plastic flow phenomenon, local welding of the asperities occurs, and as will be apparent in the next chapter, it is the shearing of these welds or the equivalent which gives rise to the adhesion component of friction. One interesting feature of the interacting profiles in Fig. 2.21(a) is that the percentage contact length given by $\Sigma L_i/L$ also represents the percentage contact area $\Sigma A_i/A$, where A is the apparent or nominal contact area between the surfaces.

For an elastomer pressed against a textured rigid base, the nature of contact is such that the more flexible surface readily adapts to the contour of the other by draping action, as shown in Fig. 2.21(b). Molecular–kinetic, stick–slip motion is permitted by a "flowing" action as the elastomer continues to drape over the rigid asperities.

In practical examples it is extremely difficult to accurately estimate actual areas of contact between pairs of surfaces. We give here a qualitative explanation of how contact occurs, but quantitative information depends on many experimental factors and will undoubtedly remain an approximation for many decades. The condition of contact between random surfaces is itself relatively complex, and to this must be added a practical familiarity with metallurgy, plasticity, chemistry, and thermodynamics before we can attempt to predict contact areas with any degree of certainty. The role of experiment in tribology studies will always remain a necessity, and, when complimented by appropriate analysis following established principles, will provide solutions to most relevant engineering problems.

Having considered the nature of surface texture and how contact is established between surfaces, we must examine the frictional mechanism which arises when relative tangential motion between the contacting surfaces is attempted. This will occupy our attention in the next three chapters.

CHAPTER 3

FRICTION OF METALS

3.1 Classic Laws of Friction

Leonardo da Vinci (1452–1519) is generally credited with being the first to develop the basic concepts of friction, and his famous sketches inspired the French scientist Amontons to conduct experiments and later formulate his laws of friction. Coulomb also conducted careful experiments, and he expanded on Amontons' work. The classic laws of friction as they evolved from these early studies may be summarized as follows:

(a) Friction force is proportional to load.
(b) Coefficient of friction independent of apparent contact area.
(c) Static coefficient is greater than the kinetic coefficient.
(d) Coefficient of friction independent of sliding speed.

The classic laws have survived the years without significant amendment until recent times, and we might therefore assume that they are reasonably valid. In the light of the latest advances, however, most of the laws have been found to be incorrect. At the same time, considering the tools available to the early workers, the classic laws reflect remarkable insight into the mechanism of dry friction, and they served to guide later and more precise investigations.

The first law is correct except at high pressure when the actual contact area approaches the apparent area in magnitude. It generally takes the form

$$F = fW, \tag{3.1}$$

where F is the friction force, f the coefficient of friction, and W the normal load. Equation (3.1) is commonly referred to as Coulomb's law, and can be regarded as a definition for the coefficient of friction. The remaining classical laws must be severely qualified. Thus the second law appears to be valid only for materials possessing a definite yield point (such as metals), and it does not apply to elastic and viscoelastic materials. The third law is not obeyed by any viscoelastic material; indeed, a controversy exists today as to whether viscoelastic materials possess any coefficient of static friction at all. The fourth law is not valid at all for any material, although in the case of metals it is not as incorrect a statement as for elastomers where viscoelastic properties are dominant. A fifth law states that the

33

coefficient of friction is material dependent. However, this can be regarded more as an observation by the early workers rather than a separate law, and it is therefore not included specifically in the original list of classical laws.

3.2 General Friction Theories

Some of the general friction theories which have been proposed to explain the nature of dry friction are summarized below:

(a) *Mechanical Interlocking*

Amontons and de la Hire in 1699 proposed that metallic friction can be attributed to the mechanical interlocking of surface roughness elements. This mechanism gives an explanation for the existence of a static coefficient of friction, and it also explains dynamic friction as the force required to lift the asperities of the upper surface over those of the lower one.

(b) *Molecular Attraction*

Tomlinson in 1929 and Hardy in 1936 attributed friction forces to energy dissipation when the atoms of one material are "plucked" out of the attractive range of their counterparts on the mating surface. Later work attributed adhesional friction to a molecular–kinetic bond rupture process in which energy is dissipated by the stretch, break, and relaxation cycle of surface and subsurface molecules.

(c) *Electrostatic Forces*

According to this theory presented as recently as 1961, stick–slip phenomena between rubbing metal surfaces can be explained by the initiation of a net flow of electrons, which produces clusters of charges of opposite polarity at the interface. These charges are assumed to hold the surfaces together by electrostatic attraction.

(d) *Welding, Shearing, and Ploughing*

This most recent theory proposed by Bowden in 1950 is now widely accepted for metal friction. High pressures developed at individual contact spots cause local welding, and the junctions thus formed are sheared subsequently by relative sliding of the surfaces. Ploughing by the asperities of the harder surface through the matrix of the softer material contributes the deformation component of friction, as shown later.

Many of the early theories and laws of friction were geometrical and mechanical in nature, as might be expected in the absence of a detailed understanding of physical, chemical, and mechanical fundamentals. This is especially true of the mechanical interlocking and early molecular attraction theories. The interlocking theory may be abandoned, since the raising and lowering of the asperities of one surface over those of the other surface in the sliding operation involves no energy dissipation, and friction is certainly a dissipation mechanism. Similarly, the electrostatic theory implies a leakage of electrons from the interface over long time intervals which would lower the coefficient of friction, yet no such observations have been made. The welding–shearing theory offers the most satisfactory physical explanation for metallic friction on a macroscopic level, whereas in the case of elastomers the most recent molecular attraction theories appear to be valid today.

3.3 Elastic Contact in Metals

Before proceeding to examine the basic friction mechanism in metals, it is first necessary to consider in detail the nature of contact between metallic surfaces. This has been touched upon in Section 2.13, but greater detail is required here so that a proper understanding of contact conditions and friction emerges. Consider the theory of Hertz[13] for the contact of two spherical bodies within the elastic limit. The pressure p at any radius r and the radius of contact a are given by

$$p = \frac{3W}{2\pi a^2}\left[1 - \left(\frac{r}{a}\right)^2\right]^{1/2}, \qquad (3.2)$$

$$a = \sqrt[3]{\left(\frac{3WR}{4E'}\right)}, \qquad (3.3)$$

where W is the applied load, R the equivalent radius of curvature given by the relationship[†]

$$1/R = 1/R_1 + 1/R_2,$$

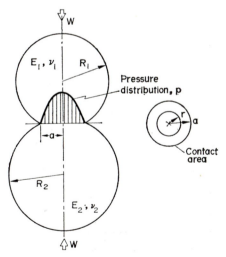

FIG. 3.1. Hertzian contact between two spheres.

† For convex–convex contact only. In the case of convex–concave contact, a minus sign is used.

E' is a composite elastic modulus defined by

$$\frac{1}{E'} = \frac{1-v_1^2}{E_1} + \frac{1-v_2^2}{E_2},$$

E_1, E_2 the Young's moduli for the elastic bodies, v_1 and v_2 their Poisson's ratios, and R_1, R_2 their radii of curvature, as shown in Fig. 3.1. When one of the contacting bodies is highly elastic, $E_1 \gg E_2$ and $E' = E_2/(1-v_2^2)$. Also, if the lower body is plane rather than spherical in profile, $R_2 \to \infty$ and $R = R_1$.

Since the contact area $A_i = \pi a^2$, it follows from eqn. (3.3) that

$$A_i = KW^{2/3}, \tag{3.4}$$

where K is a constant. This equation is most important in visualizing the nature of elastic contact between bodies. Thus if we imagine a surface to consist of many asperities of equal height, each of which has an approximate radius of curvature R at its near-spherical peak, it is clear that the total actual area of contact varies with the two-thirds power of the applied load.

Practical surfaces, however, exhibit close to a normal or Gaussian distribution of asperity heights according to the relationship

$$\phi(s) = [1/\sqrt{(2\pi)}] \exp\left(-\tfrac{1}{2}s^2\right) \tag{3.5}$$

where $\phi(s)\mathrm{d}s$ expresses the probability that a particular asperity in the texture has a height between s and $(s+\mathrm{d}s)$ above some reference plane. If this height distribution is assumed, while each individual asperity has a spherical tip of mean radius R, it can be shown that[12] in place of eqn. (3.4) the actual contact area is now directly proportional to load thus:

$$A = K_1 W, \tag{3.6}$$

where K_1 is a constant, as shown in Fig. 3.2. It is of interest to note that virtually the same curve applies for apparent or nominal contact areas which differ by a factor of 10, so that the real and apparent areas of contact are independent of each other. Since the mean

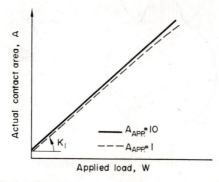

FIG. 3.2. Relationship between actual contact area and load.

actual pressure $\bar{p}$ is defined by the relationship

$$W = A\bar{p}, \tag{3.7}$$

it follows from eqns. (3.6) and (3.7) that $\bar{p}$ is constant and independent of load.

Another interesting feature of the contact of rough surfaces is that the average size of a micro-contact is constant, even within the elastic range, according as the load W is increased. This implies that the increase in area of existing contact spots is exactly matched by the continual creation of new contacts of zero or smaller area. Thus

$$A/M = \text{const} = \bar{A}_i, \tag{3.8}$$

where A is the total actual contact area and M the number of contact spots at any instant, and $\bar{A}_i$ denotes the mean or average actual contact area per asperity during the loading process. From eqns. (3.6) and (3.8) it follows that

$$W = \text{const} \times M, \tag{3.9}$$

so that the load is directly proportional to the number of asperities.

When two metallic surfaces come into contact, the probability is that plastic flow occurs at the tips of leading asperities in each surface, as described in the next section. However, this is by no means the only possibility. If the asperities are rounded and of relatively large radius[†] so that the surface is virtually smooth rather than rough, the pressures will be high but still within the elastic range. The proportionality between real area and load in accordance with eqn. (3.6) then becomes a statistical effect dependent on the varying height of asperities. Under these conditions, the mean actual pressure is virtually independent of load, and the mean size of a micro-contact is constant. The criterion for elastic contact conditions to exist is expressed by the inequality[(14)]:

$$(E'/H)\sqrt{(\sigma/R)} < 1, \tag{3.10A}$$

where H is the real hardness of the material and σ the standard deviation of the surface roughness ($= Z_2$, see Table 2.2). The term $(E'/H)\sqrt{(\sigma/R)}$ is called the "plasticity index". The hardness H in many cases is identical with the mean actual pressure $\bar{p}$.

3.4 Elastoplastic Contact

Consider the interaction of two metallic surfaces with a compressive load W, as shown in Fig. 3.3. It is assumed here that, in contrast with the previous section where elastic conditions prevail, the relative sharpness of the asperities permits plastic flow to occur at asperity peaks. When the two surfaces are initially brought together, physical contact between them is established in at least three locations. If A_i represents the area of a particular contact spot at the ith location and p_i is the local contact pressure, then

$$W = \sum_{i=1}^{M} A_i p_i, \tag{3.11}$$

[†] Typical radii are an order of magnitude greater than asperity heights.

where $M \geqslant 3$. At the instant when contact is first established, the summation of the individual contact areas is too small to carry the load W with finite pressures. As a result, the pressure on the larger asperities increases rapidly to the plastic flow or yield pressure p^* of the softer surface. The load is therefore carried by asperities where the pressure at the peaks

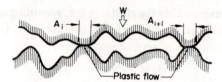

FIG. 3.3. Elastoplastic contact between metal surfaces.

has reached p^* and by lesser asperities where the pressure is still largely elastic. This condition is expressed by the following equation:

$$W = p^* \underbrace{\sum_{i=1}^{M_1} A_i}_{\substack{\text{Plastic}\\\text{contact}}} + \underbrace{\sum_{i=M_1}^{M} p_i A_i}_{\substack{\text{Elastic}\\\text{contact}}}. \tag{3.12}$$

It is assumed here, of course, that the plastic flow pressure p^* is constant. In practice, however, the effects of work hardening create an additional increment Δp^*, as shown in Fig. 3.4, so that the effective flow pressure $(p^* + \Delta p^*)$ varies with strain ε. Let us further replace the

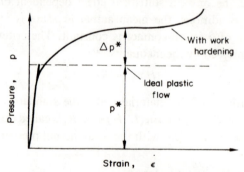

FIG. 3.4. Ideal and actual plastic flow at asperity peaks.

variable elastic pressure component p_i by some mean value $\bar{p}$, which is less than p^*. Putting $\bar{p} = p^* - \Delta\bar{p}$ and replacing p^* by $p^* + \Delta p^*$ in eqn. (3.12),

$$W = (p^* + \Delta p^*) \sum_{i=1}^{M_1} A_i + (p^* - \Delta\bar{p}) \sum_{i=M_1}^{M} A_i. \tag{3.13}$$

Assuming that the two incremental terms in eqn. (3.13) cancel, then

$$W = p^* \sum_{i=1}^{M} A_i = p^* A, \tag{3.14}$$

which is a form identical with eqn. (3.6).

The occurrence of plastic flow at asperity peaks therefore increases the size of individual contact areas from zero to finite values, and the total contact area A adjusts to match the applied load W in accordance with eqn. (3.14). As W increases, it is clear that there is a proportional increase in A just as in the case of purely elastic contact. The average contact area A/M also remains constant with increasing load, as suggested by eqn. (3.8). Since plastic flow occurs at the tips of the major contacting asperities, the metal surfaces are essentially welded together at these locations. The condition for plastic flow to occur at asperity peaks is given by the inequality

$$(E'/H) \sqrt{(\sigma/R)} > 1. \tag{3.10B}$$

Several interesting conclusions may be drawn from this brief study of elastic and elasto-plastic conditions, thus:

(a) The total actual contact area A is proportional to the applied load W irrespective of the nature of the deformation occurring at asperity peaks (i.e. elastic, elastoplastic, or purely plastic).

(b) The criterion for elastic or plastic contact conditions to occur is independent of the load W, as shown by eqns. (3.10A and B). This contradicts the earlier view that at very light loads elastic contact predominates, whereas at heavy loads, plastic flow occurs at asperity tips. In fact, the occurrence or absence of plastic flow depends to a very large extent on the radius of curvature of asperity peaks, and bears no relation to the load W.

(c) The mean elastic pressure $\bar{p}$ for elastic contact, or the mean plastic flow pressure p^* (in the absence of work hardening), are virtually independent of load.

(d) The average size of a micro-contact A/M remains constant with increasing load for both elastic and plastic contact.

Table 3.1 shows the nature of contact, the distribution of asperity heights, and the form of the relevant equation as outlined in these last two sections.

TABLE 3.1. THE NATURE OF METAL-ON-METAL CONTACT

Asperity height distribution	Nature of contact	Equation
Constant height, M = constant	Elastic	$A \sim W^{2/3}$
Normal height distribution		
$\bar{A}_i = A/M$ = constant	Elastic	$A \sim W$
Constant height, M = constant	Plastic	$A \sim W$

3.5 Basic Mechanism of Friction

Having considered in detail how the asperities of randomly rough metallic surfaces interact under load, the next logical step is the study of how frictional forces are generated when we attempt to slide one of the surfaces relative to the other. This will occupy our attention in the next section. As a prelude, we consider here the basic nature of friction.

Two main factors contribute to the friction generated between unlubricated surfaces in relative motion. The first, and usually the more important, factor is the adhesion which occurs at the regions of real contact, and the second may be described as a deformation term. If we assume no interaction between the two factors, we may write

$$F = F_{\text{adhesion}} + F_{\text{deformation}} \qquad (3.15)$$

where F is the total friction force. Dividing by the load W, the corresponding equation is written in terms of friction coefficients

$$f = f_A + f_D, \qquad (3.16)$$

where the suffixes A and D denote the adhesion and deformation terms respectively.

By a careful selection of experimental conditions, it is possible to separate the adhesion and deformation terms. Thus by choosing optically smooth surfaces the roughness features are virtually eliminated so that the contribution of the deformation coefficient is negligible. The total measured friction force is then due to adhesion alone. Alternatively, the use of a lubricant between rough surfaces in relative motion virtually eliminates the adhesion term, and the measured friction force can be attributed solely to the deformation component. In the normal case of dry sliding between rough surfaces, the coefficient of adhesional friction is generally at least twice as large as the deformation contribution. The physical mechanism responsible for the adhesion and deformation terms is widely different for metals and elastomers.

3.6 Welding–Shearing–Ploughing Theory

When the two metal surfaces are brought into contact under load, local welding occurs at the tips of the major asperities of the surfaces as described earlier until the total area matches the applied load W. If a sideforce F is now applied to the upper surface as shown in Fig.

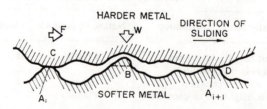

FIG. 3.5. Mechanism of welding, shearing and ploughing.

3.5, it is clear that relative sliding between the surfaces will occur only if the welded junctions are sheared. This shearing action gives rise to the adhesion component of friction.

Assuming that the total area of shear is given by A and the average shear strength of the junctions is s, we may write

$$F_{\text{adhesion}} = As. \qquad (3.17)$$

Figure 3.5 shows, however, that when the shearing of the welds at C and D has taken place, the surfaces are not entirely free to slide relative to one another. If the upper surface is made of harder material than the lower surface, it will plough or groove a path through the softer metal as indicated at the point B in the figure. This grooving effect gives rise to the deformation term in metal friction thus:

$$F_{deformation} = F_{ploughing}. \qquad (3.18)$$

We therefore rewrite eqn. (3.15) in the following form for metals:

$$F = As + F_{ploughing}. \qquad (3.19)$$

In most cases, the ploughing term is insignificant, and eqn. (3.17) can then be viewed as contributing the entire resistance F to motion. If we now divide both sides of eqn. (3.17) by W, we obtain an expression for the coefficient of sliding friction f:

$$f = F/W = As/Ap^* = s/p^*, \qquad (3.20)$$

where W has been replaced by the right-hand side of eqn. (3.14). The relationship in eqn. (3.20) defines the simple theory of adhesion for metals.[1]

There is no reason to suppose at this stage that the area of shear in the numerator of eqn. (3.20) differs from the area of plastic flow in the denominator, and hence both are designated by A. We note that if metals do not work harden appreciably, the shear stress s of the interface is approximately equal to the critical shear stress τ of the metal. The flow or yield pressure p^* is found to be about 5τ[†], and so eqn. (3.20) gives $f \doteq 0.2$. In practice, most metals in air give $f = 1.0$. This sizeable discrepancy can be explained largely by two phenomena not accounted for in the simple theory. These are *junction growth* and *work hardening*.

3.7 Junction Growth

Experiments as recent as 1957 on the contact between a hemispherical metal slider and a flat surface[15] showed a remarkable phenomenon which was later destined to become a necessary part of the widely accepted welding–shearing–ploughing theory of metallic friction. In the absence of a tangential sideforce, the area of contact is a single circular region. By now applying a slowly increasing tangential sideforce until gross sliding or slippage occurred, it was discovered that the area of contact had increased three- or fourfold prior to sliding. This is because the plastic yielding of the junction is controlled by the combined effect of the normal and tangential stresses p and s according to the following yield criterion

$$p^2 + \alpha^2 s^2 = p^{*2}, \qquad (3.21)$$

where α is a constant with a value of about 10[1]. When the normal load is first applied, the local pressure at asperity tips rapidly approaches the yield value p^* as described earlier, and at a particular location i, a contact area A_i is defined as follows:

$$A_i = W_i/p^*. \qquad (3.22)$$

† Under certain conditions, $p^* = 3\pi$[1], see page 144

The subsequent application of a sideforce F_i introduces a small tangential stress s, and eqn. (3.21) can then only be satisfied if p^* reduces to the elastic pressure p_i. Thus the area of contact A_i must increase to the value

$$A_i + \Delta A_i = W_i / p_i. \tag{3.23}$$

It may be conjectured that since p_i is now elastic, the growth of junctions in accordance with eqn. (3.23) is aborted when p drops below the value p^*. However, the material still behaves plastically since the total pressure $\sqrt{(p^2 + \alpha^2 s^2)}$ still corresponds to the case of plastic flow, and junction growth continues. If the surfaces are very clean, it is possible that ΔA_i may reach a value 9 times as large as A_i before fracture occurs. Figure 3.6 shows schematically the phenomenon of junction growth when one of the contacting metal surfaces is perfectly smooth.

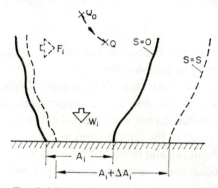

FIG. 3.6. Phenomenon of junction growth.

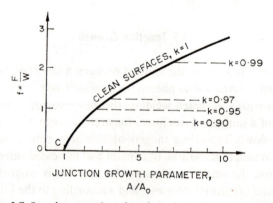

FIG. 3.7. Junction growth as function of cleanliness of surfaces.

The graph in Fig. 3.7 shows the effect of surface contamination on the magnitude of junction growth prior to shearing. If we define a cleanliness factor $k = s/\tau$, where s is the shear stress of the interface in the presence of contaminants and τ is the critical shear strength of the metal, then shearing occurs at higher values of friction for cleaner surfaces as indicated here. The junction growth parameter is defined as the ratio of total actual contact area for finite sideforce to total contact area for zero sideforce. In terms of individual junctions,

$$A = \sum_{i=1}^{M} (A_i + \Delta A_i),$$
$$A_0 = \sum_{i=1}^{M} A_i. \tag{3.24}$$

We observe that the point C corresponds to zero growth, and for this case $f = 0$. Also, it is apparent that junction growth proceeds until limited by the shear strength s of the contaminated interface. Thus for $k = 1$ or zero contamination, junction growth proceeds indefinitely, whereas a coefficient $f = 1.0$ corresponds to a cleanliness factor of 0.95 and a threefold increase in contact area.

3.8 Work Hardening

Most metals work harden appreciably when plastic flow occurs, and the strength of a welded junction is therefore often greater than that of the softer metal. The shearing action necessary to permit relative tangential motion between surfaces may therefore occur along a plane which is different from that defined by localized welding. Figure 3.4 earlier has shown that work hardening creates an additional pressure increment Δp^*, which is added to the existing yield pressure p^* of the softer material. Consequently, the critical shear stress τ is increased to $\tau + \Delta\tau$, where $\Delta\tau = 0.2\Delta p^*$. Figure 3.8 shows two possible shearing planes

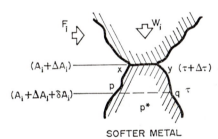

FIG. 3.8. Work hardening effects in metal contact.

xy and pq along which relative motion may take place. The criterion for failure to occur is expressed by the inequality

$$\tau (A_i + \Delta A_i + \delta A_i) \lessgtr (\tau + \Delta\tau)(A_i + \Delta A_i)$$

which simplifies to the relationship

$$\left(\frac{\delta A_i}{A_i + \Delta A_i} \right) \lessgtr \left(\frac{\Delta p^*}{p^*} \right). \tag{3.25}$$

Here, δA_i is the increase in cross-sectional area at the section within the bulk of the softer material compared with the original interface xy, and ΔA_i is the increase of the latter which has already taken place due to junction growth. Failure will occur due to shearing along pq when the left-hand side of the inequality in eqn. (3.25) is less than the right-hand side, and along xy when the opposite is true. The equality sign indicates an equal probability that failure will occur along xy and pq.

We can write the following relationships for F_i and W_i:

$$F_i = (A_i + \Delta A_i + \delta A_i)\, s, \qquad (3.26)$$
$$W_i = A_i(p^* + \Delta p^*). \qquad (3.27)$$

By using these equations and substituting from eqn. (3.25) for the ratio $\Delta p^*/p^*$, we find that for M junctions

$$f = \frac{\Sigma F_i}{\Sigma W_i} \lesseqgtr \frac{s}{p^*}\left[1 + \left(\frac{\Sigma \Delta A_i}{\Sigma A_i}\right)\right], \qquad (3.28)$$

which shows generally an increase in the coefficient of friction compared with the simple adhesion theory expressed by eqn. (3.20). If junction growth without work hardening is considered, then

$$F_i = (A_i + \Delta A_i)s, \qquad (3.29)$$
$$W_i = A_i p^*, \qquad (3.30)$$

and thus we can write

$$f = \frac{\Sigma F_i}{\Sigma W_i} = \frac{s}{p^*}\left[1 + \left(\frac{\Sigma \Delta A_i}{\Sigma A_i}\right)\right], \qquad (3.31)$$

which shows that the equality sign in eqn. (3.28) is applicable either when work hardening is non-existent and shearing occurs along xy in Fig. 3.8, or when the effect of work hardening is to create an equal probability of shearing along xy or pq. In any event, the overall effects of work hardening on the coefficient of friction f are generally small compared with the influence of junction growth.

3.9 The Ploughing Component of Friction

The last three sections have been essentially devoted to the adhesion component of metallic friction, with special emphasis on the role played by the actual contact area. In this section we examine the nature of the ploughing component for model asperities such as spheres or cylinders. Figure 3.9 shows the grooving produced by a rigid metal sphere which slides on a softer yielding metal. We can write expressions for the load-support area A_1 and the grooving area A_2 as follows:

$$A_1 = \pi d^2/8 \qquad (3.32)$$

and
$$A_2 = \tfrac{1}{2}\, R^2(2\theta - \sin 2\theta). \qquad (3.33)$$

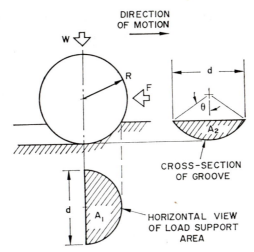

FIG. 3.9. Sliding of metal sphere on softer metal.

Assuming that the plastic-yielding metal is isotropic and that its yield pressure is p^*, then

$$W = p^*A_1 \quad \text{and} \quad F = p^*A_2, \tag{3.34}$$

where W and F are the applied load and frictional resistance respectively. It follows that the coefficient of friction due to grooving is defined as

$$
\begin{aligned}
f_{\text{grooving}} &= F/W = A_2/A_1 \\
&= (4R^2/\pi d^2)(2\theta - \sin 2\theta) = C(\theta)/\pi, \tag{3.35}
\end{aligned}
$$

where $C(\theta) = (2\theta - \sin 2\theta)/\sin^2 \theta = 4\theta/3$ for small θ. Thus, with $d = R$, $\theta = 30°$, and $f_{\text{grooving}} = 0.23$, or with $d = 2R$, $\theta = 90°$, and $f_{\text{grooving}} = 1.00$. In practice, of course, the magnitude of the grooving component of friction is quite small.

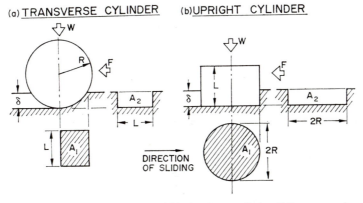

FIG. 3.10. The ploughing component of friction for a cylinder sliding on a softer metal.

In the case of a cylinder sliding on a softer metal surface, two cases must be distinguished, depending on whether the cylinder is placed in a transverse or upright position, as shown in Fig. 3.10. For the transverse case,

$$A_2 = L\delta \quad \text{and} \quad A_1 = L\sqrt{[(2R - \delta)\,\delta]},$$

so that

$$f_{\text{grooving}} = \frac{A_2}{A_1} = \sqrt{\left[\frac{1}{2(R/\delta) - 1}\right]}. \tag{3.36}$$

For the upright cylinder,

$$A_2 = 2R\delta \quad \text{and} \quad A_1 = \pi R^2,$$

so that

$$f_{\text{grooving}} = A_2/A_1 = (2/\pi)(\delta/R). \tag{3.37}$$

We observe from eqns. (3.36) and (3.37) that the upright cylinder gives lower values of f_{grooving} at all (δ/R) ratios.

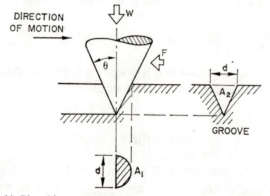

FIG. 3.11. Ploughing component for a cone sliding on a softer metal.

Finally, we consider the case of a cone sliding on a softer metal surface, as shown in Fig. 3.11. The load-support and grooving areas are given by

$$A_1 = \tfrac{1}{8}\pi d^2 \quad \text{and} \quad A_2 = \tfrac{1}{4}d^2 \cot\theta,$$

so that

$$f_{\text{grooving}} = F/W = A_2/A_1 = (2/\pi)\cot\theta. \tag{3.38}$$

For a semi-cone angle of 60°, $f_{\text{grooving}} = 0.32$, whereas for a very pointed cone with $\theta = 30°$, $f_{\text{grooving}} = 1.1$.

The calculation of the grooving component of friction for the three basic asperity shapes (sphere, cylinder, and cone) in this section neglects "pile-up" of material ahead of the slider. Such realism is noticeably absent from Figs. 3.9, 3.10, and 3.11. Figure 3.12 shows the wave pattern produced by a spherical slider as a result of the crumpling and accumulation of

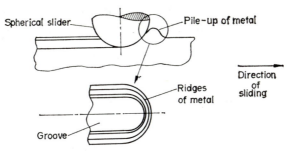

FIG. 3.12. Material pile-up ahead of a spherical slider.

material ahead of the grooving path. The energy required to form the crest of material in front of the slider increases the magnitude of the grooving component as calculated from eqns. (3.35) to (3.38) inclusive, since a substantial enlargement of the grooving area A_2 occurs. It is very difficult, however, to assess this effect quantitatively.

We note particularly that no mention of adhesional friction is made since we are concerned here with the resistance to sliding which the asperities of the harder metal experience in grooving a path through the softer material. In fact, the isotropic condition assumed in eqn. (3.34) is not entirely true, as seen from experiments.[16] To allow for this deviation, we may insert a coefficient K_p in front of the expressions for f_{grooving} in eqns. (3.35)–(3.38). Thus typical values of K_p for different materials are listed in Table 3.2. We observe in general that according as the relative hardness of the particular metal in question decreases, the coefficient K_p is increased.

TABLE 3.2. VALUES OF
COEFFICIENT K_p

Material	Coefficient
Tungsten	1.55
Steel	1.35–1.70
Iron	1.90
Copper	1.55
Tin	2.40
Lead	2.90

3.10 The Adhesion Component of Friction

Equation (3.20) has shown that the adhesion coefficient of friction in the simple theory is equal to the ratio of shear strength to flow pressure s/p^*. In the case of our three basic asperity shapes, we assume a general relationship $K_A(s/p^*)$ governs the adhesional contribution, and we now estimate the values of the coefficient K_A for sphere, cylinder, and cone.

Consider first the interface between sphere and softer metal, as shown earlier in Fig. 3.9 and reproduced in greater detail in Fig. 3.13. The semi-circumferential area dA is given by

$$dA = \pi R^2 \cos \phi \, d\phi,$$

and the adhesion force $dF' = s \, dA$, where s is the effective shear strength at the interface. The horizontal component of dF' is dF'' as shown, such that

$$dF'' = dF' \sin \phi$$
$$= \pi R^2 s \cos \phi \sin \phi \, d\phi.$$

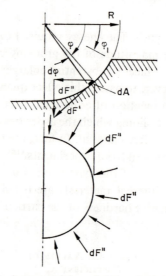

FIG. 3.13. Adhesion component of sliding friction for a sphere.

The average value of dF'' opposing the direction of travel is $(2/\pi) \, dF''$, and so by integrating from ϕ_1 to $\pi/2$,

$$F = \int (2/\pi) \, dF'' = R^2 s \int_{\phi_1}^{\pi/2} \sin 2\phi \, d\phi = \tfrac{1}{2} R^2 s \, (1 + \cos 2\phi_1), \qquad (3.39)$$

where F is the effective adhesion force opposing sliding motion. Putting $\theta = \pi/2 - \phi_1$, so that Figs. 3.9 and 3.13 have the same notation, and using the first part of eqn. (3.34), we find that

$$f_{\text{adh.}} = F/W = (2/\pi)(s/p^*), \qquad (3.40)$$

which shows that the adhesional contribution is independent of θ and that $K_A = 2/\pi$.

A similar analysis for a rigid cylinder sliding on a softer metal gives the following results:

$$\left. \begin{array}{ll} f_{\text{adh.}} = s/p^* & \text{for transverse position} \\ f_{\text{adh.}} = [(2/\pi)\,(\delta/R)+1]\,s/p^* & \text{for upright position} \end{array} \right\}. \qquad (3.41)$$

We are now in a position to combine the grooving and adhesional coefficients of friction for the cylinder and sphere thus:

Sphere

$$f = f_{\text{adh.}} + f_{\text{grooving}} = \frac{c(\theta)}{\pi} + \frac{2}{\pi}\left(\frac{s}{p^*}\right).$$

Transverse cylinder

$$f = f_{\text{adh.}} + f_{\text{grooving}} = \sqrt{\left[\frac{1}{2(R/\delta)-1}\right]} + \left(\frac{s}{p^*}\right). \qquad (3.42)$$

Upright cylinder

$$f = f_{\text{adh.}} + f_{\text{grooving}} = \frac{2}{\pi}\left(\frac{\delta}{R}\right)\left(1+\frac{s}{p^*}\right) + \frac{s}{p^*}.$$

Caution must be used in algebraiclly adding the two components of friction as in eqn. (3.42). The adhesion contribution must be very small, so that s can be more readily visualized as the shear strength of a lubricated interface. Otherwise there is evidence that the adhesion and ploughing terms interact in a complex fashion which cannot be predicted by theory. The ratio s/p^* is therefore considerably less than unity in eqns. (3.40)–(3.42).

The normal situation which prevails at a sliding interface between two metals is where adhesion is relatively high. The main factor which determines the magnitude of adhesion between metallic surfaces is the presence or absence of surface contaminants, and these are discussed in the next section.

3.11 Surface Contaminants

We have seen earlier in Fig. 3.7 that the phenomenon of junction growth is inhibited by surface cleanliness, and this has an important bearing on the frictional coefficient. Most surface contaminants consist of oxide layers which form rapidly on exposed clean metal surfaces. Figure 3.14 is a schematic diagram showing the topography and structure of a

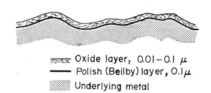

〰️ Oxide layer, 0.01–0.1 μ
— Polish (Beilby) layer, 0.1μ
▨ Underlying metal

FIG. 3.14. Topography and structure of polished metal specimen.

typical polished metal specimen. In this case, the oxide layer is 0.01 to 0.10 μ thick and the polished layer is somewhat greater (0.1 μ). Mechanically polished or abraded metal surfaces are frequently used in frictional studies, chiefly because the polishing action removes contaminating grease films in a simple and convenient manner.

The formation of oxide films on metal surfaces occurs spontaneously when the latter are exposed to air. A chemical reaction occurs at the surface of the clean metal following the adsorption of molecules of oxygen and water vapour. The kinetics of growth of the film involve the diffusion of metal ions through the oxide film. It is interesting to note that the topography of metal oxides depends on environmental conditions during their growth period.

Thus slowly grown oxides are relatively smooth, but at higher temperatures crystals and whiskers may appear and create a rough profile.

When clean metal surfaces are placed in contact, strong metal–metal bonds are formed. In fact the atoms on one of the surfaces (in the case of similar metals) approach those of the other surface until they are as close as the atoms within the bulk of the metal itself. At this stage, the interfacial atoms can no longer distinguish between their own neighbours and those of the other body. Consequently, the forces at the interface are similar in magnitude to those between metal atoms in the bulk of the bodies in contact. In practice, there will be some mis-matching between the engaging crystal lattices, so that the interface generally exhibits some imperfections which constitute weak regions. These regions may close up because of plastic flow or thermal diffusion, and in this event the interface ceases to have any physical meaning. *Adhesion then corresponds to the strength of the metal itself.* A similar mechanism applies to dissimilar metals, such as indium and gold. Here the interfacial forces are probably an average of internal atomic forces in each of the contacting members. Thus the average adhesion between dissimilar metals is stronger than within the bulk of the weaker metal (in this case, indium), and when the surfaces are pulled apart they will separate within the weaker metal rather than at the original interface.

The absence of strong interfacial adhesion between metal surfaces can be attributed primarily to two causes. The first is, of course, the presence of surface contaminants as discussed above, which reduces the effective shear strength s at the interface below the critical shear strength τ of the metal. The second factor is a lack of ductility among contacting asperities, which effectively leads to their mechanical failure under tangential stress. Surface films may be removed[1] by thorough outgassing in a high vacuum, whereas the lack of ductility of work-hardened asperity junctions may be remedied by suitable annealing before removal of the load.

Surface contamination as a result of the addition of a lubricant will not be dealt with in this chapter since the case of lubricated sliding will receive special emphasis in three subsequent chapters.

3.12 Metal Transfer

The transfer of metal from one sliding surface to the other can be regarded as part of the normal wear process, and this is discussed at length in Chapter 9. However, it is also part of the friction mechanism between metals following the welding–shearing–ploughing theory. Thus we have seen that the junctions formed by plastic flow at asperity tips very often remain even during the subsequent relative motion between the surfaces, whilst the shearing process occurs within the softer metal. This is apparent from Fig. 3.8 earlier, and we can say that a fragment of metal *xyqp* originally forming part of the softer metal has now transferred to the harder body. Russian research[16] attributes metal transfer to a second cause, namely particle removal from the softer metal as a result of fatigue, followed by adherence to the other surface. It is highly probable that both causes contribute to the metal transfer phenomenon on a macroscopic scale.

Metal may also be transferred from one surface to another on an atomic or molecular level. This is a consequence of the inter-diffusion of surface atoms at the sliding interface. Experimental studies[16] have shown that when bronze slides against steel in a lubricant consisting of a mixture of alcohol and glycerine, the steel surface is enriched with copper atoms which diffuse into it from the bronze.

The significant application of metal transfer is machining, where metal–metal contact occurs between the tool and the workpiece. Small wear fragments from the tool are inevitably transferred to the surface being machined. Not only are these fragments undesirable as impurities, but more seriously they often become the source of considerable corrosion. Common examples are transfer from hammers, clamps, screwdrivers, tool bits, riveters, and twist drills. Metal transfer is generally reduced by the use of effective lubricants, by reducing as much as possible the amount of slip between tool and workpiece and by increasing the hardness of the tool itself.

3.13 Heat Generation in Sliding Friction

The generation of frictional forces between sliding bodies is fundamentally an energy dissipation mechanism, and this gives rise to heating effects which originate at the sliding interface. It is important to obtain an approximate expression for the temperature rise which accompanies steady-state sliding in terms of material properties and experimental conditions, and in this section we apply the technique of dimensional analysis as a step towards this objective.

Consider two asperities in contact at a given instant and at the ith location, and let the temperature rise ΔT at the interface be a function of the following independent variables:

$$\Delta T = \phi(f, W_i, V, \bar{k}, A), \tag{3.43}$$

where the frictional coefficient $f = F_i/W_i$, F_i and W_i are the frictional force and load at the ith location, V is the relative sliding speed, A the area of actual contact, and $\bar{k}$ the average thermal conductivity of the two contacting bodies 1 and 2 according to the relationship

$$\bar{k} = \tfrac{1}{2} (k_1 + k_2).$$

Application of the Buckingham *Pi* theorem gives two dimensionless parameters π_1 and π_2, such that $\phi(\pi_1, \pi_2) = 0$. If we assume $\pi_1 = f$, then it can be shown simply that

$$\pi_2 = \frac{\Delta T \bar{k} \sqrt{A}}{W_i V} = \frac{\Delta T \bar{k} L}{W_i V}, \tag{3.44}$$

where L is a generalized length dimension. We then write

$$\phi \left(f, \frac{\Delta T \bar{k} L}{W_i V} \right) = 0 \quad \text{or} \quad \Delta T = \text{const} \left(\frac{f W_i V}{\bar{k} L} \right). \tag{3.45}$$

The form of eqn. (3.45) can also be obtained from the principle of similarity in dimensional analysis. Thus the heat produced in sliding is given by $\dot{Q}_p$ and the heat conducted away from the interface is $\dot{Q}_c$, where the dot notation is used to indicate time derivatives d/dt. Then

$$\dot{Q}_p = F_i V = f W_i V \tag{3.46}$$

and
$$\dot{Q}_c = \bar{k}(\Delta T/L)\, A = \bar{k}\, \Delta T L. \tag{3.47}$$

Application of the similarity principle indicates that the ratio $\dot{Q}_p/\dot{Q}_c$ must define a characteristic number N which is constant for the given experimental conditions. Accordingly,

$$\frac{\dot{Q}_p}{\dot{Q}_c} = \frac{f W_i V}{\bar{k}\, \Delta T L} = N = \text{const}, \tag{3.48}$$

from which eqn. (3.45) emerges.

The foregoing analysis is based upon two main assumptions which must be examined here. The first is the condition of steady-state sliding which need not be discussed further. The second assumption is that part of the heat generated by sliding friction is not required to heat the continuously oncoming cool surface, so that the analysis so far is valid at very slow sliding speeds.[†] When the velocity of sliding is relatively high, the interface may not reach a steady-state value because of the continuously and rapidly oncoming cool surface. Here, the thermal diffusivity α of the moving surface becomes important. The Fourier law of heat conduction in the unsteady state and in one dimension may be written as

$$\frac{DT}{Dt} = \alpha \left(\frac{\partial^2 T}{\partial x^2} \right) \tag{3.49}$$

where DT/Dt is the substantial derivative of temperature with respect to time. Equation (3.49) can be written more conveniently in dimensionless form thus:

$$\frac{DT'}{Dt'} = \left(\frac{\alpha t_0}{L^2} \right)\left(\frac{\partial^2 T'}{\partial x'^2} \right), \tag{3.50}$$

where $T' = T/T_0$, $t' = t/t_0$, $x' = x/L$, and T_0, t_0, and L are reference values for temperature, time, and length respectively. The constant of proportionality $\alpha t_0/L^2$ in eqn. (3.50) is called the Fourier number, F_o, which is also related to the Peclet number, P_e, as follows:

$$F_o = \frac{\alpha t_0}{L^2} = \frac{\alpha}{VL} = \frac{1}{P_e}.$$

Thus, for higher sliding speeds, the Fourier or Peclet number constitutes a third dimensionless quantity π_3, and we can write

$$\phi(\pi_1, \pi_2, \pi_3) = 0,$$

or
$$\left. \phi\left(f, \frac{\Delta T \bar{k} L}{W_i V}, P_e \right) = 0, \right\}$$

or
$$\Delta T = \phi'\left(\frac{f W_i V}{\bar{k} L}, P_e \right). \left. \right\} \tag{3.51}$$

[†] At slow sliding speeds there will be an initial transient to store heat in the oncoming surface, but this disappears when steady-state sliding ensues.

The application of the similarity laws shows that the ratio of the heat production rate $\dot{Q}_p$ to the sum of the rate of heat conduction away from the interface $\dot{Q}_c$ and the rate of heat storage in the moving surface $\dot{Q}_s$ defines a characteristic number N for the sliding pair, which can be assumed constant in a particular case thus:

$$\frac{\dot{Q}_p}{\dot{Q}_c + \dot{Q}_s} = N = \text{const.} \tag{3.52}$$

From thermodynamics, we write for $\dot{Q}_s$

$$\dot{Q}_s = \varrho_1 C_{p_1} \frac{DT}{Dt} L^3 = \varrho_1 C_{p_1} \Delta T V L^2, \tag{3.53}$$

where the suffix 1 refers to the moving body, ϱ_1 is the density, and C_{p_1} the specific heat at constant pressure for the moving surface. Finally by substituting for $\dot{Q}_p$, $\dot{Q}_c$ and $\dot{Q}_s$ from eqns. (3.46), (3.47), and (3.53) and putting $\alpha_1 = k_1/\varrho_1 C_{p_1}$ in eqn. (3.52),

$$\Delta T = \text{const} \left[\frac{f W_i V}{(\bar{k} + k_1 P e_1) L} \right]. \tag{3.54}$$

Equations (3.45) and (3.54) show clearly that ΔT will be larger for smaller thermal conductivities, since it is then more difficult to conduct frictional heat away from the interface. In the case of metals, it is fortunate that in this respect $\bar{k}$ has a relatively large value (compared with elastomers or other materials[†]), so that under otherwise identical conditions the heat generated by sliding friction produces a lower temperature rise. These effects are particularly significant at higher sliding velocities as will be apparent in the next section. The melting point of one or both of the sliding metals also appears as an important parameter which does not appear in eqn. (3.54). Thus if the melting point and thermal conductivity of one metallic surface are relatively low, high temperatures are generated in a thin surface layer and molten metal creates a liquid film which hydrodynamically lubricates the interface. On the other hand, high values of melting point and thermal conductivity restrict surface melting to the tips of asperities even at high sliding speeds, and this produces a rapid polishing action.

3.14 Effects of Sliding Speed

It is usually observed for metals that the coefficient of friction decreases as the speed of sliding is increased, although at the moderate velocities used in engineering practice this effect is not very pronounced. Indeed, the early investigators[‡] could not agree whether the coefficient of friction increased, remained constant, or actually decreased with increasing speed, and several empirical equations were proposed to explain in quantitative terms the

---

[†] $\bar{k}$ metals/$\bar{k}$ elastomers $\doteqdot$. 350–2000.

[‡] Coulomb, Vince, Hirn, Rennie, etc. (16).

conflicting trends. Part of the apparent contradiction is due to the limited speed range over which the early tests were conducted, and part is no doubt due to what we now recognize as the viscoelastic nature of friction, particularly when one sliding member is made of a flexible material (such as rubber, leather, or plastic).

The results of most investigations on the speed influence suggest that it is reasonable to assume the following relationship as valid:

$$f = (a+bV)\exp(-cV)+d, \tag{3.55}$$

where the constants a, b, c, and d are given for different sliding materials and at different normal pressures in Table 3.3.[16] We note that this expression allows in principle for the

TABLE 3.3. VALUES OF CONSTANTS a, b, c, AND d IN EQN. (3.55)

Materials	Pressure (kg/cm²)	Frictional constants			
		a	b	c	d
Cast iron–copper	0.19	0.006	0.114	0.94	0.226
	2.20	0.004	0.110	0.97	0.216
Cast iron–cast iron	0.83	0.022	0.054	0.55	0.125
	3.03	0.022	0.074	0.59	0.110
Fibre–steel	0.124	0.052	0.148	0.86	0.251
	0.324	0.051	0.157	0.99	0.243

coefficient of friction to increase because of the initial predominance of the bracketed term, followed later by an overall decrease as the exponential term gains the upper hand. Thus eqn. (3.55) may indicate a viscoelastic peak of adhesional friction† when at least one of the sliding surfaces is flexible. In general, however, the exponential term predominates over the entire useful frictional range of velocities as we see in Table 3.3. The influence of pressure appears in this table, but it is more obvious from Fig. 3.15. It is seen that the coefficient of

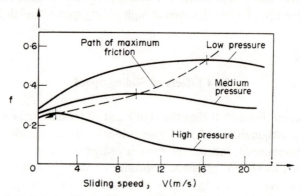

FIG. 3.15. Relationship between coefficient of friction and sliding speed.

† See Chapter 4.

friction is always smaller for higher pressures, and that the maximum coefficient at a particular pressure moves towards the origin of coordinates according as pressure increases. The
main difficulty in proposing a mathematical relationship between the coefficient of friction
and speed as in eqn. (3.55) is that during sliding the physical and mechanical properties of
the materials vary due to temperature change at the contact, and the values of the constants
in Table 3.3 can therefore only be considered as very approximate.

At low and moderate sliding velocities, friction is largely due to local adhesion and shearing at regions of contact as we have seen. Under these conditions, frictional resistance appears as surface heating which can be regarded as a transient and localized phenomenon. We
can state in general that such heating effects do not have a great effect on the resultant frictional mechanism. Consider now what happens at very high sliding speeds. In marked contrast to medium- and low-speed friction, metal surfaces are subjected to very intense frictional heating which profoundly changes the state of the surface layers in which the sliding
takes place. The surface damage associated with these rubbing processes ranges from extensive plastic flow at the lower velocities to melting on a large scale at the higher speeds. In
certain cases depending on the type of metal involved, diffusion and brittle disintegration
of the surface layer have also been observed.[1] The speeds at which large-scale melting and
other associated phenomena may occur range up to about 1000 m/s, which is several times
the speed of sound at room temperature. While such high speeds are outside the normal
range of application, they are of growing interest. Two possible applications of high-speed
friction are the motion of a shell in a gun barrel and the friction of rocket-propelled vehicles
on rails.

As might be envisaged, considerable experimental difficulties arise in producing ultra-
high speeds in the laboratory. One extremely successful method which we will describe here
is known as the deceleration technique.[1] A steel ball is suspended in the magnetic field of
a solenoid, vertical stability being achieved by a photoelectric feedback system. If the ball
sinks, more light falls onto a photocell which supplies more current to the electromagnet,
and the ball is lifted (the converse occurs if the ball is raised rather than lowered). The freely

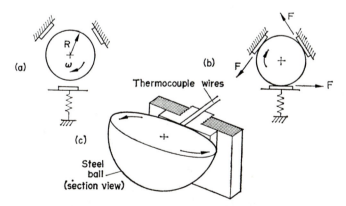

FIG. 3.16. Schematic of high-speed deceleration friction apparatus. (a) Magnetic suspension of steel
ball between three symmetrically placed friction pads. (b) Deceleration of spinning ball by activation
of friction pads. (c) Thermocouple method for estimating surface temperatures.

suspended sphere is accelerated by a rotating magnetic field of constant frequency (22,500 Hz). Figure 3.16 shows the steel ball rotating between three symmetrically placed friction pads. Because of the method of suspension and rotation of the ball (these are not shown in the figure), the surface speed which may be attained is limited only by the tensile or bursting strength of the metal. A commercial, hardened ball-bearing ball withstands a peripheral speed of about 1000 m/s without bursting. The speed of rotation of the ball is measured by marking a fine scratch on the polished surface so that a photomultiplier cell picking up reflected light from the steel surface experiences a signal once every revolution.

When the friction pads are activated by releasing a spring mechanism, the ball is slowed by the action of three symmetrical friction forces F as shown in Fig. 3.16(b). Let $\dot{\omega}$ represent the rotational deceleration at any speed, and I the moment of inertia of the ball about its spin axis. We now equate the total deceleration torque $3FR$ to the rate of change of angular momentum $I\dot{\omega}$ of the ball. Putting $f = F/W$, where W is the normal pad force acting on the ball per pad,

$$f = \frac{I\dot{\omega}}{3WR}.$$

(3.56)

The temperature at the sliding interface is measured using a thermoelectric method, as shown in Fig. 3.16(c). Two fine thermocouple wires of platinum and platinum–rhodium (diameter 25 μ) are embedded in one of the friction pads so that the wire ends were just level with the surface of the pad when the spinning ball touched it. At high speeds the friction causes the metal to flow so that a hot junction is formed which completes the thermoelectric circuit. Suitable calibration gives the interface temperature during deceleration of the ball.

Figure 3.17 shows the results obtained when the friction pads are made of copper. The curves are typical of many metals and alloys which possess a lower melting point than the

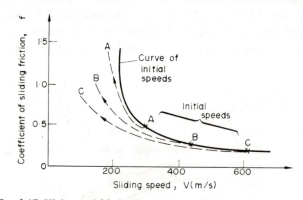

FIG. 3.17. High-speed friction between steel ball and copper pads.

steel ball used. When the friction pads are activated at any speed in the range 400–800 m/s, the ball continues to rotate smoothly between the copper pads for about 10 s. The frictional coefficient f is initially low (about 0.2 at 600 m/s) but increases steadily during the deceleration until the surfaces seize violently at a speed of about 120–140 m/s.[1] The dashed curves in Fig. 3.17 each represent a deceleration test, starting from different initial speeds A, B, and *C*. The coefficient of friction is calculated from the rate of change of angular speed according to eqn. (3.56). The curve of initial speeds in Fig. 3.17 shows a very steep rise in *f* as *V* is reduced.

Several interesting observations may be made from the results in Fig. 3.17 together with magnified images of the pad surfaces during the tests. Neither the roughness nor the state of cleanliness of the sliding surfaces proved to be of importance. Furthermore, at peripheral ball speeds in excess of 100 m/s a thin ribbon of copper was formed about the ball equator after a few revolutions. Thus the high-speed friction of copper-on-copper was, in fact, being studied. We conclude that the familiar localized abrasion marks which typify low sliding speeds give rise to a new form of surface damage (due to flow of the copper) at high sliding speeds (>100 m/s). There is evidence that at these speeds the copper behaves as a viscous lubricant, and the relatively large actual contact area which is present gives relatively high coefficients of friction not exceeding 1.5. At still higher sliding speeds (600 m/s) the area of contact receives a smooth shiny polish and the coefficient of friction is reduced to about 0.2. It is thought that in the latter case, the sliding speed is too rapid to permit the welding–shearing–ploughing theory of friction to apply, and the surface asperities are subjected to shock loading and rapid polishing.

3.15 Hardening of Metals

One of the most important properties of metals designed to resist friction forces, wear, erosion, or plastic deformation is their hardness. Several methods of hardness testing are available, and the hardness numbers in greatest use are the Brinell, Vickers, Rockwell, Shore, and Mohs hardness numbers. Usually, a hard indenter is pressed into the metal surface with a known load, and the size of the resulting indentation is measured.

The Vickers indenter is a square-based pyramid made of diamond in which the angle subtended by opposite faces is exactly 136°, and the Vickers hardness number H_V is defined as the ratio of load to the pyramidal area of indentation. We note that H_V is not too different in magnitude from the mean pressure $\bar{p}$ during the indentation process. In industrial practice, hardness measurements are often made with a Rockwell tester, which measures on a dial gauge the depth of penetration under a standard load. In this case, a conical indenter with a minute hemispherical tip is used. The mean pressure $\bar{p}$ for a cone pressed into a softer material can be shown to be independent of load. Furthermore,

$$\bar{p} \sim 1/d^2 \quad \text{and} \quad t = \tfrac{1}{2} d \cot \theta,$$

where *t* is the depth of penetration of the indenter, *d* the diameter of the indentation formed, and *θ* the semi-angle of the cone. By eliminating *d* from these equations, we find that

$$t \sim 1/\sqrt{\bar{p}},$$

so that the depth of penetration is smaller for harder materials. To establish the Rockwell hardness number H_R, we subtract the depth of penetration *t* from a fixed reference value *C* thus:

$$H_R = C - t = C - (K/\sqrt{\bar{p}}), \tag{3.57A}$$

where K is a constant. Replacing $\bar{p}$ by the Vickers hardness number H_V, a relationship between H_R and H_V emerges thus:

$$H_R = C - (K/\sqrt{H_V}). \tag{3.57B}$$

This relationship is plotted in Fig. 3.18, which also relates shore hardness and the tensile strength of metals as shown.

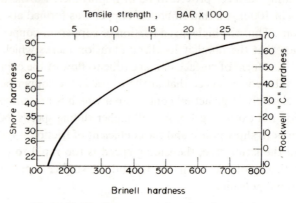

FIG. 3.18. Hardness conversion chart for metals.

An older type of hardness measurement due to Brinell involves the use of a hard, spherical indenter. This is usually a standard steel ball such as a ball-bearing which is both cheap and accurate in dimensions. When a spherical indenter is pressed into the surface of a fully work-hardened metal, the indentation pressure is roughly 3 times the yield stress of the metal (as with the Vickers indenter), so that Brinell and Vickers hardness numbers are almost identical. Differences arise if the metalwork hardens during indentation.

Another measure of hardness substitutes a scratch–hardness test for the indentation hardness experiments described above. The Mohs scratch test is based upon the general observation that one mineral will scratch a second mineral only if the hardness of the first is at least 20% greater than that of the second. Mohs proposed a scale of ten minerals ranging from talc to diamond such that each mineral will scratch the one on the scale below it but will not scratch the one above it. The scratch–hardness number M is related to indentation hardness H by a relation of the form

$$H = k(1.2)^M. \tag{3.58}$$

In general, both indentation hardness and scratch–hardness measurements on metals (and on brittle solids such as minerals) are essentially a measure of the plastic yield properties of the materials under examination.

It is often found with the micro-hardness tests described above that at small loads the hardness is greater than at large loads. This is due to the fact that the surface layers of the specimen are harder than the underlying material. Thus at sufficiently small loads, we are in fact obtaining a measure which approaches the true hardness of the surface layer, whereas for large loads and deep indentations the hardness approaches the bulk hardness of the metal.

3.16 Vacuum Conditions

It has been shown earlier in this chapter that surface contaminants have a very marked effect on friction between metals. In fact it has been proved conclusively that sliding friction is impossible without the presence of surface films, because otherwise the two surfaces would weld rigidly together. Since most surface films are oxides of one form or another, friction experiments are often conducted in a vacuum to eliminate the contaminating effect of air and thus preserve the cleanliness of exposed metal surfaces.

Remarkable increases in the coefficient of friction between different metal combinations can be obtained by conducting experiments in air and then in a vacuum after removal of the film. Table 3.4 shows a comparison of coefficients of friction after the removal of sur-

TABLE 3.4. COEFFICIENTS OF FRICTION IN AIR AND VACUUM

Combination of metals	Coefficient of friction in	
	Air	Vacuum (after removal of film)
Nickel–tungsten	0.3	0.6
Nickel–nickel	0.6	4.6
Copper–copper	0.5	4.8
Gold–gold	0.6	4.5

face films. In this case, the film-covered surfaces were produced by roughening with fine emery paper in air. Removal of the film was accomplished by heating the surfaces in a vacuum to various high temperatures depending on the particular surfaces used. The vacuum used was 10^{-6} mmHg.

If we assume that instead of removing surface films we somehow vary their thickness, some interesting results emerge. For very thin absorbed surface oxide films there is a decrease in the coefficient of friction with increasing film thickness as shown in Fig. 3.19, this

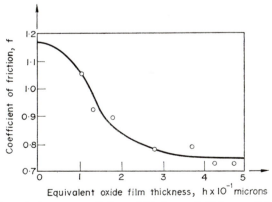

FIG. 3.19. Relationship between coefficient of friction and thickness of thin oxide films. [16]

being attributable to a decrease in the forces of molecular attraction between surfaces. On the other hand, thicker oxide films than those represented by the abscissa in Fig. 3.19 have a well-defined molecular structure which may produce either a decrease or increase in friction with increasing film thickness.

The effects of vacuum conditions on the coefficient of sliding friction between metals have particular relevance to space exploration, as we may visualize. In fact, conditions are more severe than we would anticipate by conducting experiments in a vacuum. This is because in addition to the absence of air which creates contaminating oxide films, desorption or evaporation of existing films is expedited in space. Thus exposed surfaces in the environment of space are robbed of lubricants and oxides which normally provide lubricating action and cold welding of surfaces becomes a real possibility.

CHAPTER 4

FRICTION OF ELASTOMERS

4.1 Fundamental Friction Mechanism

Consider the condition of contact between a rough, rigid surface and an elastomer which is loaded against it with load W, as we see in Fig. 4.1. The elastomer generally assumes the contour of the base surface by physically draping about its major asperities,

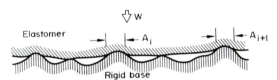

FIG. 4.1. Contact between elastomer and rigid base.

and elastic pressures develop in the elastomer over the areas of contact. The load is supported by these individual contact areas according to the relationship

$$W = \sum_{i=1}^{M} A_i p_i = \bar{p} A, \qquad (4.1)$$

where the mean pressure per asperity is $\bar{p}$, and the total actual contact area A is given by

$$A = KM(\bar{p}/E)^n. \qquad (4.2)$$

Here, K is a constant, E the Young's modulus for the elastomeric material, and the index $n \to 1$ in most practical cases.[†]

If we now attempt to dislodge the upper surface laterally by applying a tangential force F, adhesive forces develop at the contact locations, and if no net relative motion occurs between the surfaces, we can write for the equilibrium of forces

$$\underbrace{F}_{\substack{\text{Applied} \\ \text{force}}} = \sum_{i=1}^{M} \underbrace{F_i}_{\substack{\text{Resistive adhesion} \\ \text{forces}}}. \qquad (4.3)$$

† See Chapter 9, ref. 2.

61

This condition is identified by Fig. 4.2, and the adhesive forces F_i are due to the interaction of surface molecules at the contact interface.

Let us now increase the force F until gross sliding at steady speed V occurs at the elastomer–rigid boundary. The resistive force F now comprises adhesion and deformation terms in accordance with eqn. (3.15) for the general case of sliding. The deformation term is due

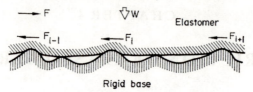

FIG. 4.2. Static friction forces at areas of contact.

to a delayed recovery of the elastomer after indentation by a particular asperity, and gives rise to what is generally called the hysteresis component of friction. We may therefore write

$$F_{\text{deformation}} = F_{\text{hyst}},\qquad(4.4)$$

so that eqn. (3.15) becomes

$$F = F_{\text{adhesion}} + F_{\text{hyst}}.\qquad(4.5)$$

Figure 4.3 gives some insight into the general nature of the adhesion and hysteresis terms in eqn. (4.5). It is seen that adhesion is distinctly a surface effect, whereas hysteresis is a bulk

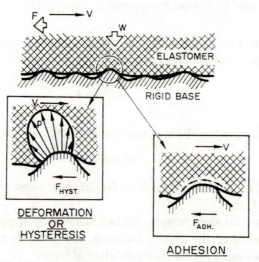

FIG. 4.3. Principal components of elastomeric friction.

phenomenon which depends on the elastic or viscoelastic properties of the elastomer. Each of these terms will be described and explained fully in succeeding sections.

If we divide both sides of eqn. (4.5) by the load W, we obtain the following:

$$f = f_A + f_H,\qquad(4.6)$$

which is analogous to eqn. (3.16) earlier for metals. Here, $f_A(=F_{adhesion}/W)$ is the coefficient of adhesional friction, and $f_H(=F_{hyst}/W)$ the coefficient of hysteresis friction. Both coefficients are, of course, the result of energy dissipation which gives rise to substantial temperature increase both at the sliding interface and within the bulk of the sliding bodies.

As in the case of metals, the magnitudes of f_A and f_H can often be adjusted by changing experimental conditions. The deformation (hysteresis) component is virtually eliminated on very smooth surfaces, and the adhesion term is substantially reduced when the sliding interface is lubricated. In the latter case, the lubricant effectively suppresses the molecular–kinetic interchange responsible for adhesion between the sliding surfaces.

4.2 The Adhesion Term

Modern theories of adhesion in a given velocity and temperature range have described adhesion as a thermally activated molecular stick–slip process. Unlike a hard material, the elastomer structure is composed of flexible chains which are in a constant state of thermal motion. During relative sliding between an elastomer and a hard surface, the separate chains in the surface layer attempt to link with molecules in the hard base, thus forming local junctions. Sliding action causes these bonds to stretch, rupture, and relax before new bonds are made, so that effectively the elastomer molecules jump a molecular distance to their new equilibrium position. Thus a dissipative stick–slip process on a molecular level is fundamentally responsible for adhesion, and several theories exist to explain this phenomenon.

The making and breaking of molecular junctions can be viewed as a causative mechanism for adhesion, since we concern ourselves with the fundamental behaviour of individual molecules in the sliding surfaces. We may also view adhesion as a resulting or macroscopic mechanism by considering that each body participating in sliding behaves as a continuum,[†] the properties of which may be simulated using mechanical models. From the macroscopic viewpoint, we can write for the total adhesional force:

$$F_{adh} = \sum_{i=1}^{M} F_i = \sum_{i=1}^{M} A_i s_i, \qquad (4.7)$$

where s_i is the local effective shear strength of the interface. On a molecular level, we can write at each location i:

$$F_i = n_i j_i, \qquad (4.8)$$

where n_i is the number of molecular junctions or bonds between the elastomer and base at location i, and j_i is the local effective junction strength. From the last two equations, it follows that

$$s_i = \left(\frac{n_i}{A_i}\right) j_i, \qquad (4.9)$$

[†] A continuum exhibits homogeneous properties that can be regarded here as the statistical resultant of molecular motion.

so that the local shear strength of the macroscopic model is a function of the molecular junction strength j_i. The coefficient of adhesional friction f_A is obtained by dividing F_{adh} by the applied load W thus:

$$f_A = \frac{F_{adh}}{W} = \sum_{i=1}^{M} F_i/W \qquad \text{[from eqn. (4.3)]}$$

$$= \sum_{i=1}^{M} n_i j_i/W \qquad \text{[from eqn. (4.8)]}$$

$$= Nj/W, \qquad (4.10)$$

where $N = \sum_{i=1}^{M} n_i = Mn$, M is the total number of macroscopic sites, n the mean number of molecular junctions per site, and N the total number of junctions of mean strength j. If we assume in the macroscopic model that each contact location has mean area $\bar{A}_i$ and shear strength s, then we can obtain an alternative expression for f_A:

$$f_A = \frac{F_{adh}}{W} = \sum_{i=1}^{M} A_i s_i/W \qquad \text{[from eqn. (4.7)]}$$

$$= \frac{M\bar{A}_i s}{W}. \qquad (4.11)$$

The various molecular–kinetic stick–slip theories of adhesion for elastomers[2] essentially attempt to establish how N and j vary with loading, speed, temperature, and lubrication. On the other hand, mechanical model theories examine in a less precise but simplified manner the variation of A_i and s with the same parameters. As we shall see, both approaches indicate the distinct viscoelastic properties of adhesion, and in most practical cases the coefficient of adhesional friction exhibits a peak value at creep speeds in the range 0–3 cm/s at room temperature. Before examining in detail some of the theories of adhesion, it is first of all necessary to obtain an understanding of the hysteresis component of friction.

4.3 The Hysteresis Term

The hysteresis component of friction can be visualized by Fig. 4.4, which shows the distribution of pressure about a single symmetrical asperity of the surface (a) when no relative motion exists at the contact interface (this is the condition of static friction shown in Fig. 4.2), and (b) in the presence of relative sliding.

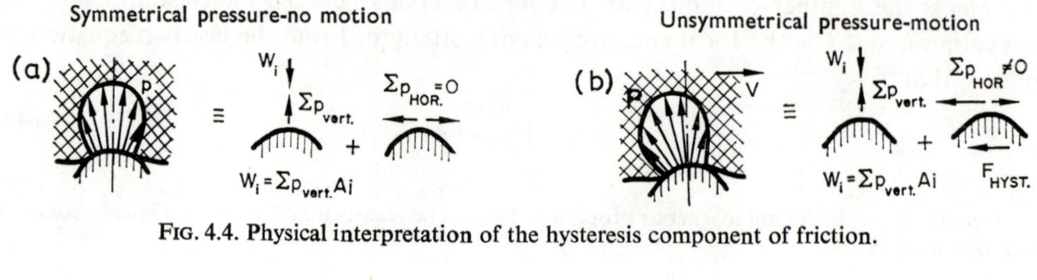

FIG. 4.4. Physical interpretation of the hysteresis component of friction.

We see clearly that the absence of relative motion produces a symmetrical draping of the elastomer material about the contour of the asperity. If the pressure distribution which is normal to the contour of the asperity is resolved in this case into vertical and horizontal components, it is seen that the summation of vertical pressures must be in equilibrium with the load W_i and the horizontal components cancel. If, however, the elastomer is moving with a finite velocity V relative to the base surface, it tends to accumulate or "pile up" at the leading edge of the asperity and to break contact at a higher point on the downward slope. Thus the contact arc moves forward compared with the static case, and this effect creates an unsymmetrical pressure distribution where the horizontal pressure components give rise to a net force (F_{hyst}) which opposes the sliding motion.

The friction of an elastomer sliding under load on a textured base is characterized by a "flowing" action, as the elasticity of the elastomer permits it to conform to the roughness elements in the support surface. The deformation or hysteresis component can be viewed as the "inertia" associated with the portion of draped elastomer between asperities at any instant during the sliding action. This inertia causes a delayed reaction within the bulk of the elastomer, which causes material to accumulate on the positive flanks of asperities and to fail to follow the contour of the latter on the negative slopes as indicated in Fig. 4.4.

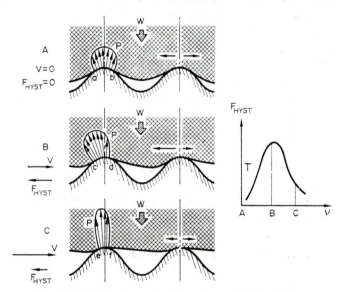

FIG. 4.5. Effects of asperity interaction on hysteresis force.

The hysteresis mechanism physically described above requires the existence of a finite slip speed at the contact interface. It remains to examine in more detail the speed effect and how it gives rise to pronounced viscoelastic properties. We will complicate the simplistic picture in Fig. 4.4 by including two symmetrical asperities rather than one and by considering two widely differing sliding speeds. Figure 4.5 illustrates the effects of asperity interaction on the magnitude of hysteresis forces during sliding. At zero speed we have a symmetrical draping and pressure generation as shown in Fig. 4.5A, so that $F_{hyst} = 0$.

When relative motion at speed V exists at the contact interface, the pressure asymmetry effect described earlier causes the contact arc ab to move forward (relative to the direction of sliding of the elastomer) to a new position cd, as shown in Fig. 4.5B. There is a corresponding rise in the value of the hysteresis force F_{hyst} as indicated in the insert in the same figure. As speed is further increased, the contact arc restores partly to the original symmetrical position as shown in Fig. 4.5C, and there is a drop in the value of F_{hyst}. We also note carefully that

$$|ef| < |cd| < |ab|,$$

which can be regarded as an effective "stiffening" of the elastomer which occurs as sliding speed is raised. The stiffening effect increases the mean pressure $\bar{p}$ with speed over each asperity in the base, and there is a corresponding reduction in the mean area of contact per asperity according to the relationship

$$\bar{A}_i = W/M\bar{p}, \tag{4.12}$$

where M is the total number of asperities over which the load W acts.

At the higher speeds of sliding (this is indicated by the relatively small contact arc ef in Fig. 4.5C), the delayed recovery of the elastomer on the downward slope of one asperity will decrease the accumulation on the positive slope of the succeeding asperity. Thus the point c (which perhaps had moved to the left relative to the static position a at modest sliding speeds), now moves to the right to the position e at higher speeds, as shown in Fig. 4.5C. The points c and d therefore approach each other, and the contact arc at very high sliding speeds has not only a minimum value, but it is virtually symmetrical. The hysteresis force due to pressure asymmetry is therefore reduced at very high speeds of sliding.

The physical mechanism of hysteresis described above accounts for the viscoelastic peak shown in the insert in Fig. 4.5, and it occurs at intermediate velocities of sliding. It has been tacitly assumed, of course, that the interface temperature T is constant at all sliding speeds. This assumption is made more plausible by the neglect of adhesion, which in turn implies the existence of an interfacial lubricant film. Let us now assume that the speed of sliding V is increased to the value $(V+\Delta V)$, so that a new temperature $(T+\Delta T)$ prevails at the sliding interface. Figure 4.6 shows that the effect of increasing temperature is to move the hysteresis peak to a higher velocity. The actual magnitude of ΔV for a given ΔT may be calculated from the Williams–Landel–Ferry transform method described later (Section 4.12). The

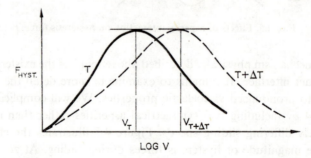

FIG. 4.6. Effects of temperature and sliding speed on hysteresis force.

curves in Fig. 4.6 have practical significance in a number of applications, as seen in Part II. If we choose to increase sliding speed in a given example without making any other changes in experimental conditions, we find generally that f_A reduces and f_H increases. Adjustment of the values of the adhesional and hysteresis contributions to elastomeric friction is an extremely useful design technique as the later applications will demonstrate.

4.4 Viscoelasticity

Before examining any theories of adhesion or hysteresis it **is** important to obtain a grasp of the fundamentals of viscoelasticity. The need for this is simply that adhesion and hysteresis are recognized to be viscoelastic phenomena which occur at different scales of magnitude (i.e. macro- or micro-) within elastomers during sliding. The term viscoelasticity is commonly applied to materials which are neither ideal elastic solids nor viscous liquids, but in fact possess characteristics which are typical of both.

We may, of course, represent elastic behaviour by an ideal spring and viscosity by an ideal dashpot, so that it is not surprising to find that viscoelastic behaviour is commonly represented by different assemblies of springs and dashpots. Such assemblies are called mechanical models, and they are widely used in different combinations and orders of complexity to simulate viscoelastic behaviour. The two basic mechanical models are known as Voigt and

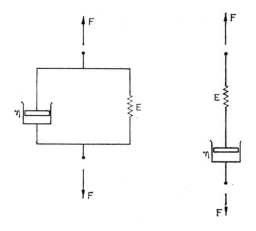

(a) VOIGT MODEL   (b) MAXWELL MODEL

FIG. 4.7. Simple (a) Voigt, and (b) Maxwell models.

Maxwell models, as shown in Fig. 4.7, and they consist simply of a spring and dashpot in parallel and series respectively. These models are generally too simplified to represent dynamic performance in viscoelastic materials, and they are therefore somewhat modified, as shown in Fig. 4.8. Thus by placing a spring in series with a simple Voigt model and in parallel with a simple Maxwell model, we obtain modified Voigt and modified Maxwell models respectively. These are far more representative of actual viscoelastic behaviour in

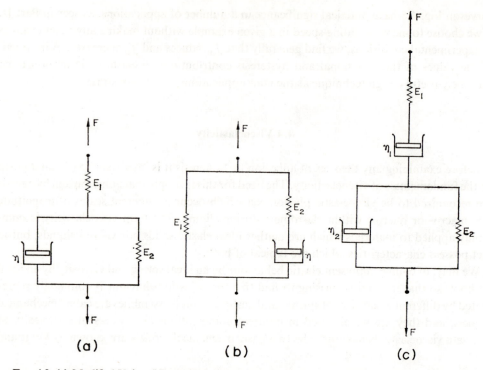

FIG. 4.8. (a) Modified Voigt, (b) modified Maxwell, and (c) Maxwell–Voigt models of viscoelastic behaviour.

materials, primarily because one of the springs assumes the role of retarded elasticity and the other that of instantaneous elasticity—these being typical characteristics of viscoelastic performance. By combining the simple Maxwell and Voigt models in series, the Maxwell–Voigt model shown in Fig. 4.8(c) offers even more realism. For a true representation of actual material behaviour, generalized Voigt and generalized Maxwell models offer arrays of simple Voigt and Maxwell models in different combinations.

Each of the five models in Figs. 4.7 and 4.8 establishes a particular relationship between applied force F and resulting extension δ.[2] These relationships can also be expressed in terms of stress σ and strain ε, where $\sigma = F/A$ and $\varepsilon = \delta/L$. We will not develop the various relevant equations here, but we observe that the stress σ is generally out of phase with the resulting strain ε, and we may express the ratio of stress to strain as

$$\sigma/\varepsilon = K' + jK'' = K^*, \qquad (4.13)$$

where K^* is a complex modulus comprising real and imaginary parts K' and K'' respectively, which are in phase and 90° out of phase with the strain ε, as shown in Fig. 4.9. Let the angle between K' and K^* be denoted by δ as indicated in the figure, then, writing

$$\tan \delta = K''/K', \qquad (4.14)$$

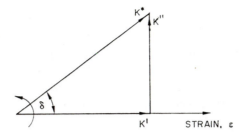

FIG. 4.9. Complex modulus and component parts.

we define the left-hand term in eqn. (4.14) as the tangent modulus or damping factor for the viscoelastic material. The complex modulus K^* becomes the complex shear modulus G^* (with corresponding components G' and G'') for a material in shear, or the complex Young's modulus E^* (with components E' and E'') for a material in tension or bending. Each of these modes of deformation has a corresponding tangent modulus, and we therefore append the subscript s or E to δ as appropriate. In general, $\tan \delta_s = \tan \delta_E$. For a material in compression, K^* becomes the complex bulk modulus B^*, and the corresponding $\tan \delta_B \to 0$.

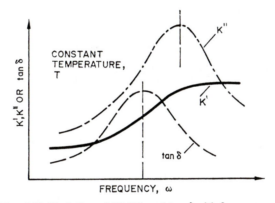

FIG. 4.10. Variation of K', K'', and $\tan \delta$ with frequency ω.

The variation in each of the parameters K', K'', and $\tan \delta$ according as frequency is raised is shown in Fig. 4.10 for a viscoelastic material at constant temperature T. We can interpret the variation of K' with speed as an effective stiffening of the material. The characteristic peak of the $\tan \delta$ curve has very important implications for friction forces as we shall see in the following sections.

4.5 Theories of Adhesion

Several molecular–kinetic and mechanical model theories of adhesion exist to describe the frictional behaviour of elastomeric materials.[2] Because of their relative complexity and for reasons of conciseness, we will deal here with two relatively simplified theories.

(a) Mixed Theory

Part of the physical model considers a simplified stick–slip event on a molecular level, and part uses information from a mechanical model—hence the name "mixed" theory. Consider an elastomer sliding on a rigid surface with velocity V as shown in Fig. 4.11. Let "adhesion" persist for a time during which the body moves a distance λ and then release takes place. An associated strain develops in the material causing energy to be stored elastically in the element. When the elastic stress exceeds the adhesive force, failure of the adhesive bond takes place and the element relaxes. If the maximum stress on an area δA is σ_0, then the

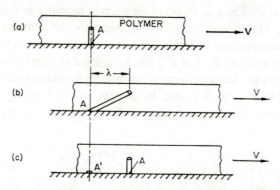

FIG. 4.11. A simple mechanism of adhesion.[17] (a) "Adhesion" takes place at points A. (b) Elastomer sample moves a distance λ at velocity V and frictional drag is developed. Elastic energy is stored in element. (c) Adhesion at A' fails. Energy stored in element is returned in part to system. New point of attachment at A.

work done in stretching the element is $\frac{1}{2}\sigma_0 \lambda\, \delta A$. Upon release, the elastic energy returned to the system is

$$\tfrac{1}{2}\sigma_0\,\lambda\,\delta A(1-\alpha),$$

where the energy loss fraction is α. Computation of α is made by assuming (for example) a Voigt model representation of viscoelastic behaviour [Fig. 4.7(a)]. During a quarter-cycle of sinusoidal vibration applied to one end of this model, the work required to extend the spring (equal to the elastic energy stored in the spring in a quarter-cycle of vibration) can be shown to be[†]

$$\tfrac{1}{2}E'L\delta_0^2,$$

whereas the energy dissipated during a quarter-cycle of vibration in the dashpot is[†]

$$(\pi/4)\,\eta\omega L\delta_0^2.$$

Here E and η denote the spring modulus and dashpot viscosity, L is a characteristic length, and δ_0 is the maximum spring extension. The ratio of energy dissipated in the dashpot to

[†] See section 10.3.

that stored in the spring is the loss fraction α thus:

$$\alpha = \pi \left(\frac{\eta \omega}{2E'} \right) = \frac{1}{2} \pi \tau \omega$$

$$= \tfrac{1}{2} \pi \tan \delta \qquad \text{[see ref. 2 and eqn. (4.24A)],}$$

where τ is the relaxation time of the Voigt element, ω the applied frequency of vibration, and tan δ the tangent modulus or damping factor. Using this expression for α, the energy loss is given by

$$\Delta E_{loss} = (\pi/4)\sigma_0 \lambda \, \delta A \tan \delta.$$

For N bonds or adhering elements, the total energy loss is

$$E_{loss} = (\pi/4)N\sigma_0 \lambda \, \delta A \tan \delta,$$

and this must be equated to the external work of friction $F\lambda$ (where F is the frictional force). Thus

$$F = (\pi/4)N\sigma_0 \, \delta A \tan \delta. \qquad (4.15)$$

The adhering area is $N \, \delta A$, and this is obviously proportional to load W and inversely proportional to hardness H:

$$N \, \delta A \sim W/H. \qquad (4.16)$$

By substituting eqn. (4.16) in eqn. (4.15),

$$F = K\sigma_0(W/H) \tan \delta, \qquad (4.17)$$

where K is a proportionality constant. Also, since the coefficient of adhesional friction f_A is defined by F/W,

$$f_A = K(\sigma_0/H) \tan \delta. \qquad (4.18)$$

Equations (4.17) and (4.18) show all the observed properties of measured frictional behaviour, thus:

(a) Frictional force F proportional to load W (Amonton's law).
(b) Coefficient of friction f_A decreases as hardness increases.
(c) Viscoelastic properties contained in tan δ term (Fig. 4.10).

Indeed, for a given elastomer, eqn. (4.17) shows that the frictional force F is directly proportional to tan δ. Figure 4.12 shows (a) separate plots of tan δ and F for SBR[†] tread rubber as a function of temperature, and (b) the same curves plotted by transferring the tan δ curve horizontally until the two peaks occur at approximately the same temperature. The equivalent frequency corresponding to the new position of the tan δ curve is about 10^6 Hz, this having been obtained by the Williams–Landel–Ferry transform method described in Section 4.12. But the sliding speed at which the frictional data has been recorded is 1 cm/s. We can compute the mean distance λ of a molecular jump:

$$\lambda = \frac{1 \text{ cm/s}}{10^6 \text{ Hz}} \doteq 100 \text{ Å}.$$

[†] Styrene butadiene.

6*

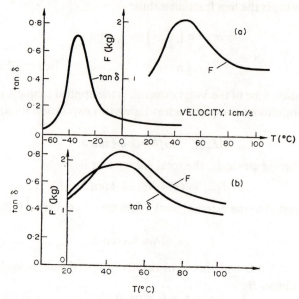

FIG. 4.12. (a) Friction force F and tan δ vs. temperature for SBR rubber. (b) Same curves with tan δ transformed to equivalent friction frequency.

Other plots of both F and tan δ for various rubbers and elastomers give essentially the same result as the SBR example shown in Fig. 4.12. Thus experiments and the theory resulting in eqn. (4.17) indicate that the frictional force F is proportional to tan δ, and so the visco-elastic nature of F is assured.

(b) *Simple Theory*

The simple theory of adhesion described here represents an ingenious and convincing application of the basic adhesion equation for solids to the case of viscoelastic materials. The frictional force in sliding is proportional to the product of actual contact area and effective shear strength s according to the simple relationship[18]

$$F = As. \tag{4.19}$$

For slow-speed sliding of a rigid sphere on an elastomeric surface, the area of contact can be shown from Hertzian theory† to be proportional to $E^{-2/3}$, where E or E' is the storage modulus of the elastomer. The variation of E' with frequency ω or sliding speed V is well known from viscoelastic theory, and can be interpreted[19] as an effective stiffening of the elastomer at higher deformation frequencies (see K' in Fig. 4.10). It follows that the contact area A must decrease in an inverse manner to the E' curve, as shown in Fig. 4.13. The variation of shear strength s with frequency or speed can be obtained as follows.

† See eqn. (3.3).

For the simple Voigt model in Fig. 4.7(a) we can write both shear and pressure equations thus:

$$p = \frac{E'z}{L} + \frac{\eta \dot{z}}{L} \qquad (4.20A)$$

and

$$s = G'\theta + \eta\dot{\theta}, \qquad (4.20B)$$

where p and s denote pressure and shear stress, E' and G' are the corresponding real moduli n compression and shear as defined earlier, η is the dashpot viscosity, z and $\dot{z}$ are the resulting normal deflection and its rate of change, and θ and $\dot{\theta}$ signify shear strain and rate of strain in the thin surface layer at a frictional interface. We can also write these equations in the form

$$p = (E^*/L)z \qquad (4.21A)$$

and

$$s = G^*\theta, \qquad (4.21B)$$

where the complex moduli E^* and G^* are defined by the following:

$$E^* = E' + jE'' \qquad (4.22A)$$

and

$$G^* = G' + jG''. \qquad (4.22B)$$

Here E'' and G'' are the loss moduli in compression and shear respectively. Let us now assume that the cyclic compression and shear deformation correspond to a sinusoid of equivalent frequency ω thus:

$$z = z_0 e^{j\omega t} \quad \text{and} \quad \theta = \theta_0 e^{j\omega t}. \qquad (4.23)$$

By substituting eqn. (4.23) into the appropriate forms of eqns. (4.20) and comparing the result with that obtained when eqns. (4.22) have been inserted into eqns. (4.21), the conditions for complete equivalence are:

$$E'' = \omega\eta \quad \text{for compression} \quad \text{and} \quad G'' = \omega\eta \quad \text{for shear.}$$

Now, since

$$\tan \delta_E = E''/E' \quad \text{and} \quad \tan \delta_G = G''/G',$$

it follows that

$$\tan \delta_E = \frac{\omega\eta}{E'} \qquad (4.24A)$$

and

$$\tan \delta_G = \frac{\omega\eta}{G'}. \qquad (4.24B)$$

Since $\tan \delta_E \doteq \tan \delta_G$, it is common practice to neglect the suffixes E and G, as indicated earlier. From eqns. (4.20B), (4.23), and (4.24B) we find that

$$s = (1 + j\tan \delta) G'\theta \quad \text{or} \quad |s| = G'\sqrt{(1 + \tan^2 \delta)} |\theta|.$$

Now, the variations of both G' and $\tan \delta$ with frequency ω are given in Fig. 4.10, so that the variation in $|s|$ with frequency (for constant amplitude) clearly has the form shown in Fig. 4.13.

Both s and A now vary in such a manner in Fig. 4.13 that their product may either give the viscoelastic peak we require or possibly a flat characteristic. The latter will occur if the drop in A with frequency matches the initial rise in s, and the viscoelastic peak will obviously occur if the fall in A takes place substantially later than the rise in s on the frequency or

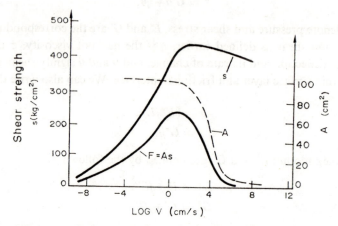

FIG. 4.13. Explanation of viscoelastic nature of F by simple product solution.

speed scale. Let the rate of loading r_L of the elastomer in the contact region be defined as the sliding speed divided by the contact diameter. Thus from experimental results,[18] we find that

$$r_L = \frac{V}{L} \doteq \frac{1 \text{ cm/s}}{1 \text{ mm}} \doteq 10 \text{ s}^{-1}.$$

Let us also assume that frictional shear occurs in a very thin surface layer t of the elastomer (about 100 Å). The shear rate r_s is then defined by

$$r_s = \frac{V}{t} \doteq \frac{1 \text{ cm/s}}{10^{-6} \text{ cm}} \doteq 10^6 \text{ s}^{-1}.$$

Thus we conclude that

$$r_s/r_L \doteq 10^5,$$

so that the s-curve is effectively shifted 5 decades to the left relative to the A-curve. This ensures that the product solution $F = As$ gives the viscoelastic peak evident both in Figs. 4.12 and 4.13.

The above simple theory differs from the mixed and other theories of adhesion[2] in that the frictional peak is explained in physical terms rather than by showing a proportionality between frictional force F and tan δ. Using this approach, both F and tan δ may be shown to possess similar viscoelastic peaks, and therefore their proportionality follows. It is a matter of preference as to which order of logic we finally select, but the important conclusion is that $F \sim \tan \delta$ in accordance with eqn. (4.17). More exhaustive studies of adhesion have shown the following relationship between the coefficient of adhesional friction f_A and $\tan \delta$:

$$f_A = B\phi'(E/p^r)\tan \delta, \tag{4.25}$$

where the exponent $r < 1$, B is a constant, and ϕ' a function which depends on the adhesion-generating ability of the interface.

4.6 Adhesion as a Contact Problem

The theories of adhesion discussed above and those appearing elsewhere[2] have assumed nominally flat sliding surfaces, although we do not specify whether a fine scale of micro-roughness exists at the interface. Practical engineering surfaces, however, exhibit both macro- and micro-roughness effects, as shown in Fig. 4.14 (see also Fig. 2.1). These different scales of roughness jointly determine the actual area of physical contact when an elastomer

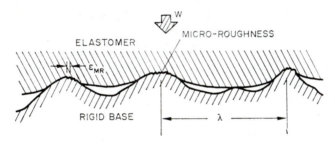

FIG. 4.14. Macro- and micro-roughness effects on contact area.

is draped on the surface under the action of an applied load W. The macro-roughness has some mean wavelength λ which, in general, varies between the limits of 0.25 mm on machined metal surfaces to 10 mm on coarse asphaltic road surfaces. For a given load W the mean actual pressure p on each asperity of the macro-roughness greatly exceeds the apparent or nominal pressure p_{app} according to the relationship

$$p = \left(\frac{A_{app}}{A_{act}}\right)p_{app}. \tag{4.26}$$

Here the apparent area A_{app} can be visualized as the area of contact for completely smooth surfaces, and A_{act} is the summation of individual areas at the summits of the asperities for the actual surfaces shown in Chapter 2. Almost all of the theories of adhesion indicate that f_A is inversely proportional to p [see eqn. (4.25)], and p, of course, exceeds p_{app} as shown above; thus the very existence of a macro-roughness reduces the coefficient of adhesion. The total frictional force F is also reduced, since $A_{act} < A_{app}$. The exact effects of the micro-roughness cannot be predicted at this stage, although it is certain that a further reduction in A_{act} occurs. The wavelength ε_{MR} of the micro-roughness varies between the limits of 0.25 μ[†] (extra-fine honing operations) to 100 μ (coarse road surfaces).

It is normally desirable that a surface should exhibit a distinct texture pattern to achieve optimum non-skid properties, and we may wonder how this fact can be reconciled with the reduction in adhesion. The solution to this apparent contradiction lies, of course, in the distinction between wet and dry conditions of sliding. For the dry case there is no doubt that the maximum adhesion is attained on smooth surfaces, and since the interfacial area has a maximum value the mechanism of molecular–kinetic bonding is most widespread. Under wet conditions, however, an interfacial film of liquid is spread uniformly, and it effectively suppresses intermolecular bonding at the surfaces; thus the adhesion falls to a very low value. When the rigid surface has a distinct macro-roughness, the voids between asperities act as reservoirs for the liquid under wet conditions,[‡] and the pressure distribution at each asperity summit promotes local drainage effects. There is therefore a greater probability of suitable conditions existing for adequate adhesion under wet conditions when the surface exhibits a macro-roughness compared with the completely smooth case. This probability is greatly enhanced by providing a distinct micro-roughness at asperity peaks, as discussed in a later chapter. The overall conclusion is that the combined effects of the macro-roughness and the micro-roughness under wet conditions are to minimize the decrease in the coefficient of adhesional friction below the dry value.

We emphasize that irrespective of whichever theory of adhesion seems most acceptable in a given situation, the adhesion itself remains essentially a contact problem. Only suitable experimentation will determine exact areas of real contact, and it appears to be extremely difficult today to accomplish this objective since the frictional interface by its very nature is removed from sight. Analysis is useful in determining contact areas only in the case of single and multiple asperities of idealized shape (usually rods, cylinders, or spheres) which often bear no resemblance whatever to a typical random surface texture. Apart from this, the combined effects of speed, temperature, viscoelastic draping, and elastohydrodynamic inter-action complicate any analysis to the extent that satisfactory practical solutions may never be feasible. It is therefore highly desirable to develop reliable methods of experimentally determining actual contact areas despite the difficulties involved.

4.7 Theory of Hysteresis Friction

There are two commonly accepted and recent theories of hysteresis friction for elastomers and rubbers.[2] The *unified theory* proposes semi-empirical equations by analogy with a corresponding theory for adhesion, whereas the *relaxation theory* uses a Maxwell model of viscoelastic behaviour[*] to define a coefficient of hysteresis friction for ideal asperity shapes. Since the form of the resulting equations is similar in both theories, we will here describe the simpler of the two:[20]

[†] $1\,\mu = 1$ micron $= 10^{-6}$ m.
[‡] See Chapter 8.
[*] Refer to Fig. 4.7.

Unified Theory

We can express the hysteresis component of frictional force as follows:

$$F_{\text{hyst}} = MJ, \tag{4.27}$$

where M is the total number of macroscopic sites or contacting asperities and J is the frictional resistance at each site. Equation (4.27) is analogous to eqn. (4.9) for the adhesional force on an individual asperity. The number of asperities M encountered by the sliding elastomer is equal to

$$M = \gamma M_0, \tag{4.28}$$

where M_0 is the maximum possible number of asperities in a given surface, and $\gamma \leqslant 1$ is an asperity density factor accounting for the degree of "packing" of the asperities on the surface. We can define M_0 in turn as

$$M_0 = \frac{A}{(\frac{1}{2}\lambda)^2}, \tag{4.29}$$

where A is the actual contact area, λ the mean wavelength of the surface, and $\lambda/2$ the mean linear dimension of a typical asperity at its base. From eqns. (4.28) and (4.29), we can write

$$M/A = 4\gamma/\lambda^2. \tag{4.30}$$

The frictional resistance J can be expressed as the energy E_d dissipated per site divided by the mean sliding distance between asperities (which is the mean wavelength λ) thus:

$$J = E_d/\lambda. \tag{4.31}$$

Since the total applied load W can be written as

$$W = pA, \tag{4.32}$$

we can define the coefficient of hysteresis friction f_H as follows:

$$f_H = F_{\text{hyst}}/W = (M/A)\,(J/p). \tag{4.33}$$

By substituting from eqns. (4.30) and (4.31) into eqn. (4.33), the following expression for f_H is obtained:

$$f_H = (4\gamma/\lambda^3)(E_d/p). \tag{4.34}$$

This equation must be further modified to obtain an expression for the energy dissipated per asperity E_d. This is identical with the product of the volume of the elastomer Q which is deformed over each asperity during sliding and the loss modulus E'' thus:

$$E_d = QE''. \tag{4.35}$$

Figure 4.15 indicates by shading the projection of Q in the direction of motion for a sliding cone, sphere, or cylinder. For these three basic shapes, the draping height w_0 can be represented by an equation of the form

$$w_0 = K(p/E)^m, \qquad (4.36)$$

where the values of the constants K and m are listed in Table 4.1. We note that K assumes a constant value for cones only when the applied load W is constant. If we now express the deformed volume Q in terms of w_0, and then substitute for w_0 from eqn. (4.36), the resulting

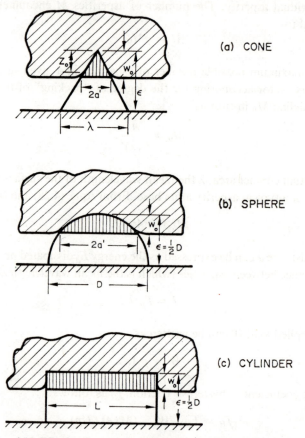

FIG. 4.15. Effective flow cross-sections for various asperity shapes.

expression can be used in eqns. (4.34) and (4.35). The coefficient of hysteresis friction then assumes the following general form

$$f_H = 4C\gamma(p/E)^n \tan \delta, \qquad (4.37)$$

where the constants C and n are listed in Table 4.2 for spheres, cylinders, and cones.

The index n in eqn. (4.37) according to this theory would then appear to lie between the values 2 and 3 for a random asperity, assuming that the latter exhibits some shape that lies

TABLE 4.1. K AND m FOR SPHERES, CYLINDERS, AND CONES

Asperity shape	K	m
Sphere	$\dfrac{9\pi^2 R(1-\nu^2)^2}{16}$	2
Cylinder	$\dfrac{64R(1-\nu^2)^2}{\pi^2}$	2
Cone	$\left\|\dfrac{2\sqrt{2}}{\sqrt{\pi}}(1-\nu^2)^{3/2}\sqrt{\dfrac{R}{\varepsilon}}\sqrt{\dfrac{W}{E}}\sqrt{\dfrac{w_0}{Z_0}}\right\|$	1

E = Young's modulus. ν = Poisson's ratio.

between the extremes of infinite and zero sharpness. To account for finite speeds of sliding, the Young's modulus E in the denominator should be replaced by the complex modulus E^*, which assumes a different numerical value at each speed. However, the calculation of the deformed volume Q neglects contact asymmetry in the direction of sliding (see Fig. 4.5), and this effect may increase the value of f_H in eqn. (4.37). This may be regarded as a minor discrepancy in the theory—yet the form of eqn. (4.37) is relatively precise. Undoubtedly, the loss modulus E'' in eqn. (4.35) accounts for the asymmetry effect in the area of contact. The later and more rigorous relaxation theory gives the following equation for f_H:[21]

$$f_H = \zeta A \phi(\bar{p}/E) \tan \delta, \qquad (4.38)$$

where A is a numerical factor dependent on the shape of the asperity, ζ is a length ratio, and ϕ a geometric factor < 1. The similarity of eqns. (4.37) and (4.38) is apparent despite

TABLE 4.2. C AND n FOR SPHERES, CYLINDERS, AND CONES

Asperity shape	C	n
Sphere	$\dfrac{81\pi^5(1-\nu^2)^4}{2048}$	3
Cylinder	$\dfrac{256\sqrt{2}}{3\pi^3}\left(\dfrac{L}{R}\right)(1-\nu^2)^3$	2
Cone	$\left\|\dfrac{2}{3}\sqrt{\left(\dfrac{2}{\pi}\right)}(1-\nu^2)^{\frac{9}{2}}\left(\dfrac{W}{ER^2}\right)^{\frac{3}{2}}\left(\dfrac{w_0}{Z_0}\right)\left(\dfrac{R}{\varepsilon}\right)^{\frac{15}{2}}\right\|$	2

the radically different approaches taken in the underlying theories. It is believed more realistic to assume a unity index for the ratio $(\bar{p}/E)$ in accordance with the more elaborate theory. By comparing eqns. (4.25) and (4.38), we observe that both the adhesion and hysteresis coefficients of friction are proportional to tan δ. *It is remarkable that f_A is proportional in some way to $(E/\bar{p})$ and f_H to its reciprocal.* This inverse dependence on pressure poses interesting design problems in practical applications, and ultimately a compromise solution must be agreed upon between the adhesion and hysteresis requirements. It is more correct to substitute the complex Young's modulus E^* for E in the hysteresis equations (4.37) and (4.38) to allow for temperature and frequency variations.

4.8 Generalized Coefficient of Hysteresis Friction

For a given lower surface, we may rewrite eqn. (4.38) in the form

$$\left(\frac{f_H}{\tan \delta}\right) = \text{const} \left(\frac{\bar{p}}{E}\right), \qquad (4.39)$$

where the ratio $(f_H/\tan \delta)$ is called a generalized coefficient of hysteresis friction. For constant speed and temperature, this theoretical relationship is shown in Fig. 4.16 to be in close agreement with experimental results. The latter were obtained using a cylindrical slider moving with a velocity of 0.06 m/s on both natural and butyl rubber. If temperature or sliding speed are varied, the viscoelastic nature of the frictional force F_{hyst} or coefficient f_H will be apparent, as in Fig. 4.6. In this case, the generalized coefficient of hysteresis

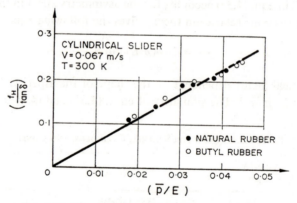

Fig. 4.16. Generalized coefficient of hysteresis friction.

friction increases slightly with sliding speed for constant temperature because of the stiffening effect described earlier.[2] It has also been shown that whereas both f_H and $\tan \delta$ exhibit viscoelastic peaks when plotted separately versus speed, the sliding velocity at which f_H attains this peak is substantially lower than that at which the $\tan \delta$ maximum occurs. Thus a typical speed value for the coefficient of hysteresis friction maximum to take place is 55 m/s for natural rubber on a surface having a mean wavelength $\lambda = 12$ mm. We will see in Part II how the hysteresis peak of friction can be adjusted in practical applications to correspond to the most advantageous sliding speeds.

A generalized coefficient of adhesional friction $(f_A/\tan \delta)$ may also be defined from eqn. (4.25). It follows from Fig. 4.12 that this ratio increases slightly with temperature for a fixed speed of sliding.

4.9 Mean Pressure Effects

The mean pressure $\bar{p}$ per asperity increases with sliding speed V because of the stiffening effect. If the speed of sliding is held constant and the mean wavelength λ of the surface asperities is reduced by using a finer scale of texture, it can also be shown that $\bar{p}$ increases.

Let M be the number of asperities in a given surface and W the total load which presses a viscoelastic plane or elastomer on to this surface. Assuming for simplicity that the asperities are hemispheres, then M is given by

$$M = W/\pi a^2 \bar{p}, \tag{4.40}$$

where the radius of contact a is obtained from Hertzian theory[†] as follows:

$$a \sim \bar{p}(R/E) \tag{4.41}$$

with R representing the radius of any sphere or hemisphere in the base surface. By substituting eqn. (4.41) into eqn. (4.40),

$$M \sim \frac{E^2}{\bar{p}^3 R^2} \tag{4.42}$$

assuming W is constant. Since R^2 is proportional to the projected area of each asperity, it follows that

$$MR^2 = \text{const} \tag{4.43}$$

for a given total surface area according as the scale of texture is varied. From the last two equations, we conclude that

$$E^2/\bar{p}^3 = \text{const}. \tag{4.44}$$

Now we define the operating "frequency" experienced by the elastomer (during sliding at speed V on a base surface with mean asperity wavelength λ) by the relationship $\omega = V/\lambda$. Thus for a fixed sliding speed the "frequency" of indentation ω increases according as λ is reduced on finer surfaces. We know from Fig. 4.10 that E' or E increases with ω; thus it follows from eqn. (4.44) that $\bar{p}$ also must increase on finer surfaces for the same speed of sliding and operating temperature.

4.10 Macro- and Micro-hysteresis

The hysteresis mechanism which develops in sliding over the broad roughness features of a given surface can also be described as macro-hysteresis. There also exists a lesser contribution to hysteresis friction which results from the fine-scale or micro-roughness existing at asperity tips, and the term micro-hysteresis is often used to describe this component. It is certain that the micro-roughness grooves or deforms a thin surface layer of the elastomer during sliding, but it is not at all sure whether to classify this effect as a hysteresis or ad-

[†] See eqns. (3.2) and (3.3).

hesion mechanism. Indeed, if the scale of the macro-roughness is progressively diminished to a value less than the micro-roughness (perhaps approaching molecular dimensions), then it is clear that the hysteresis component of friction has been replaced at some stage by adhesional friction.

The concept of micro-hysteresis implies therefore that there is some overlap or ambiguity in the separation of frictional effects into distinct adhesional and hysteresis components. *Both mechanisms may, in fact, be attributed to the same viscoelastic dissipation of energy,*[20] *with a different scale of events for identification purposes.* Thus the average molecular jump distance which is typical for rubber adhesion is about 100 Å, whereas the mean wavelength λ of the macro-roughness which determines the hysteresis component is several millimetres. The adhesion of elastomers must be viewed as taking place within a microscopically thin surface layer of the elastomer rather than at an interface having no dimensions. We can then classify the micro-hysteresis effect by observing the amplitude of the micro-roughness of asperity peaks relative to this surface layer. Thus the smallest micro-roughness of significance that we measure will confine the deformation of the elastomer to the adhesion layer— and the corresponding micro-hysteresis effect is an adhesional mechanism. On the other hand, larger micro-roughnesses will begin to deform the bulk of the sliding elastomer in addition to the surface layers, and in this case the micro-hysteresis phenomenon is truly hysteretic.

4.11 Separation of Adhesion and Hysteresis Terms

In most practical examples, the friction of elastomeric materials comprises both adhesion and hysteresis as explained earlier. If we are free to adjust experimental conditions, we can separate the two components and thus quantify the contribution of each. The most common

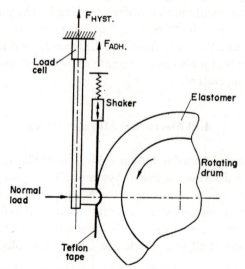

FIG. 4.17. Schematic of tape and shaker mounting.

methods are the use of a lubricant to minimize or eliminate the adhesion component, and the use of smooth, dry surfaces to eliminate hysteresis (see Sections 3.5 and 4.1). We are therefore capable of estimating the magnitude of each term in a given example.

One of the difficulties, however, of isolating the hysteresis component of friction is that the adhesion term can never be entirely eliminated[21] even when a lubricant is introduced at the interface. At very high sliding speeds, of course, the hysteresis contribution to friction dominates, and the use of a lubricant will reduce the adhesion to a level which (although not zero) is negligible by comparison. However, if the sliding speed is low, the adhesion may well exceed the hysteresis contribution to friction even in the presence of a lubricant.

The adhesion term can be effectively suppressed by placing a vibrating, teflon-covered tape between the slider and the elastomer. Figure 4.17 shows a schematic of the tape and shaker mounting. The elastomer is mounted on the surface of a rotating circular drum, and the slider penetrates radially under the action of a normal load. The slider is restrained by a force cell which records only the hysteretic friction force, whereas the tape is restrained separately. Ideally, the tape should be completely rigid in tension and at the same time perfectly flexible to wrap around the slider profile. Such a tape, if obtainable, would eliminate the adhesive component of friction entirely from the force cell reading and permit hysteresis losses to be measured. Teflon tape (which has been prestretched to eliminate further extension) fulfils the requirements, but it is extremely difficult to specify the amount of prestretch which is necessary, and further complications develop from load cell displacement. A pneumatic shaker interposed between the tape and its restraint was found to completely eliminate the adhesive friction. The average force on the tape is now zero, and the force cell records bulk losses only, because of the indentation of the elastomer. In the case of rubber,[2] a frequency of vibration of about 250 Hz applied to the tape was found to be most effective.

At higher sliding speeds, the use of a lubricant may be sufficient to reduce the adhesion coefficient to a value which is negligible compared with the hysteresis contribution, as indicated earlier, and no special apparatus is required. However, caution should be exercised in interpreting the results, since elastohydrodynamic events oppose the hysteresis effects when a lubricant is present, as shown clearly in Chapter 8. Furthermore, the total effect of the elastohydrodynamic separating force increases rapidly with sliding velocity, so that it is difficult to identify the separate contributions at high sliding speeds. Temperature rise at the interface is another factor which is difficult to account for or measure precisely.

4.12 The Williams–Landel–Ferry Transform

Both the adhesion and hysteresis components of elastomeric friction have been clearly shown to be viscoelastic phenomena, and their magnitudes vary with both temperature and frequency (or speed). We shall see that temperature change produces an equivalent effect at constant speed, as frequency (or speed) change at constant temperature. The interchangeability of temperature and frequency provides a valuable tool in experimental design, and thus we desire to know if possible the amount of change in the one variable which is necessary to compensate for a known or given change in the second variable. The Williams–

Landel–Ferry equation[22] accomplishes this objective by providing an expression for the horizontal frequency shift factor a_T as related to operating temperature T thus:

$$\log_{10} a_T = \frac{-8.86(T-T_0)}{101.5+T-T_0},\tag{4.45A}$$

where T_0 is a reference temperature chosen as characteristic of a given polymer. In general, $T_0 \doteq T_g+50°C$, where T_g is the glass-transition temperature[†] for the polymer. By applying eqn. (4.45A) in its present form, the value of any viscoelastic property (notably friction) obtained at temperature T and frequency ω can be related to the reference temperature T_0

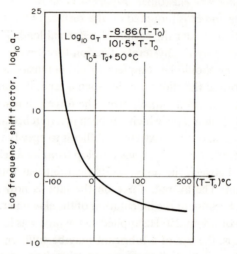

FIG. 4.18. Frequency shift factor of the Williams–Landel–Ferry transformation.

by a frequency shift of amount $a_T\omega$, which is readily computed. This is an extremely useful tool in assessing data from tests conducted at various temperatures T, since they can be reduced to the equivalent data at a fixed and common reference temperature T_0. Figure 4.18 shows the variation of the frequency shift factor a_T with temperature difference $(T-T_0)°C$. We note that a_T is very sensitive to temperature changes in the region of T_0 but less dependent at higher temperatures.

If we wish to reduce experimental data recorded at (ω, T) to a common temperature T_c which is distinct from T_0, we must first find the frequency shift factor a_T for any data relative to the reference temperature T_0 according to eqn. (4.45A). We then find the frequency shift factor a_{T_c} for data at temperature T_c relative to T_0 by substituting $T=T_c$ in eqn. (4.45A). Thus

$$\log_{10} a_{T_c} = \frac{-8.86(T_c-T_0)}{101.5+T_c-T_0}.\tag{4.45B}$$

[†] This is the temperature at which transition to the "glassy" or brittle state occurs.

The frequency shift Q in decades is then obtained by subtracting eqn. (4.45A) from (4.45B):

$$Q = \log_{10} a_T - \log_{10} a_{T_c} = 8.86 \left(\frac{T_c - T_0}{101.5 + T_c - T_0} - \frac{T - T_0}{101.5 + T - T_0} \right).$$

Each data point recorded at (ωT) with $\log_{10} \omega$ (or $\log_{10} V$) as abscissa must therefore be shifted by an amount Q to correspond to the newly selected reference temperature T_c.

Many factors affect the magnitude of the frictional forces generated between an elastomer and a textured base, and we can therefore give only rounded estimates of frictional performance in a given example. It is true that many of the equations in this chapter yield relatively precise data particularly in experiments, but the conditions and assumptions made are often limiting and unrealistic in terms of practical utility. Thus we may know precisely from equations the magnitude of hysteresis and adhesion, but we are not sure about phenomena such as humidity effects, surface absorption, dust deposits, lubricity, etc. Indeed, there may be as many as 100 variables to be accounted for in determining tyre–road interaction in rolling, sliding, or cornering, and we must undoubtedly await further developments in research which will quantify as many of these as possible. Nevertheless, we can make accurate qualitative predictions (and often provide a good deal of quantitative information) on the basis of today's knowledge of tribological principles. Our lack of complete knowledge in this field may be compensated for by judicious and timely experiments, which guide the theoretical investigations (and vice versa) and provide a link with practical experience.

CHAPTER 5

FRICTION OF VARIOUS MATERIALS

THE previous two chapters have summarized the principles of friction as they apply to the two most basic categories of engineering materials, namely metals and elastomers. This chapter includes a qualitative and experimental treatment of materials that do not fall under either of these groups. Such include lamellar solids, diamonds, ice and brittle solids. Other polymeric materials such as wood, yarns, and fibres are also included because of their particular engineering interest.

5.1 Friction of Lamellar Solids

It is well known that solids having a plate-like or lamellar structure have a low friction due to their very noticeable anisotropic or directional properties. The principal materials in this category are graphite, molybdenum disulphide, talc, mica, and boron nitride.

Graphite

The structure of graphite shows a series of parallel planes or sheets which are relatively far apart, as illustrated in Fig. 5.1. The carbon atoms in the individual planes are in hexagonal array, and they exhibit a strong inter-atomic bond strength

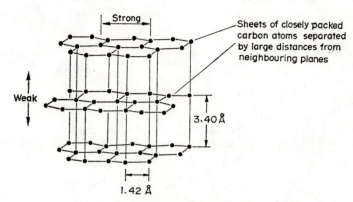

FIG. 5.1. The molecular structure of graphite.

(approximately 100 kcal/g-atom). By contrast, the bond strength between carbon atoms in different planes is relatively low (about a few kcal/g-atom). It is therefore not surprising to find that cleavage always occurs between sheets. The friction of steel-on-graphite or graphite-on-graphite in air is of the order $f_A \doteq 0.1$, and electron diffraction studies of the graphite surface after rubbing show that many of the basal planes of atoms acquire an angle of tilt of about 5° to the plane of rubbing, as shown in Fig. 5.2. Thus $\tan \theta = 0.1$, which is about the same value as the coefficient of friction itself.

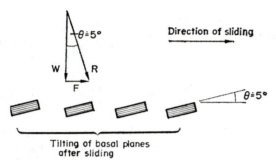

FIG. 5.2. Tilting of basal planes of polycrystalline graphite after sliding.

This implies that the basal planes after rubbing are normal to the resultant of the load and frictional force. This is referred to as compressional texture,[1] but the detailed process by which such a structure is achieved is not yet known. Graphite appears to be strong in compression (giving a small contact area) and weak in shear (providing a low interfacial shear strength) so that the coefficient of friction in air is understandably low.

If we now thoroughly outgas graphite by heating in a vacuum and then measure the friction at room temperature, we find that $f_A \rightarrow 0.5$ or 0.6. There appears to be a correspondingly dramatic increase in wear, although gross seizure does not occur because of the absence of

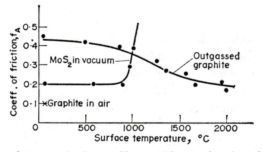

FIG. 5.3. Friction of outgassed polycrystalline graphite as a function of temperature.

junction growth (see Chapter 3). The high coefficient of friction obtained by outgassing decreases with increasing surface temperature, as seen in Fig. 5.3. Here, again, the reason for this phenomenon is not known. If surface films of oxygen or water vapour are introduced on clean or outgassed graphite, the friction and wear return to the normally low "atmospheric" values. Only minute amounts of O_2 or H_2O vapour are required to produce this effect, and they have been shown to cover only the edges of the graphite crystallites. Hydrogen and heptane vapours produce similar results.

The cleavage faces of graphite are low-energy surfaces, and this can be shown by the inability of water to wet or spread over them. On the other hand, the edges of graphite crystals react readily with oxygen or water vapour to produce a surface containing various oxygenated groups. In air, the surface friction will be small—irrespective of whether we have face–face, edge–face or edge–edge interactions, as shown in Fig. 5.4. In this case, the

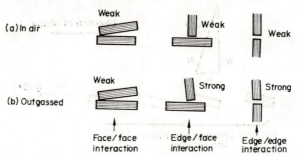

FIG. 5.4. Interaction of edges and cleavage faces in sliding of graphite-on-graphite.

edge faces have greatly reduced surface energy, so that there is no combination of the crystal faces which can give a large bond strength. After outgassing in a vacuum, the cleavage faces are scarcely affected, but the edge–edge and edge–face interactions are greatly increased. This mechanism accounts for the dramatic increase in friction and wear which occur when vacuum conditions are substituted for air, and it also indicates that we need only saturate the edges of the graphite crystallites with vapour to restore friction and wear to their normally low values.

Molybdenum Disulphide

Molybdenum disulphide (MoS_2) has a similar layered molecular structure to that of graphite. The individual sheets are either pure molybdenum or pure sulphur. Cleavage occurs readily between neighbouring planes of sulphur, since the bond strength between them is relatively low. For single crystals of MoS_2 in air, the coefficient of friction for cleavage faces on steel is about 0.1, whereas for crystal edges on steel it rises to the value 0.26. In contrast to the behaviour of graphite, the friction of molybdenum disulphide in a vacuum is little different from that in air[†] even at temperatures up to 800°C. This is shown clearly in Fig. 5.3. It is not practical to raise the temperature above 800°C since marked decomposition of the MoS_2 occurs. This in turn leaves solid molybdenum, which gives a very high friction as shown, approaching that of metals. Again, the friction mechanism for MoS_2 is very similar to that of graphite. The friction remains low even in a vacuum because temperature can never be increased to the point where the surface oxides are removed.

[†] $f_A \doteq 0.2$ for MoS_2 on MoS_2 in air.[(1)]

The low friction of molybdenum disulphide *in vacuo* has made it particularly appealing in space applications. Thus MoS_2 does not require the presence of gases and vapours to maintain a lubricity condition. It can withstand exceedingly high pressures when bonded to metal (as high as 30,000 bar) in cases where graphite and titanium disulphide have been known to fail. MoS_2 generally appears in one of three forms:

  (a) As a powder which is normally added to a carrier material.
  (b) As a grease mixture containing 40–70% MoS_2 in a semifluid consistency.
  (c) As an aerosol or dispersion spray.

Another useful technique is the inclusion of MoS_2 during the manufacture of sintered metal parts. It has been found that electric motor brushes made of an MoS_2–silver composite have a wear rate as low as a hundredth to a thousandth that of graphite brushes. The following bar charts in Fig. 5.5(a) and (b) compare the relative performance of molybdenum disulphide and other additives to a non-polar mineral oil at low sliding speeds. Of the three solids

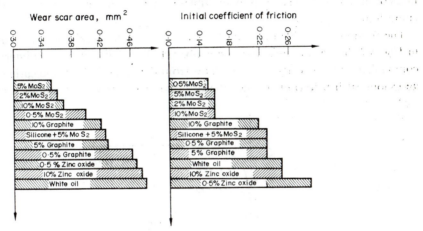

FIG. 5.5. Comparison of frictional and wear performance of MoS_2 and other additives to mineral oil.

used (MoS_2, zinc oxide, and graphite), only MoS_2 has the ability to suppress stick–slip behaviour. Furthermore, concentrations of graphite in excess of 5% are not as effective in reducing wear as MoS_2 in concentrations as low as 0.5%. It is little surprise that molybdenum disulphide has been acclaimed today as a wonder sliding material.

The applications of MoS_2 are diverse, ranging from the effective lubrication of heavy doors to sophisticated uses in the environment of space. When the first lunar spaceship carrying astronauts landed on the moon in July 1969, a molybdenum disulphide phenolic resin permitted ready extension of its adjustable legs at a temperature of –157°C. As a solid lubricant, MoS_2 has been successfully used in oil-free internal combustion engines, in the lubrication of roller chain drives, in heavy duty roll extrusion, and oil pipeline valves. Its performance is superior to that of graphite and teflon (PTFE), and it may be applied either by conventional methods (such as in powder, grease, or spray form) or by sputtering or plasma spraying.

Mica

Mica is a material which has long been known to have an extremely well-defined cleavage. The cleavage plane may be considered to consist of a matrix of silicone oxide in which are embedded polarized potassium atoms. When cleavage occurs, about one half of the potassium atoms remain in each surface, leaving a corresponding number of negative "holes" in the opposite face, as shown schematically in Fig. 5.6. The surface energy of the cleavage

FIG. 5.6. Schematic of cleavage plane in mica.

plane of mica is approximately 20 times higher than that of graphite. Indeed, the cleavage faces of mica have been shown by multiple-beam interferometry methods to be molecularly smooth over large areas.

The friction versus load relationship for crossed cylindrical mica surfaces shows clearly in Fig. 5.7 that Amonton's law is not obeyed. The sliding process is intermittent and accompanied by the tearing of small flakes from the mica surface, indicating a very strong adhesion. Such behaviour in the case of clean mica surfaces is modified after prolonged

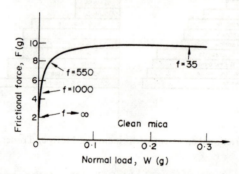

FIG. 5.7. Coefficient of friction of crossed mica shell surfaces as function of load.[1]

exposure to the atmosphere—in this case the friction falls by a factor of 10 or more. The strong adhesion between cleavage faces can be greatly reduced by adding talc to the sliding surfaces. The action of the talc in substituting a weak bonding of a van der Waals type across the cleavage planes in place of the existing high surface energy, is to reduce the coefficient of friction to the low value of about 0.1.

Boron Nitride

The structure of lamellar boron nitride on a molecular level resembles that of graphite in Fig. 4.1 except that the natural cleavage face contains two different elements. With graphite the cleavage face is all carbon, with MoS_2 it is all sulphur—but with boron nitride it consists of boron and nitrogen. Typical values of the coefficient of friction in air range from $f = 0.2$ to $f = 0.4$. Under vacuum conditions, the variation of f with temperature is surprisingly similar to that of outgassed graphite in Fig. 5.3.

It is known that graphite films in air provide good lubrication up to 500°C, whilst films of boron nitride give a low friction from 500°C up to almost 900°C. This suggests that a mixed film of graphite and boron nitride should provide effective lubrication up to 900°C. The results of superimposing layers of boron nitride and graphite on a base surface are shown in Fig. 5.8, and indeed we observe a reasonably constant and low coefficient of friction up to

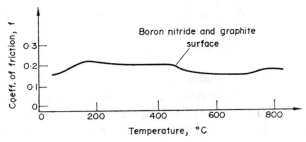

Fig. 5.8. Friction vs. temperature characteristic in air for platinum surface with superimposed layers of boron nitride and graphite. [1]

800°C. However, at the higher temperatures, only the boron nitride remains on the surface, since the graphite has oxidized. The low friction observed in Fig. 5.8 is therefore a single event phenomenon, and will not repeat as temperature is lowered from the higher values below 500°C.

Having dealt briefly with the molecular structure and frictional performance of some common lamellar solids, we will now deal briefly in the next section with the behaviour of polytetrafluoroethylene (teflon). PTFE is commonly classified as a polymeric material, and it has important desirable engineering properties which deserve emphasis. It is introduced here rather than in the previous Chapter 4 so that a comparison with molybdenum disulphide may be readily made.

5.2 Friction of Teflon (PTFE)

Polytetrafluoroethylene can be described as a relatively hard plastic material with an unusually low coefficient of friction. Table 5.1 compares the bulk yield pressure, bulk shear strength, surface shear strength, and range of friction coefficients of various polymers including PTFE. Also included in most cases is the contact angle[†] measured by placing a drop of water on the surfaces in question. We see that PTFE is unique in having a particularly low coefficient of friction in the range 0.05–0.10. It is also observed that, in contrast with the other polymeric materials, the surface shear strength s is substantially lower than

† Contact angle is related to surface energy and adhesion, as explained in Chapter 7.

TABLE 5.1. PROPERTIES OF PTFE AND OTHER POLYMERS

Polymer	Usual name	Bulk yield pressure $p(\text{kg/mm}^2)$	Bulk shear strength $\tau(\text{kg/mm}^2)$	Surface shear strength $s(\text{kg/mm}^2)$	Coefficient of friction	Contact angle
Polyvinyl chloride	PVC	15	5	7	0.4–0.5	—
Polystyrene	Polystyrene	20	4	8	0.4–0.5	—
Polymethyl methacrylate	Perspex	20	6	10	0.4–0.5	Zero
Nylon 66	Nylon	10	6	5	0.3	Zero
Polyethylene	Polythene	2	1.4	1	0.6–0.8	89°
Polytetrafluoroethylene	{Teflon {PTFE	2	2	0.4	0.05–0.1	126°

the bulk shear strength τ. The coefficient of friction remains low except as temperature approaches the softening point of teflon (*ca.* 320°C). As yet, no satisfactory explanation of this behaviour has been proposed. At low temperatures, the coefficient of friction rises to about 0.15 at -40°C as shown in Fig. 5.9. The ratio s/p is found to follow the same general

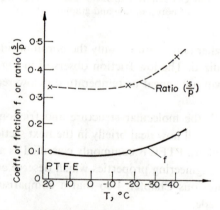

FIG. 5.9. Effect of temperature on coefficient of friction and (s/p) ratio for teflon.

trend as indicated by the dotted line in the figure. The curves in Fig. 5.9 show that there is qualitative (although not quantitative) agreement with the simple theory of adhesion for metals outlined in Chapter 3. At the same time, the variation in f with temperature shows clearly the viscoelastic nature of the frictional mechanism. The generally low coefficient of friction for PTFE can be explained by the predominance of anisotropic (directional) strength properties, so that certain long-chain molecules of the polymer network permit ready shearing in one direction and not another. Thus a mechanism not unlike that observed with lamellar solids (such as graphite) is responsible.

If PTFE is stretched or drawn, the molecular chains are aligned in the direction of extension and the structure becomes very highly oriented. Figure 5.10 shows the variation in the adhesional component of friction with applied load W for bulk teflon prestretched 280%

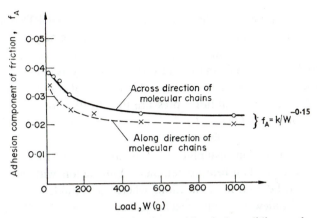

FIG. 5.10. Adhesion component of friction for steel hemisphere sliding on drawn PTFE.

at room temperature. The adhesion component f_A may be obtained by subtracting the hysteresis or deformation component f_H from the total coefficient f, following eqn. (4.6). We observe that f_A is slightly higher when sliding occurs across the oriented molecules than along them. Furthermore, both curves follow a mathematical relationship of the form $f_A = k/W^{0.15}$, with average values in the range $0.02 \rightarrow 0.03$.

Despite the very low friction of teflon up to about 300°C, it has four major defects as a practical bearing material:

(a) it is not strong enough mechanically;
(b) it is a poor conductor of heat;
(c) it has a high coefficient of thermal expansion; and
(d) at high speeds, $f \rightarrow 0.3$.

Accordingly, if PTFE were used directly as a bearing material, it would get hot, expand and stick. These difficulties can be overcome by *incorporating it in the surface of a porous metal (such as sintered copper). In this way, we obtain a material which has the bulk mechanical and thermal properties of copper—but has the surface frictional properties of teflon.* High temperatures are most commonly produced by high sliding speeds, and experiments have

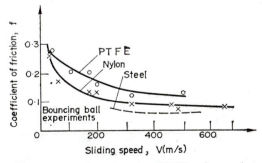

FIG. 5.11. High-speed friction of steel on PTFE, nylon, and steel.[1]

shown that rapid sliding destroys the smoothness of a polished PTFE surface, leaving a wear mark with a coarse texture. Figure 5.11 shows that at extremely high sliding speeds, the friction of PTFE-on-steel is greater than nylon-on-steel or steel-on-steel. Under these conditions, material is removed from the PTFE surface partly as a fine powder and partly in the form of fibrous debris. Excessive wear of this type is also obtained if a 100% PTFE surface is used under vacuum conditions (pressure 10^{-6} to 10^{-7} torr).

It appears that we must use fillers extensively with teflon to obtain long life and low friction. Usually copper, bronze, or glass-fibres are used as fillers in the proportion 15–20%. The addition of 5–15% molybdenum disulphide to glass-fibre-filled PTFE gives close to optimum frictional performance for ball-bearing retainers. Pure teflon with no additives will undoubtedly continue to be used in applications where appreciable frictional heating can be avoided, e.g. machine slideways, simple mechanical components, etc. With these limitations, its overall performance is comparable with that of MoS_2. When vacuum conditions apply, however, there is no known material which can give a performance comparable with that of molybdenum disulphide.

5.3 Friction and Wear of Diamond

Diamond is the hardest material known to man. It will indent and scratch the hardest metals and ceramics without suffering permanent deformation. Its yield pressure is in excess of 2000 kg/mm², and it has a Young's modulus approximately 4 or 5 times that of steel. The molecular structure of diamond shows two interpenetrating face-centred cubic lattices with a natural cleavage face along the octahedral or oblique planes. All the atoms are carbon apart from approximately 0.1–1% impurities. Diamond exhibits distinct anisotropic properties because of the directed nature of the carbon–carbon bonds. Thus the octahedral plane is considered "hard" because it is extremely difficult to polish, grind, or saw along this face, whereas other planes are easier to machine (these are termed "soft" directions). There appears to be a direct correlation between the ease of grinding and the coefficient of sliding friction, "hard" directions showing a much lower coefficient than "soft" directions.

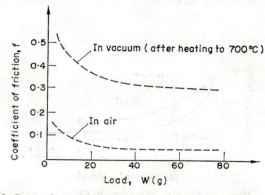

FIG. 5.12. Comparison of friction of diamond surfaces in air and vacuum.

The friction of a diamond stylus on an oblique diamond face is shown as a function of normal load in Fig. 5.12. The lower curve indicates the frictional behaviour in air, and we observe four characteristic features:

(a) The coefficient of friction is generally low ($f = 0.05$).
(b) The friction is practically the same whether the diamond surfaces are thoroughly cleaned and degreased or covered with mineral oil or fatty acid.
(c) The coefficient of friction rises as the load diminishes.
(d) The friction shows a marked dependence on orientation.

As an approximation we may write for the load range from 5–70 g

$$f = kW^{-1/3}. \tag{5.1}$$

This expression can also be deduced from Hertzian theory which appears valid for diamond-on-diamond contact. Since the actual area of contact is proportional to $W^{2/3}$ from eqn. (3.4), then the friction force F is given by

$$F = sA = skW^{2/3} \tag{5.2}$$

and the coefficient of friction $f = F/W$ then assumes the form shown in eqn. (5.1). The value of s for contaminated diamond surfaces is about 100 kg/mm².

Thorough cleaning of the diamond surfaces is perhaps not feasible, but much of the contaminant film can be removed by heating to about 700°C in a high vacuum. The upper curve in Fig. 5.12 shows that after such cleaning, there is a dramatic rise in the value of f. At the same time, the variation in frictional coefficient with load can still be described by an equation of the form shown in eqn. (5.1). The value of the shear strength s of the cleaned interface is approximately 1000 kg/mm², which is comparable with the bulk strength of diamond. At heavier loads than those shown in Fig. 5.12, the coefficient of friction rises with increasing load, thus reversing the trend which is evident at low loads. This is believed to be due to cracking of the diamond surface which increases the resistance to sliding.

We have so far considered the friction of diamond-on-diamond, and we must now concern ourselves with the friction generated between diamond and metal surfaces. Figure 5.13

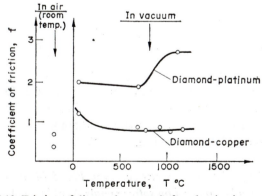

Fig. 5.13. Friction of diamond-on-metal after cleaning in a vacuum.

shows that the friction of diamond on clean platinum in a vacuum is extremely high ($f \doteq 2$ to 3) even at room temperature. This contrasts with the relatively low friction of diamond-on-diamond in Fig. 5.12.

The reason for the latter effect is that diamond is not ductile like a metal, so that junction growth cannot occur. On the other hand, metals flow plastically and permit much larger areas of contact with accompanying high coefficients of friction. The trends evident in Fig. 5.13 apply for a variety of *clean* metals; thus at room temperature

$$\begin{cases} f \doteq 0.4 \text{ to } 0.6 & \text{in air,} \\ f \doteq 1 \text{ to } 3 & \text{in vacuum,} \end{cases}$$

for diamond–metal sliding pairs.

Sliding velocity is a critical parameter in determining the friction and wear of metals on diamond. Figure 5.14 compares the friction of chromium, steel, and copper on diamond at speeds as high as 800 m/s. The high-speed apparatus described in Section 3.14 and shown schematically in **Fig. 3.16** was used to determine the curves in Fig. 5.14. The

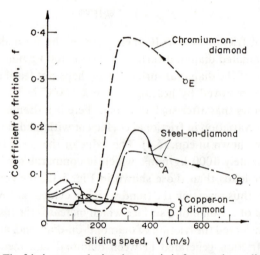

Fig. 5.14. The friction vs. velocity characteristic for metals-on-diamond.

points *A*, *B*, *C*, *D*, and *E* identify the initial speed of the rotating metal ball prior to deceleration by the frictional pads. The following trends are apparent from Fig. 5.14:

(a) All the curves shown exhibit a similar pattern. Thus the friction increases after the initial deceleration and reaches its peak value. As speed is further reduced, a sharp drop in *f* occurs, followed by small variations in friction in the low-speed range.

(b) The coefficient of friction and accompanying diamond wear are greater for metals with higher melting points, as seen from Table 5.2.

Indeed, *the characteristic friction vs. speed curves in Fig. 5.14 are of interest because they explain how diamonds can be polished by softer metals.* Within the high-speed, high-friction

TABLE 5.2. WEAR AND FRICTION OF DIAMOND BY METALS

Metal surface	Melting point (°C)	Diamond wear (10^{-9} g/1000 (m)	Friction coefficient (100–200 (m/s))
Copper	1083	1–2	0.05
Ball-bearing Steel	1500	30–60	0.08
Chromium	1615	200–500	0.10

range, the diamond is covered with a thin transferred film of metal due to large-scale melting of the latter; thus metal is met by metal leaving the diamond surface untouched. Below a critical sliding speed, however, the frictional heating is insufficient to cause large-scale melting of the metal. Instead, hot spots are formed on the diamond surface, the hot-spot temperature being limited by the melting point of the sliding metal. The diamond is transformed by high temperatures into amorphous carbon, the process being slow at 1000°C but very rapid above 1600°C. *Thus the polishing of diamond is due in large measure to a high-temperature change of carbon form (graphitization)*, and the effectiveness of the different metals which bring about diamond wear increases rapidly with their melting point.

One final word must be said about the use of diamond dust in polishing. Diamonds may be abraded at relatively low speed by polishing with porous cast-iron impregnated with diamond dust. Large directional effects may be observed during this process because of the existence of "hard" and "soft" directions as described earlier. This anisotropic behaviour can be explained by imagining diamonds to have a mosaic structure of little octahedral and tetrahedral blocks. During abrasion, some of these are knocked out of the diamond surface by the oncoming diamond dust. The directional effect then arises from the ease of displacing the elemental blocks in one direction rather than another.

5.4 The Adhesion of Ice

The adhesion of ice to metals appears to be quite distinct from its adhesion to polymers. When water is frozen on to clean metal surfaces, the interface is stronger than the ice itself and fracture occurs within the ice. Also, the presence of surface contaminants on metals reduces the adhesion appreciably. With polymeric materials, the interfacial forces are generally smaller than the cohesive forces within the bulk of the ice over a wide temperature range, so that the break occurs truly at the interface. For both metals and polymers, the actual adhesion recorded depends on temperature, as shown in Fig. 5.15. The dependence on temperature appears to disappear for polymers in the range $-15°$ to $-30°C$. The contact angle of water on the various surfaces indicates on a relative scale the adhesion strength of the interface with smaller contact angles corresponding to larger adhesion values, and vice versa.

From a practical viewpoint it makes little sense to change from one metal to another, since for all ice-to-metal combinations the interface is stronger than the bulk of the ice. Thus for ships in arctic waters or aircraft deicing surfaces, removal of the ice cannot be effected by overcoming the interfacial adhesion, but only by breaking the ice near the interface. If the ice is unconfined, it will tend to fracture in a brittle manner, this effect being more pronounced at lower temperatures. If, on the other hand, the ice is confined, it will tend to flow plastically; this effect requires greater forces to shear the ice according as the temperature is

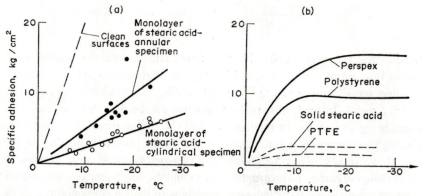

FIG. 5.15. The adhesion of ice to (a) stainless steel, and (b) various polymeric surfaces. [1]

lowered. Surface layers of polymeric materials appear to give a very low adhesion, particularly if they are hydrophobic. This is particularly true of PTFE.

The friction of various sliders (stainless steel, polystyrene, PTFE) on ice at low speeds of sliding shows similar trends to the adhesion measurements in Fig. 5.15. Figure 5.16 shows a general increase in the coefficient of friction as temperature is reduced, this effect being more pronounced for metals. Such results can be simply explained by the mechanism of

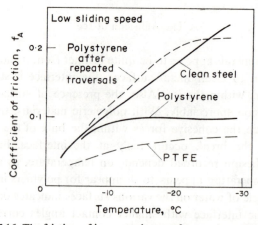

FIG. 5.16. The friction of ice on various surfaces at low sliding speeds.

pressure melting. There is evidence that the low friction is due to the formation of a thin surface film of water at the regions of contact, the actual contact area being only a small fraction of the apparent area. This is obviously easier to achieve at temperatures nearer 0°C, and progressively more difficult as temperature is lowered; hence the broad trends in Fig. 5.16. The polystyrene and PTFE surfaces have lower coefficients than steel at low temperatures because they are essentially hydrophobic or water-repellent. After repeated traversals with the polystyrene slider, the coefficient of friction increased to the value for clean stainless steel; this may be attributed to a diminution in the hydrophobic effect due to prolonged rubbing.

At high speeds of sliding, the friction of practically all materials falls to a very low value. This effect can be explained by *frictional melting.* Again, a thin surface film of water is formed at the sliding interface. Figure 5.17 shows the friction versus speed characteristic of

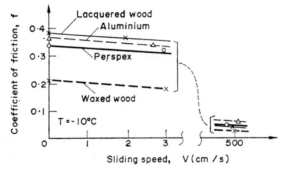

FIG. 5.17. Friction of miniature skis on ice as function of sliding speed.

miniature skis on snow. The skis are made of waxed wood, lacquered wood, perspex, and aluminium as shown. Whereas differences in the value of the coefficient of friction for the different materials appear to exist at low sliding speeds, these largely disappear at high speeds. Furthermore, the friction at high speeds is fractionally greater for a good thermal conductor because the frictional heat necessary to form the water film is more readily conducted away from the interface. At the higher speed of 5 m/s, we observe that the range of friction coefficients is from 0.02 to 0.04. This represents a tenfold drop for the aluminium ski.

One interesting feature of the characteristic friction versus speed curves in Fig. 5.17 is that if speed is reduced from the high value, a sudden increase in the friction occurs (this is not shown in the figure). The surfaces appear to seize together, and the adhesion became sufficiently great to break the measuring device.[1] The water film formed by frictional heating at the higher speed appears to freeze when speed is reduced, and the surfaces adhere firmly together. This behaviour was observed for all sliding materials except PTFE. In the latter case, the friction remained low and the sliding was smooth at all speeds. Further details of the friction of skis on ice are given in Part II.

5.5 Friction of Brittle Solids

There are a number of solids such as rocksalt, sulphur, lead sulphide, and sapphire which can be described as brittle. Unlike metals, these materials are not ductile—they crack and fragment at very small tensile strains. Friction experiments on rocksalt have shown that for a steel hemispherical slider, microscopic fragmentation of the rocksalt surface causes tilting of the rocksalt blocks relative to the original level.[1] Despite this fragmentation and minute cracking, however, the rocksalt surface has the typical grooved appearance characteristic of metals or other ductile materials. Thus the region of sliding is dominated by plastic flow, so that the overall frictional mechanism is very similar to that of metals. Large-scale junction growth, however, which is characteristic of clean metals, does not occur with brittle materials, and the coefficient of friction does not therefore exceed the value 0.7 to 1.0 even after vigorous cleaning. The friction of other brittle materials such as lead sulphide, lead selenide, sulphur, sapphire, and glass is similar to that of rocksalt.

5.6 The Friction of Wood

The physical structure of wood is surprisingly complex. Indeed, wood was one of the materials employed by Amontons in his classical experiments of 1699, and its internal composition was then believed to be relatively simple. In softwoods, about two-thirds of the woody material consists of polysaccharide and hemi-celluloses, the remainder being non-cellulosic polyaromatic lignin.[1] Wood has a remarkable affinity for water, which substantially affects its mechanical properties and subsequent frictional behaviour. Two stages can be identified in the absorption of water:

(a) Water is first absorbed on the surface of wood fibres and within the fibre walls. This process is called molecular sorption and accounts for approximately the first 30–40% by weight of water.
(b) The remainder is then absorbed into the bulk of the wood and held in capillary spaces.

Wood can be classified as a viscoelastic material, and its frictional performance can therefore be understood in terms of adhesion and hysteresis as developed in the previous chapter. When the total water content is less than 40% we speak of "dry" wood, whereas the term "wet" wood will refer to percentages in excess of 66%.

Figure 5.18 shows (a) the coefficient of friction as a function of normal load for hemispherical and steel sliders on wet extracted balsam wood, and (b) the corresponding coefficient of friction also as a function of load for wooden sliders on both steel and PTFE surfaces. The moisture content of the wood was maintained during the experiments by spraying water at intervals, and the friction equipment was kept in an atmosphere of 95% relative humidity at a temperature of 18°C. Small amounts of lubricating fatty acid esters present in the surface of the balsam wood were initially removed by treating the wooden specimens with ethanol and water; this procedure ensured that repeatable frictional coefficients would be obtained in the experiments.

For a given load the coefficient of friction of the steel and teflon sliders on the wood depended on the diameter D of the sliders used as shown. This is most clearly illustrated by crossing sliders of extreme diameter—1.5 and 25 mm respectively. We observe from Fig. 5.18(a) that the coefficient of friction f increases with load, the rate of increase being greater for the smaller diameter slider. The results in this figure suggest a hysteresis mechanism of friction exists, due to the time-dependent deformation of the relatively soft balsam track by

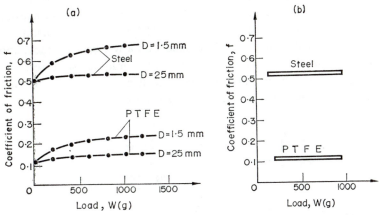

FIG. 5.18. Friction of (a) steel and PTFE sliders on wet wood and (b) of wooden sliders on steel and PTFE.

the harder sliders. To eliminate this effect, the materials of slider and track were interchanged. The results in Fig. 5.18(b) show that for a curved balsam slider on both steel and teflon smooth tracks, the coefficient of friction is now independent of load and slider diameter. With this arrangement it is clear that no grooving of the flat surface (whether steel or teflon) can occur, and therefore the hysteresis contribution to friction is zero. *We conclude that the frictional coefficient in Fig. 5.18(a) includes both adhesional and hysteresis terms, whereas that in Fig. 5.18(b) is due to the adhesional mechanism only.*

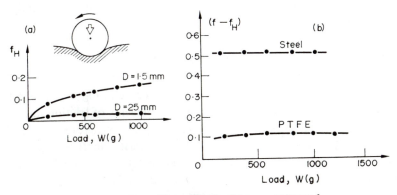

FIG. 5.19. The rolling of hard spheres on wet wood.

This result can be confirmed by an additional ingenious experiment in which a hard sphere rolls over the wood at the same loads as those used in the first set of sliding experiments described above. It has been shown clearly elsewhere that during rolling there is practically no contribution to friction from interfacial adhesion,[23] and the total frictional force measured is due solely to hysteresis, provided the sliding speed is reasonably low.

Figure 5.19(a) shows the measured hysteresis friction coefficient obtained in the rolling experiments. If the ordinates of these curves are subtracted at each load from the corresponding ordinates in Fig. 5.18(a), the result appears as in Fig. 5.19(b). The coefficient of friction $(f-f_H)$ obtained in this manner is due almost entirely to adhesion, and we observe the remarkable similarity between Fig. 5.18(b) and Fig. 5.19(b). We are now reasonably certain that the friction of wet wood on smooth steel or PTFE is due solely to adhesion, and that for the inverse experiment of steel or PTFE sliders on a wet balsam track, approximately 20% of the total measured friction is due to deformation or hysteresis losses.

The coefficient of friction on balsam wood also depends critically on the percentage moisture content, as shown in Fig. 5.20. Sliding friction experiments using a PTFE ball

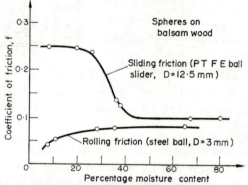

FIG. 5.20. Variation of rolling and sliding friction for spheres on balsam wood as function of moisture content.

slider show a distinct drop in f from an initial value of 0.25 on "dry" wood to 0.10 on wet wood. Since most of the coefficient of friction in the sliding experiments is due to adhesion, we conclude that *adhesion decreases with increasing moisture content*. The results of rolling friction experiments using a steel ball are also shown in Fig. 5.20. Since rolling friction is due almost entirely to hysteresis, we conclude that *hysteresis increases with increasing moisture content*. The curves in Fig. 5.20 apply for different slider materials and diameters, and they cannot be superimposed or subtracted for any purpose. We observe, however, that the adhesion contribution to the upper curve must decrease with increasing moisture content even more dramatically than the total friction curve in Fig. 5.20 suggests. It is also to be noted that whereas the hysteresis losses and coefficients of friction must be identical for both PTFE and steel spheres of the same diameter (since hysteresis is determined by the bulk properties of the wood), the adhesion term by contrast is markedly dependent on surface conditions at the sliding interface. Permanent rather than viscoelastic deformations are produced in both rolling and sliding over dry wood ($< 30\%$ moisture content). Microscopic examination of the deformed fibres shows that the central voids are completely closed from the action of the normal load. The addition of water to the deformed fibres causes them to return instantaneously to their original shape.

The friction characteristics of lignum vitae deserve mention briefly. This is an extremely hard and dense wood having a complex molecular structure and containing a large amount of natural waxes. These waxes are extremely difficult to remove, so that the coefficient of friction is low (about 0.1). This property, in fact, had made lignum vitae a popular bearing material before the advent of solid lubricants and even liquid lubrication. The friction value remains low after repeated traversals because the natural waxes continue to be exuded at the sliding interface. Removal of the wax material causes an immediate rise in friction to the value $f \doteq 0.45$. In contrast to balsam, lignum vitae is very little affected by moisture. Even after prolonged immersion in water, its friction and hardness are scarcely changed. It is therefore still popular as a marine bearing material.

We conclude this section by observing in summary that although the molecular structure of wood is relatively complex and its physical properties are directional or non-isotropic, the frictional mechanism is relatively simple. This arises in general from adhesion and deformation at the areas of real contact. Although the importance of wood as an engineering material has declined in the last few decades with the introduction of new synthetic and plastic materials, there are still applications where its properties and performance are the most suitable to be found. Thus wood offers the best long-life, high-friction performance as a brake-shoe material on steel wheels in railway carriages and rapid-transit systems. When impregnated with a suitable lubricant, wooden brake blocks ensure a smooth, vibration-free, high-friction braking action on metallic rims.

5.7 Friction of Yarns and Fibres

Yarns and fibres can be regarded as individual elements or strands within a bulk fibrous material, and a knowledge of their frictional performance is essential in the weaving, hosiery, or cloth fabrication industries. The friction between crossed fibres or cylinders of polymer is commonly measured over a wide load range using a crossed-cylinder apparatus, as described in Chapter 11. This apparatus consists of a lower cylinder or fibre held in a fixture and capable of travelling at a speed of about 0.2 cm/min in a direction coincident with the longitudinal axis of the fibre. An upper fibre at right angles to the first is pressed against the latter and therefore flexed in a vertical plane. When the lower fibre is set in motion, the upper fibre is dragged along with it and the horizontal deflection is a measure of the frictional force. In general, the motion is intermittent, and in any one run a large number of stick–slip cycles can be observed. The upper fibre is, in fact, a cantilever, and fibres of different thicknesses can be used to extend the load range from about 10^{-6} to 10 g. Most of the fibres (except PTFE) are in the "undrawn" state, and their frictional properties can therefore be compared directly with those of the bulk polymer.

The results for PTFE, nylon, and polythene appear to obey the following friction versus load relationship:

$$f = K'/W^\beta,$$

where $0.2 < \beta < 0.3$. Their friction versus sliding speed and friction versus temperature characteristics are distinctly viscoelastic since they can be classified as polymeric materials. Also, the friction is markedly dependent on the fibre diameter, and electrostatic charging and adhesion are of negligible proportions.

The range of materials covered in this chapter is necessarily limited, and only those with important engineering applications have therefore been included. Most materials can fortunately be classified as metals or elastomers, and the mechanism of friction in each case has been dealt with in previous chapters. Some important materials (such as nylon) will receive treatment later when the lubrication section of Part I is being presented. Having discussed in some detail the mechanism of dry friction in the last three chapters, we must now consider what complications arise from introducing a lubricant at the sliding interface.

CHAPTER 6

HYDRODYNAMIC LUBRICATION

6.1 Brief History

The art of lubrication is at least as ancient as the greasing of Egyptian and Roman chariot wheels and axles with tallow. Until recent times, the slideways for launching ships were greased with animal fats, and lard oil was used widely in machining operations as a cutting fluid. Castor oil was used as a lubricant in early airplane engines, and sperm oil was highly valued by clockmakers. Thus by making use of the natural oiliness characteristics of animal and vegetable oils, an empirical art for minimizing solid friction and lubricating primitive machinery has evolved over the years.

The first quantitative investigations of hydrodynamic behaviour were carried out towards the end of the nineteenth century. As early as 1874, Stefan in Germany investigated the squeeze behaviour of a lubricant confined between approaching plane surfaces.[24] The original work on lubricating films between sliding surfaces appears to have originated a few years later independently in England and Russia. Beauchamp Tower conducted experiments in England to determine the friction of the journal of a railway-car wheel, and observed the generation of hydrodynamic pressure in the first partial journal bearing.[25] The pressures generated even in these early experiments of 1885 were sufficiently large to displace plugs or stoppers placed in oil holes in the bearing casing. Osborne Reynolds became intrigued with Tower's results and showed that they could be well explained by applying the principles of fluid mechanics. In 1886 Reynolds published his classical theory of hydrodynamic lubrication, which is widely used in the design of modern machinery.[26]

About the same time in Russia, N. Petroff was independently experimenting with bearings lubricated with mineral oil, and showed how the friction of a bearing could be entirely accounted for by shearing forces in the lubricant film. Petroff applied Newton's law of viscosity (see page 108) to a rotating concentric journal, and expressed frictional shear in terms of lubricant viscosity, speed, and physical dimensions. The Petroff equation is still applied today in the case of lightly loaded journal bearings.

Reynolds's theory of the journal bearing was subsequently developed further by Sommerfeld in Germany in 1904. The Sommerfeld number is widely recognized today as the most significant design parameter for bearings. Later, Michell in Australia and Kingsbury in the United States independently extended Reynolds's theory of hydrodynamic pressure generation in a converging fluid wedge to the design of thrust bearings. So many other contributions to the state of the art have been made during the last three decades that it would require a separate volume to list the various authors. The general trend, however, appears to be the gradual elimination of many of the original assumptions due to Reynolds.

6.2 The Generalized Reynolds Equation

The introduction of a fluid film between two moving and rough surfaces provides a lubricating effect which substantially reduces the coefficient of friction between them. Such films are usually sufficiently thin that viscous forces are large in comparison with inertia forces, so that the latter may be neglected. The Navier–Stokes equation in fluid mechanics can be shown to reduce to the Reynolds equation in lubrication theory when inertia effects are negligible.[26] In its most general form, the Reynolds equation has the following form:

$$\frac{\partial}{\partial x}\left(\frac{\varrho h^3}{\mu}\frac{\partial p}{\partial x}\right) + \frac{\partial}{\partial y}\left(\frac{\varrho h^3}{\mu}\frac{\partial p}{\partial y}\right)$$

$$\equiv \underbrace{6(U_1 - U_2)\frac{\partial(\varrho h)}{\partial x}}_{\text{Wedge term}} + \underbrace{6\varrho h\frac{\partial}{\partial x}(U_1 + U_2)}_{\text{Stretch term}} + 12\overbrace{\frac{\partial(\varrho h)}{\partial t}}^{\text{Squeeze term}}, \qquad (6.1)$$

where x, y are mutually perpendicular coordinates within the fluid, p is the local pressure within the film, ϱ and μ are the density and absolute viscosity of the lubricant, h is the film thickness, and U_1 and U_2 are the velocities of sliding of upper and lower surfaces respectively, as shown in Fig. 6.1. The physical significance of eqn. (6.1) is that the generation of

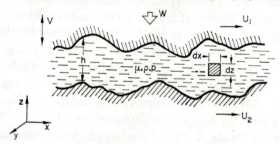

FIG. 6.1. The general case of lubricated sliding.

hydrodynamic pressure within the film is a function of three terms which have been called the wedge, stretch, and squeeze contributions to load support.

The wedge term appears to be the most important of the three terms, particularly with regard to journal and pivoted-pad bearing design and other common applications. It can be separated into two principal components, one due to density variation in the x-direction and the second as a consequence of changes in h with increasing x thus:

$$\text{Wedge term} = 6h(U_1 - U_2)\frac{\partial \varrho}{\partial x} + 6\varrho(U_1 - U_2)\frac{\partial h}{\partial x}. \qquad (6.2)$$

For liquids, density variations are small or negligible, so that the first component disappears and ϱ can be eliminated entirely from eqn. (6.1). Gas bearings, on the other hand, require that eqn. (6.1) contain the density term as shown. Most lubrication examples assume smooth and rigid bounding surfaces for the lubricant, and in such cases the stretch term has no significance. If one or both of the interacting surfaces are elastomeric in nature or exhibit flexible properties (perhaps because of a solid lubricant lining), then relative stretching of the surface material causes an additional pressure increment because of variations in either U_1 or U_2 or both with x, as seen from eqn. (6.1). The magnitude of the stretch effect, however, is small compared with the wedge or squeeze terms. The squeeze contribution is due to normal approach of the surfaces, whether or not sliding motion occurs between them. This effect can be caused by impact, slow-speed squeezing action, or more commonly relative normal vibration between the surfaces.

Since the horizontal relative velocity of both surfaces is sufficient and not the absolute magnitude of each velocity component, it is convenient to consider $U_2 = 0$ and to allow the upper surface to move with velocity U. Also, although viscosity variations occur in the lubricant film, we commonly assume an average viscosity $\bar{\mu}$ in any application[†], where

$$\bar{\mu} = \frac{1}{L^2} \int_0^L \int_0^L \mu \, dx \, dy. \tag{6.3}$$

With these substitutions, the generalized Reynolds equation takes the following form:

$$\frac{\partial}{\partial x}\left(h^3 \frac{\partial p}{\partial x}\right) + \frac{\partial}{\partial y}\left(h^3 \frac{\partial p}{\partial y}\right) \equiv 6\bar{\mu} U \frac{\partial h}{\partial x} + 6\bar{\mu} h \frac{\partial U}{\partial x} + 12\bar{\mu} V. \tag{6.4}$$

In the absence of squeeze motion, the last term on the right-hand side of eqn. (6.4) disappears. If the stretch term $6\bar{\mu}h(\partial U/\partial x)$ is also neglected, the only remaining source of hydrodynamic pressure generation is the wedge term, due to consistent variations in film thickness in the x-direction. As indicated earlier, this term is responsible for the load support feature of most conventional bearings (Michell, journal, pivoted-pad, and slipper bearings). The specific design and operating principles of these and other bearing types are dealt with in detail in Chapter 15.

The assumptions underlying the theory of hydrodynamic lubrication as originally proposed by Reynolds[26] in 1886, are as follows:

(a) Neglect of gravitation and inertia terms.
(b) Assumption of Newtonian[‡] behaviour.
(c) Constant viscosity.
(d) Incompressibility of fluid.
(e) Film thickness small compared with other dimensions.
(f) Zero slip at liquid–solid or gas–solid boundaries.
(g) No surface tension effects.

---

† The notation $\bar{\mu}$ will be continued in this section, but subsequently the more usual symbol μ is used.
‡ See Section 6.3.

The (f) and (g) assumptions are usual ones in fluid mechanics, and the neglect of inertia in the first assumption is made possible by (e). As a consequence of the latter, velocity components in the z-direction may be neglected. The conditions of isoviscosity, incompressibility, and Newtonian behaviour have been removed by subsequent refinements and extensions to the original theory.

6.3 Velocity Distribution

Consider an element of fluid in the film between solid boundaries as shown in Fig. 6.2,

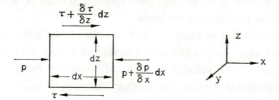

FIG. 6.2. Equilibrium of a fluid element in the lubricant film.

and let a balance of pressure and viscous forces be written for the x-direction. Thus

$$\left(-\frac{\partial p}{\partial x}\,\mathrm{d}x\right)\mathrm{d}z + \left(\frac{\partial \tau}{\partial z}\,\mathrm{d}z\right)\mathrm{d}x = 0,$$

where dy is the thickness or width of the element and τ the shear stress. Since $p = p(x)$ only and $\tau = \tau(z)$ only, this equation becomes

$$\frac{\mathrm{d}\tau}{\mathrm{d}z} = \frac{\mathrm{d}p}{\mathrm{d}x}.$$

Substituting in place of τ the condition of Newtonian behaviour[†]

$$\tau = \mu\frac{\mathrm{d}u}{\mathrm{d}z}$$

we obtain the relationship

$$\frac{\mathrm{d}^2u}{\mathrm{d}z^2} = \frac{1}{\mu}\left(\frac{\mathrm{d}p}{\mathrm{d}x}\right), \tag{6.5}$$

which can be integrated twice as follows to obtain the x-component of velocity for any point in the lubricant film

$$u = \frac{1}{2\mu}\left(\frac{\mathrm{d}p}{\mathrm{d}x}\right)z^2 + C_1z + C_2.$$

[†] A Newtonian liquid is one wherein stress is proportional to rate of strain. This condition is exactly analogous to Hooke's law for solid or elastic materials, which states that stress is proportional to strain. Thus Newtonian liquids and Hookean solids show a simple proportionality between applied stress and resulting deformation (or its rate of change).

Here, C_1 and C_2 are constants obtained from the boundary conditions. At $z = 0$, $u = U_2$ and at $z = h$, $u = U_1$ (see Fig. 6.1), so that

$$C_1 = \frac{U_1}{h} - \frac{1}{2\mu}\left(\frac{dp}{dx}\right)h \quad \text{and} \quad C_2 = U_2.$$

Thus we may write:

$$u = \underbrace{\frac{1}{2\mu}\left(\frac{dp}{dx}\right)z(z-h)} + \underbrace{U_2\left(\frac{h-z}{h}\right) + \frac{z}{h}U_1}. \tag{6.6}$$

Hydraulic flow com-   Shear flow component
ponent of velocity   of velocity

Similarly, in the transverse or y-direction, the velocity of flow v is given by

$$v = \frac{1}{2\mu}\left(\frac{dp}{dy}\right)z(z-h), \tag{6.7}$$

where (dp/dy) is the pressure gradient in the y-direction.

For the particular case of a thrust slider moving at speed U relative to a fixed base (Fig. 6.3), we may sketch the velocity distribution according to eqn. (6.6) at three points a, b and c, where the pressure gradient is positive, zero, and negative respectively. It is apparent

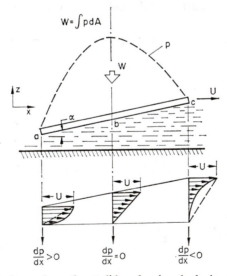

FIG. 6.3. Velocity distribution under a thrust slider, showing the hydraulic and shear components.

from these distributions that the hydraulic and shear flow components reinforce one another from a to b and that they oppose each other algebraically in the region b to c. This effect is necessary from the continuity relationship which requires that the flow past any section is constant in time. For an incompressible liquid it is therefore apparent that the hydraulic and shear components of flow should be algebraically additive at the smaller cross-section a and subtractive at c.

Let us now suppose that we wish to support a heavier load W than Fig. 6.3 suggests. If the speed U is to remain unchanged, we must obviously increase the angle of inclination α of the slider. At the point a in the figure, it is apparent that dp/dx has increased; thus there is an increase in the hydraulic flow component of velocity and perhaps a decrease in the shear component. At the point b, conditions will remain largely unchanged, but at the point c the increase in the shear contribution (due to the larger area of flow at this point) is compensated for by an increase in the negative hydraulic component. The increase in W may be sufficient to create a zero or even negative velocity of flow at the fixed lower surface at location c, thus initiating a reverse flow or separation. In the absence of strong viscosity effects, this might cause eddy and vortex formation.

6.4 Mechanism of Load Support

It is well to tabulate the various causes of load support due to hydrodynamic lubrication. and these are classified according to whether the interacting surfaces are rough or smooth,

Smooth Surfaces

Table 6.1 shows how the wedge, stretch, and squeeze terms contribute separately to pres-

TABLE 6.1. SOURCES OF LOAD SUPPORT USING SMOOTH SURFACES

Description of system[a]	Schematic representation	Hydrodynamic equation
1. *Plane, smooth, rigid, inclined* surfaces. No vertical motion		*Wedge term:* $$\frac{d}{dx}\left(h^3\frac{dp}{dx}\right) = 6\mu U \frac{dh}{dx}$$
2. *Plane, smooth, parallel* surfaces. Lower surface rigid and fixed. Upper surface flexible and held at one end		*Stretch term:* $$\frac{d}{dx}\left(h^3\frac{dp}{dx}\right) = 6h\mu \frac{dU}{dx}$$
3. *Plane, smooth, parallel, rigid* surfaces. No side motion, lower surface fixed. Upper surface reciprocates vertically		*Squeeze term:* $$\frac{d}{dx}\left(h^3\frac{dp}{dx}\right) = 12\mu V$$

[a] Incompressible, isoviscous liquid. Two-dimensional models.

sure generation and hence load support. Practical examples may incorporate these terms individually or in combinations. Thus a sliding tyre obviously exhibits the stretch term, whereas a plane slider bearing with one or more elastomeric surfaces combines the wedge and stretch effects. As will be seen later, a journal bearing may combine the wedge and squeeze contributions to load support. The list of such examples is lengthy.

Rough Surfaces

By introducing the concept of surface roughness, at least four additional contributions to load support on a macroscopic scale become apparent. These are listed and described in Table 6.2. The directional effect is obvious, and gives rise to positive pressure increments which exceed the contribution from negative pressures, giving a net load support. This type of surface effect may be deliberately introduced in machining operations to assist in lubricating shafts and spindles. More commonly, the directional effect is caused by non-uniform

TABLE 6.2. SOURCES OF LOAD SUPPORT USING ROUGH SURFACES

Classification of support mechanism	Remarks	Schematic representation
1. *Directional effect*	Directional parameter Z_7 negative. *Rigid, parallel* surfaces, lower surface fixed	
2. *Macro-elasto-hydrodynamic*	Upper surface *plane, smooth* and *rigid*. Lower surface *flexible* with sinusoidal or symmetrical roughness. Elastohydrodynamic distortion produces net load support	
3. *Cavitation*	*Rigid, parallel* surfaces. Upper surface *smooth*, lower has sinusoidal roughness. Cavitation destroys negative pressures, giving net load support	
4. *Viscosity effects*	For sinusoidal roughness in lower surface (same conditions as 3 above), pressure effect increases viscosity and load support, temperature has opposite effect	In all cases: $$W \equiv \sum p_{\text{pos.}} \, \delta A - \sum p_{\text{neg.}} \, \delta A$$

wear in a preferred direction. One typical example is pavement wear at the approach to a traffic signal, stop sign, or similar speed-reducing obstacle. The predominantly braking mode of vehicle and tyre behaviour creates eventually a positive directional effect Z_7 (see eqn. (2.10)) in the pavement profile. In wet weather this will produce a small, negative load support, assuming that brake application has caused the tyres to slide relative to the roadway. This is, of course, beneficial in this instance by improving traction.

The macro-elastohydrodynamic effect is most important and will be dealt with in detail in Chapter 8. The generation of pressure on the positive slopes of the flexible asperities tends to push down and distort the asperities, while the negative pressures on the down slopes produce a distortion of the pull or suction type. This is illustrated in Fig. 6.4 for the case of

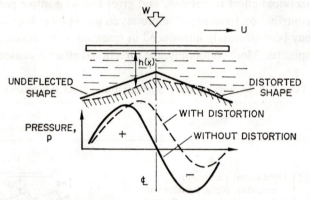

FIG. 6.4. Macro-elastohydrodynamic distortion of flexible sawtooth asperity.

a simplistic sawtooth flexible asperity shape which is covered by a lubricant over which a plane loaded surface slides as shown. In the absence of distortion, the undeflected asperity shape produces approximately a sinusoidal pressure distribution, so that the net load support is zero. On the other hand, a flexible asperity is distorted in such a manner that an effective directional parameter is introduced, and the resulting pressure distribution as shown gives a net uplifting force W.

Cavitation effects[†] destroy the contribution of pressure to negative load support, since a liquid can sustain only a modest state of tension before bubble formation or "striation" occurs. In this case, only the positive pressure increments shown in Table 6.2 contribute to load support, so that a net uplifting force is created. The absolute viscosity of the lubricant is frequently far from constant, and varies with both temperature and pressure according to the following relationships:

$$\mu = K/T^{\beta} \qquad (1.5 \leqslant \beta \leqslant 3.0),$$
$$\mu = \mu_0 \, e^{mp} \tag{6.8}$$

where K is a constant. Since m has a very small value,[‡] pressures of the order of hundreds of atmospheres are required to significantly alter the value of μ. In contrast to this, viscosity

---

[†] See Section 6.7.
[‡] $m = 10^{-8}$ m²/kg for most oils and lubricants.

is extremely sensitive to temperature change. It can be shown that the pressure-dependence of μ produces an additional load support, however minute, and the temperature sensitivity has the opposite effect.

6.5 Squeeze Films

The term squeeze film applies to the case of approaching surfaces (usually planar) which attempt to displace a viscous liquid between them. For an incompressible fluid, we neglect the wedge and stretch terms in eqn. (6.4) and write

$$\nabla^2 p = \frac{12\mu}{h^3}\left(\frac{\mathrm{d}h}{\mathrm{d}t}\right),$$ (6.9)

where h is the squeeze-film thickness and a function of time only, and the speed of descent of the upper surface[†] V is written as $(\mathrm{d}h/\mathrm{d}t)$. Figure 6.5 shows the general case of parallel

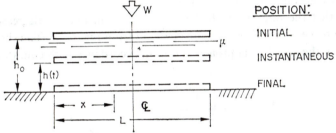

FIG. 6.5. Parallel squeeze-film effect: smooth surfaces.

sinkage of a flat plate on a smooth surface for which eqn. (6.9) applies. Since the load W must obviously equal the pressure integral taken over the plate area, we may write

$$W = \int p \, \mathrm{d}A.$$ (6.10)

By combining eqns. (6.9) and (6.10), the following basic squeeze-film equation is obtained:

$$t = \frac{K\mu L_T^4}{W}\left(\frac{1}{h^2} - \frac{1}{h_0^2}\right),$$ (6.11)

where h_0 is the initial squeeze-film thickness, L_T a typical length dimension of the plate, t the time taken for the film thickness to decrease to the value h, and K a constant[‡] determined by the shape of the plate.[(2)] For a circular plate of diameter D, $L_T = D$ and $K = 3\pi/64$. Taking the time derivative of eqn. (6.11),

$$\frac{\mathrm{d}h}{\mathrm{d}t} = -\frac{Wh^3}{2K\mu L_T^4},$$ (6.12)

[†] The upper descending surface when planar is frequently called a plate.
[‡] See Table A in Appendix.

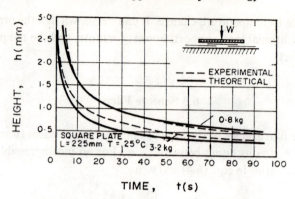

FIG. 6.6. Height vs. time relationship for squeeze film.

which shows that the speed of descent of the upper surface is proportional to the cube of film thickness. Figure 6.6 shows a plot of eqn. (6.11) for the case of a flat, square plate squeezing through a viscous oil on to a smooth base surface.

We observe that the rate at which the squeeze-film thickness h decreases with time is slower as time proceeds. This is even more apparent from Fig. 6.7, which shows the proportionality between the speed of approach and h^3 in accordance with eqn. (6.12).

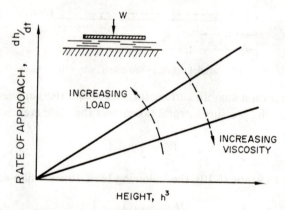

FIG. 6.7. Speed of descent as function of film thickness in parallel squeeze film.

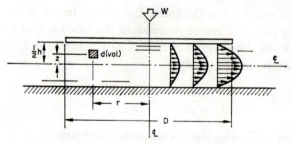

FIG. 6.8. Coordinate system for circular plate.

Velocity Profile

The velocity distribution in the squeeze film between a flat, round plate and a smooth base can be derived as follows. Consider the coordinate system shown in Fig. 6.8, and in particular the equilibrium of a small element dv within the film at radius r and height h above the instantaneous horizontal centre line. The pressure and viscous forces acting on the faces

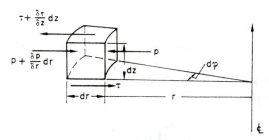

FIG. 6.9. An element dv within the squeeze film.

of the element are shown in Fig. 6.9. We note that from symmetry conditions there can be no viscous forces acting on the $dr\,dz$ faces.

By summing and equating to zero all radial forces acting on d (vol.),

$$\frac{dp}{dr} = \frac{\partial \tau}{\partial z} = \mu\,\frac{\partial^2 v}{\partial z^2} \qquad (6.13)$$

where v is the local velocity of the lubricant in the r-direction, and $\tau = \mu\,\partial v/\partial z$ from Newton's viscosity law. By integrating eqn. (6.13) twice to find v, and using the boundary conditions that $v = 0$ at $z = \pm\frac{1}{2}h$,

$$v = -\frac{h^2}{8\mu}\left(\frac{dp}{dr}\right)\left(1 - \frac{4z^2}{h^2}\right). \qquad (6.14)$$

The maximum velocity v_0 is obtained by putting $z = 0$ in this equation, and the mean velocity $\bar{v}$ can then be shown to be $2v_0/3$ thus:

$$\bar{v} = \frac{2}{3}v_0 = -\frac{h^2}{12\mu}\left(\frac{dp}{dr}\right). \qquad (6.15)$$

By now applying the continuity equation to the squeezing process, we obtain another expression for mean velocity:

$$\underbrace{\pi r^2\,dh}_{\substack{\text{Volume of liquid swept}\\\text{out in time } dt \text{ by}\\\text{descending plate}}} \equiv \underbrace{(2\pi rh)\,\bar{v}\,dt}_{\substack{\text{Volume of liquid passing}\\\text{through periphery of squeeze}\\\text{film in time } dt}}$$

$$\bar{v} = \frac{r}{2h}\left(\frac{dh}{dt}\right). \qquad (6.16)$$

Equating the right-hand sides of eqns. (6.15) and (6.16) and substituting from eqn. (6.12) with $K = 3\pi/64$ gives

$$\frac{dp}{dr} = -\frac{64Wr}{\pi D^4} \tag{6.17}$$

and substitution of dp/dr from eqn. (6.17) into eqn. (6.14) gives the final velocity distribution

$$v = \frac{8Wrh^2}{\pi\mu D^4}\left(1 - \frac{4z^2}{h^2}\right). \tag{6.18}$$

We note that v is a function of *three* coordinates, r, h, and z, of which only h is time-dependent. This equation shows clearly that whereas at any fixed height h and radius r the velocity distribution is parabolic over the cross-section of flow, its mean value at a fixed height h increases linearly with r.

Impact and Dynamic Loading

In place of the slow, natural descent of the plate through the squeeze film as depicted in Fig. 6.5, consider now that the plate (assumed circular and plane) impinges suddenly upon the squeeze film under the action of a load W. At the instant of impact, the liquid is suddenly put in motion and there is a corresponding impulsive pressure throughout the liquid which decelerates the plate. Let V_i be the plate velocity immediately before and V_0 the velocity immediately after impact. It has been shown clearly[1] that $V_0 \geqslant 0.97V_i$, so that the liquid surface starts moving with a velocity substantially equal to the initial velocity V_i of the plate.

For dynamically loaded squeeze films it is obviously incorrect to equate the integral of the pressure in the film over the plate area to the load. The velocity of the hammer is progressively reduced as it penetrates through the film, and in place of eqn. (6.10) we equate the loss of kinetic energy of the plate to the work expended against the viscous resistance in the liquid:

$$\underbrace{\frac{W}{2g}\,[V^2-(V-dV)^2]}_{\substack{\text{Loss of kinetic}\\ \text{energy of plate}}} = -dh\underbrace{\int_0^{1/2D} p\,2\pi r\,dr.}_{\substack{\text{Work against}\\ \text{viscous forces}}} \tag{6.19}$$

If we regard the plate and liquid as a single system, there is no potential energy change during the instant dt in which the changes dV and dh occur. By simplifying the left-hand side of eqn. (6.19) and dividing both sides by $V = dh/dt$, we find

$$\underbrace{\frac{W}{g}\,dV}_{\substack{\text{Change in momentum}\\ \text{of plate}}} = dt\underbrace{\int_0^{1/2D} p\,2\pi r\,dr,}_{\substack{\text{Impulse created}\\ \text{in liquid}}} \tag{6.20}$$

which shows that considerations of momentum yield the same results as energy change.

The equation which shows a balance between pressure and viscous forces in the liquid is equally valid for static or dynamic loading, so that to find p it is convenient to equate the right-hand sides of eqns. (6.15) and (6.16) remembering that $p = 0$ at $r = \pm D/2$ and $V = dh/dt$:

$$p = \frac{3\mu D^2 V}{4h^3}\left(1 - \frac{4r^2}{D^2}\right) \tag{6.21}$$

By substituting this value of p in eqn. (6.19) and integrating,

$$\frac{W}{g}\,dV = \frac{3\pi\mu D^4}{32}\left(\frac{dh}{h^3}\right),$$

which upon integration gives

$$V = V_0 - \frac{3\pi\mu D^4 g}{64W}\left(\frac{1}{h^2} - \frac{1}{h_0^2}\right), \tag{6.22}$$

where at $h = h_0$ (the initial film thickness), $V = V_0$. We note that in deriving this equation, eqn. (6.20) cannot be used in place of (6.19) since $V = dh/dt$ remains unknown and the solution for V appears as a function of time rather than h. In most cases, $V \to 0$ at a value of h which is small compared with h_0, so that $1/h_0^2$ is negligible in comparison with $1/h^2$. The plate is therefore brought to rest at a height given by

$$h = \left(\frac{3\pi\mu D^4 g}{64\,WV_0}\right)^{1/2}. \tag{6.23}$$

The maximum film pressure p_{max} in the film can be obtained by first setting $r = 0$ in eqn. (6.21), then substituting for V from eqn. (6.22), and finally setting the differential $dp/dh = 0^{(1)}$. Indeed, the film thicknesses at which the maximum velocity of flow, the maximum pressure and the maximum rate of shear occur are in the ratio $3:\sqrt{5}:\sqrt{3}$. Figure 6.10 shows how the

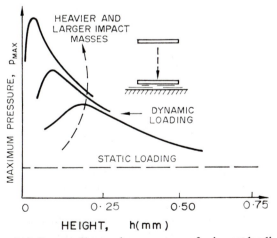

FIG. 6.10. Squeeze film maximum pressures for impact loading.

maximum pressure in the squeeze film increases with decreasing h to a maximum value which is determined by the size and weight of the plate and the viscosity of the squeeze film. We observe that for static loading, the maximum pressure is independent of h. The film thicknesses at which the velocity of flow, pressure, and rate of shear are a maximum have an order of magnitude comparable with the dimensions of the surface roughness of either approaching member, when the latter are assumed to be rigid and non-deforming. When one or both of the members are elastomers, complex elastohydrodynamic effects occur as $h \to 0$, and these will be considered in detail in Chapter 8.

Vibrational Performance

We now consider the vibration of a flat plate in a lubricant such that a parallel squeeze film exists beneath the plate and a stationary flat surface, as sketched in Fig. 6.11. The squeeze film resists the vibratory motion by providing spring and damping forces. Let a

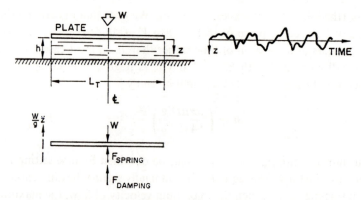

FIG. 6.11. Squeeze film subjected to vibrational loading.

spring constant k for the lubricant be defined as the ratio of spring force to displacement, whereas the damping constant c is the ratio of damping force to velocity thus:

$$k = \frac{F_{\text{spring}}}{z} ; \quad c = \frac{F_{\text{damping}}}{\dot{z}}. \tag{6.24}$$

Writing the equation of motion for the plate and making use of eqn. (6.24),

$$\frac{W}{g}\ddot{z} + c\dot{z} + kz = W. \tag{6.25}$$

The left-hand side of this equation shows clearly that the lubricant behaves as a viscoelastic material in a vibratory environment. We must find next how k and c can be expressed in terms of known properties of the liquid.

It is of interest to note that k and c can be closely approximated by considering the non-inertial or slow speed descent of the plate, even though these constants are subsequently used in a dynamic equation of the type (6.25). The spring constant k can be simply expressed as follows in terms of the bulk modulus B for the liquid, where

$$B = p/(\varDelta v/v)$$

Here p is the absolute pressure within the liquid, and $(\varDelta v/v)$ the dimensionless change in volume. If A is the area of the plate, then

$$\frac{\varDelta v}{v} = \frac{Az}{Ah} = \frac{z}{h},$$

and hence $p = Bz/h$. Thus, since $F_{\text{spring}} = pA$, we can write for k from eqn. (6.24):

$$k = \frac{pA}{z} = \frac{B(z/h)\,A}{z} = \frac{BA}{h}. \tag{6.26}$$

The damping constant c is obtained from eqn. (6.12), assuming that $F_{\text{damping}} = W$ thus:

$$c = \frac{W}{(\mathrm{d}h/\mathrm{d}t)} = \frac{2K\mu L_T^4}{h^3}. \tag{6.27}$$

We observe that both k and c diminish rapidly as the film thickness h increases.

The approach time for non-planar surfaces can be obtained in a manner similar to that described earlier for plane parallel squeeze films. The results for various configurations take the form shown in Table B of the Appendix, whereas Table C lists the corresponding spring and damping constants. These constants can be used to predict the dynamic behaviour of squeeze films when subjected to a random or sinusoidal vibratory input. In the case of journal or spherical bearings, it is often convenient to define a *generalized damping coefficient* as follows:

$$D' = c\left[\frac{(\varDelta/R)^3}{6\mu L_p}\right], \tag{6.28}$$

where $\varDelta$ is the average squeeze-film thickness or maximum displacement of journal or sphere and L_p a preferred length dimension defined for each particular system. Figure 6.12 shows a plot of D' versus the dimensionless displacement[†] H for the cases of a half-journal bearing, a complete journal bearing, and a ball-in-socket system.[‡] We observe clearly the similarity in the curves. For smaller film thicknesses (e.g. as $H \to 1$), it is seen that the damping constant increases very rapidly.

When the approaching surfaces have viscoelastic properties (this corresponds to elastomer coatings, or plastic surfaces, or solid lubricant linings in place of rigid surfaces as previously dealt with), we effectively introduce additional spring and damping terms, which must be combined with those of the squeeze film. This is shown in Fig. 6.13, where (a) depicts the

---

[†] $H = a/\varDelta$, where a is the instantaneous displacement of journal or sphere (see Table B in Appendix).
[‡] All non-rotating.

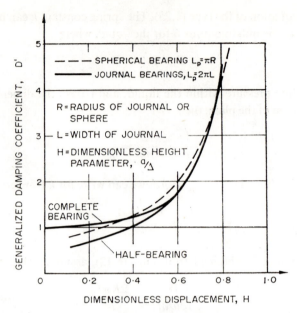

FIG. 6.12. Comparison of damping constants for cylindrical and spherical bearings.

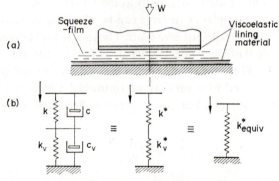

FIG. 6.13. Squeeze film with viscoelastic approach surfaces.

actual physical model, and (b) its mathematical equivalent. The subscript v denotes viscoelastic, and the asterisk indicates complex quantities thus:

$$k^* = k + j\omega c$$

for the squeeze film, and

$$k_v^* = k_v + j\omega c_v$$

for the elastomeric or viscoelastic material. By combining these two complex spring quantities in series, the effective spring constant for the combination becomes

$$k_{\text{equiv.}}^* = \frac{k^* k_v^*}{k^* + k_v^*} = k_{\text{equiv.}} + j\omega c_{\text{equiv.}}, \tag{6.29}$$

where $k_{\text{equiv.}}$ and $c_{\text{equiv.}}$ can be expressed in terms of the original parameters k, k_v, c, and c_v.

Non-parallel Approach

The speed of approach of the plate or upper surface in Fig. 6.5 can be varied by changing any of the five parameters on the right-hand side of eqn. (6.12). If these are held constant, however, only two possibilities exist for changing the rate of approach, dh/dt: (a) inclining the plate to the surface during squeeze action, and (b) replacing the smooth by a rough surface.

Intuitively, both of these methods either separately or together will increase the sinkage rate, and they will decrease the film thickness attained in a specified time. The method of inclining the plate provides, as it were, an angle of attack, and the substitution of a rough surface permits an effective drainage of the film.

Considering the first possibility, the general case of inclined sinkage[27] is where the angle $\theta(t)$ between plate and surface varies with time. For the particular case where $\theta(t)$ decreases linearly with h as the squeezing process takes place, we can write for a square plate of side L:

$$\frac{H(t)}{H_0} = \frac{\theta(t) - \beta}{\alpha - \beta},$$

(6.30)

where the initial and final values of $\theta(t)$ are α and β respectively (Fig. 6.14). The coordinate x is taken parallel to the plate and in the plane of the inclination, y is perpendicular to x and along the leading edge of the plate, and the origin of coordinates coincides with a corner of the plate. From Fig. 6.14, it is seen that

$$h(x, t) = H(t) + \theta(t) x.$$

(6.31)

If we assume an approximate pressure distribution[25] in eqn. (6.4), neglect the wedge and stretch terms, and put $dh/dt = V$, we finally obtain the following:

$$\frac{dh}{dt} = -2.37 \frac{W H^3}{\mu L^4} \left[1 + 1.5 \left(\frac{\theta L}{H} \right) + 1.2 \left(\frac{\theta L}{H} \right)^2 + 0.1 \left(\frac{\theta L}{H} \right)^3 \right].$$

(6.32)

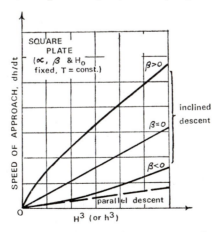

FIG. 6.14. Inclined squeeze film: smooth surfaces.

This equation shows that the approach speed is greater when a finite angle of inclination exists between plate and lower surface. For $\beta = 0$, the dimensionless angle $(\theta L/H)$ can be replaced by $(\alpha L/H_0)$ which is constant [see eqn. (6.30)]. Figure 6.15 shows a qualitative plot of eqn. (6.32) for a given square plate and lubricant. We conclude that the approach speed

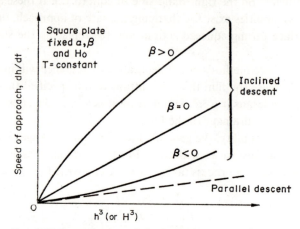

FIG. 6.15. Rate of approach for inclined and parallel plates.

of the plate is increased many times by inclining its plane to that of the underlying surface, this effect being more pronounced when the final angle of inclination β has the same sense as the initial angle α. The increase of approach speed is also particularly dominant as α itself increases, or as $h \to 0$.

Surface Roughness

If the lower surface for the case of parallel sinkage (see Fig. 6.5) exhibits a random roughness profile, the approach time of a smooth plate or approach surface is considerably reduced. The nature of the surface roughness is such that exact methods of representation are impossible, and we must therefore rely on approximate solutions to the squeeze-film equation. The most obvious method is to postulate an equation for the speed of approach which is similar in form to eqn. (6.32) except that the polynomial is expressed in terms of the dimensionless roughness (ε/h) for the lower surface thus:

$$\frac{dh}{dt} = -2.37 \frac{Wh^3}{\mu L^4}\left[1 \pm C_1\left(\frac{\varepsilon}{h}\right) \pm C_2\left(\frac{\varepsilon}{h}\right)^2 \pm C_3\left(\frac{\varepsilon}{h}\right)^3 \pm \dots\right] \qquad (6.33)$$

for a square plate of side L. Here ε is the root-mean-square or average peak-to-trough measurement of asperity size in the rough surface, and h is the film thickness. Figure 6.16 shows a comparison of the speed of approach for smooth and rough surfaces, and Fig. 6.17 shows the corresponding height versus time relationships. We observe from the first figure that the approach velocity as h reaches values comparable with ε is, indeed, considerable because of drainage in the void spacing between roughness elements, and the time of

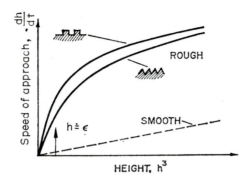

FIG. 6.16. Speed of approach on smooth and rough surfaces.[2]

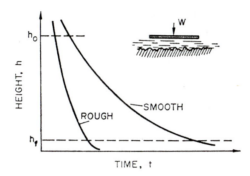

FIG. 6.17. Height vs. time curves for smooth and rough surfaces.

approach is greatly reduced. Equation (6.33) can be shown to produce the qualitative trends indicated in Fig. 6.16 (and, indeed, Fig. 6.17), but the difficulty is that no known method exists of relating the C_i constants for a particular surface to its roughness geometry. This equation is therefore of little practical use in estimating squeeze-film performance from profile or roughness measurements.

One practical method of quantifying the effects of macro-geometry on squeeze behaviour is to separate the escaping lubricant beneath the plate into bulk and channel components, as illustrated in Fig. 6.18. It is apparent that there is a bulk flow of lubricant between the plate underside and the asperity tips, and an open channel flow between the asperities.

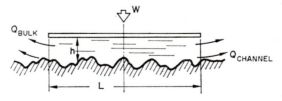

FIG. 6.18. Bulk and channel flow components on rough surfaces.

The rate of approach of the plate is dependent on both of these contributions as follows:

$$\frac{dh}{dt} = \frac{1}{A}(Q_{\text{bulk}} + Q_{\text{channel}}), \tag{6.34}$$

where A is the plate area. It has been shown that for asperities of uniform height, Q_{channel} may be evaluated as $h \to 0$ in terms of the geometry of the rough surface[28] thus:

$$Q_{\text{channel}} = \frac{3A^4 P}{80 I_p} \left(\frac{NW}{\mu L^5}\right) \text{MHR}(1 + Fh^m), \tag{6.35}$$

where A is the mean cross-sectional area of a typical channel, P the equivalent perimeter of flow per channel, I_p the polar moment of inertia of A about its geometric centre, N the number of channels appearing beneath the plate periphery, MHR the mean hydraulic radius of a typical channel ($= A/P$), F an "openness" parameter for the surface, h the film thickness measured from the plate underside to the asperity tips, and m a numerical index > 1. The openness of a surface is defined as the rate at which a horizontal cross-sectional area through the asperities decreases as the cross-sectional plane approaches the tips of the asperities at uniform speed.

Predicting the bulk flow contribution Q_{bulk} for $h > 0$ is much more difficult to achieve, and a satisfactory expression has not yet been obtained. Fortunately, the case for $h \to 0$ is of greater importance, and the bulk component is then small or negligible. If the underside of the plate has an elastomer coating, however, the larger asperities of the surface distort the elastomer as $h \to 0$, and elastohydrodynamic effects occur at localized spots as seen in Chapter 8. The form of eqn. (6.35) is still valid in such cases, but the establishment of A and P is somewhat more complex because they are directly influenced by the draping of the elastomer between asperities.

6.6 Hydrostatic Lubrication

Hydrostatic lubrication is distinct from hydrodynamic lubrication in that no relative motion[†] between surfaces occurs, although the fluid continues to flow freely in the separation between surface pairs. Figure 6.19 shows (a) a schematic diagram of a circular step bearing, and (b) the radial pressure distribution in the lubricant. As we see, hydrostatic lubrication permits a thrust load W to be applied to the upper member, which may or may not rotate about a vertical axis as shown. A lubricant with pressure p_0 enters the centre of the lower member and fills the recess (radius R_0) before being discharged radially. We may apply eqn. (6.14) for the velocity distribution v within the film thus:

$$v = -\frac{h^2}{8\mu}\left(\frac{dp}{dr}\right)\left(1 - \frac{4z^2}{h^2}\right), \tag{6.14}$$

[†] Excepting rotation.

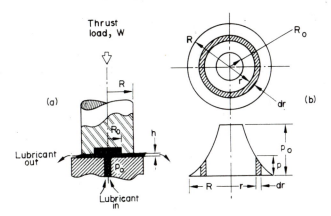

FIG. 6.19. (a) Hydrostatic step bearing and (b) radial pressure distribution.

where z has the same meaning as in Fig. 6.8, and (dp/dr) is the pressure gradient in the radial direction. Equation (6.14) applies both for the case of a squeeze film where $v = v(h, r, z)$ and for a hydrostatic bearing where $v = v(r, z)$ is a function only of position. The mean velocity of flow $\bar{v} = 2v_0/3$ where v_0 is the maximum value obtained by putting $z = 0$ in eqn. (6.14). Thus the rate of outflow of the lubricant at radial distance r is

$$Q = 2\pi rh\bar{v} = -\frac{\pi rh^3}{6\mu} \left(\frac{dp}{dr} \right).$$

By integrating this expression with respect to r and making use of the boundary conditions that $p = 0$ at $r = R$, and $p = p_0$ (the supply pressure) at $r = R_0$, we obtain the following:

$$p = \frac{6\mu Q}{\pi h^3} \ln \left(\frac{R}{r} \right) \qquad (\text{for } R_0 \leqslant r \leqslant R) \tag{6.36}$$

and

$$Q = \frac{p_0 \pi h^3}{6\mu \ln (R/R_0)}. \tag{6.37}$$

Equation (6.36) gives the pressure distribution at any radius r in the annular ring, and eqn. (6.37) gives the flow rate required to sustain a hydrostatic film of thickness h when the supply pressure of the lubricant is p_0. Now the load W is supported by the supply pressure p_0 acting over the recess area of radius R_0 and by the variable pressure p acting over the sill area of the bearing, as shown in Fig. 6.19. We may therefore write

$$W = \pi R_0^2 \, p_0 + \int_{R_0}^{R} p \, 2\pi r \, dr$$

and by substituting for p from eqn. (6.36) we finally obtain for the load capacity:

$$W = \frac{p_0 \pi}{2} \left[\frac{R^2 - R_0^2}{\ln (R/R_0)} \right]. \tag{6.38}$$

Equations (6.37) and (6.38) decide the general size of the bearing design since a certain thrust load W must be supported and some limit must be placed on the volume flow rate Q of the lubricant. However, it is then necessary to check carefully the vertical stability of the upper member, and this places an upper limit on the depth of recess $\varDelta$.[29] It is also customary to place an orifice between the supply line and the point where the lubricant enters the gap beneath the recess area. Thus a pressure drop (p_s-p_0) exists across the orifice, where p_0 is the pressure of the lubricant at the entrance point and p_s is the lubricant supply pressure. The function of the orifice is to dampen oscillations in the sill film thickness h as a result of variations in p_s and also to permit the parallel operation of two or more bearings from a single supply manifold.

The hydrostatic step-bearing in Fig. 6.19 requires an external source of pressurized lubricant for its operation, in contrast with hydrodynamic bearings which develop their load support from internal action.† In most cases, the lubricating fluid is a gas (normally air), and the overall design is commonly referred to in this case as an externally pressurized, gas-lubricated bearing. As we might anticipate, the question of positional or vertical stability is of much greater significance for gas or air bearings than for liquid bearings. This can be attributed both to the smaller values of film thickness h for gas bearings, and to the virtual absence of internal damping in gases compared with liquids. Small deviations in h from the operational or design value have a far greater probability of amplification (with the ultimate catastrophe of bearing seizure) if the working fluid is a gas rather than a viscous liquid.

The applications of hydrostatic lubrication are many and diverse. Usually heavy equipment, where the speed of relative motion between the components is small, depends on some form of externally pressurized lubricant to separate the parts in the manner of Fig. 6.19. Radio and astronomical telescopes, water-wheel generators, large-scale thrust bearings, and vertical turbo-generators are typical examples.

6.7 Cavitation

The phenomenon of cavitation has been referred to briefly in Section 6.4 and Table 6.2, since it gives rise to a net load support action particularly on rough surfaces. It can be attributed directly to a reduction of local pressure in a liquid to a level approaching its vapour pressure. In such circumstances, tiny bubbles filled with gas or vapour appear and grow rapidly. When they move subsequently to a region of increased pressure, the bubbles collapse or implode, thereby releasing tremendous energy and causing severe erosion in submerged surfaces. The phenomenon of cavitation was anticipated by Euler as early as 1754 in his theory on hydraulic turbines, and it was first observed experimentally in connection with ships' propellers in 1895. The pitting of gear teeth and the wear of bearing materials are examples of how cavitation may be produced where thin lubricant films exist.

One of the most intriguing visual observations of cavitation effects can be obtained by considering the rolling of a cylinder on a rigid, lubricated plane surface.[30] Figure 6.20

† This arises from the wedge, stretch, or squeeze terms treated in Section 6.2.

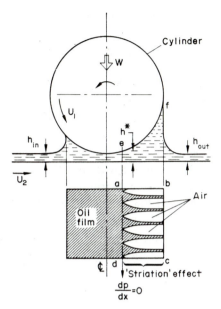

FIG. 6.20. Rolling with cavitation—cylinder on flat, lubricated plane.

shows that behind the rolling cylinder, cavitation streamers or filaments are formed when the hydrodynamic pressure reaches a negative value. The liquid thereupon contracts or "striates" because of surface tension effects, and the average thickness of the remaining lubricant filaments is very small in comparison with the original width, *ad*. The gradient of pressure and the pressure itself are both zero at point *e*, where the thickness of the film is h^*. If Q is the flow rate passing between the surfaces, then

$$Q = U_2 h_{in} = U_2 h_{out} = \tfrac{1}{2}(U_1 + U_2)\,h^*$$

and, therefore,

$$h^* = \left(\frac{2}{1+k}\right) h_{in} \tag{6.39}$$

where $k = U_1/U_2$. We note that the assumption of zero pressure gradient at the point *e* implies that this location defines the start of the cavitation regime. Also, the condition of pure sliding ($k = 0$) has no validity here because under these conditions not only does the liquid tend to accumulate in front of the body rather than behind it (thereby changing the entire pressure distribution), but in addition there is no cavitation regime whatever of the type shown in Fig. 6.20. We therefore assume that eqn. (6.39) applies only for values of k close to unity. For pure rolling, $k = 1$ and $h^* = h_{in}$.

Indeed, the cavitation streamers can also be regarded as a necessary condition for the validity of the continuity equation in the region *ef* behind the cylinder. The large increase in film thickness above the original value h_{in} must be compensated for by a reduction in the effective width of the lubricant, so that the striation effect is a logical development. In journal bearings, the lower surface is curved rather than plane, so that the rate of convergence and divergence of the film is correspondingly smaller. By using a transparent bearing structure, cavitation streamers similar to those in Fig. 6.20 have been observed.

The most significant cavitation effects have been observed as a result of etching micro-irregularities on metal surfaces which are then lubricated and permitted to operate at high velocities of relative sliding.[31] A typical example is a rotary-shaft face seal. Having photo-etched asperities of approximately 2.5 μ in height and of cylindrical cross-section on a nickel-plated steel rotor, tests were conducted using a transparent and smooth stator through which cavitation streamers were photographed, and a distinct reduction in torque was observed to accompany the onset of cavitation. Figure 6.21(a) shows the shape of a typical asperity and the resulting pressure distribution at lower sliding speeds when cavitation effects were not significant, and Fig. 6.21(b) shows the effects of cavitation at higher speeds. In the latter case, the positive pressure increments outweigh the negative pressure effects, and a net load supporting force is created. Thus the lubricating effect is due to the cavitation contribution of hundreds of such asperities. A similar argument applies if depres-

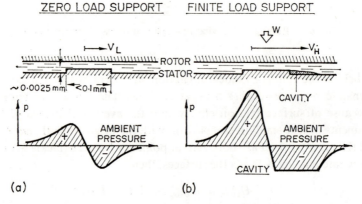

FIG. 6.21. Load support due to lubricant cavitation in sliding. (a) Low-speed sliding—no cavitation. (b) High-speed sliding—with cavitation.

sions or pits are considered instead of asperities. A combined theoretical and experimental investigation has shown[31] that for each asperity diameter d there exists an optimum density spacing for which the load per unit area $\bar{p}$ has a maximum value, as shown in Fig. 6.22. If N' is the asperity density (or number of asperities per unit area), the area fraction is $\pi N' d^2/4$, and maximum load capacity appears to occur at a constant value of this parameter ($=0.4$) irrespective of the value of N' as shown in Fig. 6.23. The value of 0.4 for the area fraction corresponds to a wavelength of approximately $2d$, so that maximum load can be achieved due to the cavitation effect for a given value of N' when the spacing between asperities has approximately the same value as the diameter of the asperities. Increasing the asperity density, of course, will also increase $\bar{p}$, as shown clearly in Fig. 6.23, and it has also been shown that a linear increase in relative sliding speed produces a corresponding linear increase in load capacity. Photo-etched asperities or micro-irregularities are therefore

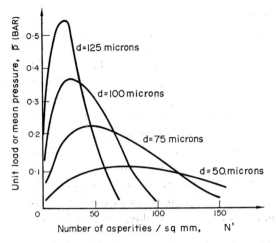

FIG. 6.22. Variation of load support with number of asperities.

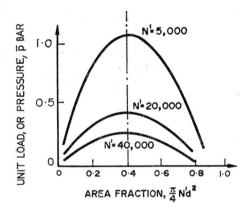

FIG. 6.23. Variation of load support with area fraction for different numbers of asperities.

intimately involved in the hydrodynamic lubrication process between two parallel surfaces through the mechanism of local cavitation behind each asperity.

The asperities considered above are, of course, rigid, so that in the absence of cavitation no load support would develop between the sliding surfaces. For flexible asperities, on the other hand, the elastohydrodynamic distortion of their otherwise symmetrical profiles (see Fig. 6.4 and Chapter 8) produces an additional load capacity, as illustrated in Table 6.2. At still higher speeds of sliding than those considered above, there is evidence that the formation of eddies behind flexible asperities is responsible for values of load support which far exceed those predicted from elastohydrodynamic theory alone. Such eddy formation on the leeward side of asperities is but another manifestation of cavitation.

Cavitation also occurs in squeeze films subjected to vibration. This is particularly noticeable according as the frequency of vibration is raised into the kHz range. In simple terms, the liquid in the squeeze film is unable to follow the rarefaction or cyclic pressure reductions which occur at high frequencies. As a consequence, the lubricant foams or cavitates, creating small bubbles which in turn are expanded and compressed. The effect of this bubble formation is a drastic reduction[32] in the damping coefficients given by eqn. (6.27) and listed in Table C of the Appendix for various squeeze-film configurations. By considering density variations produced by cavitation in the lubricant, the order of magnitude of the discrepancy between theoretical and actual values of c at higher frequencies can be accounted for. Table 6.3 shows the extent of the reduction in damping capacity as a result of cavitation in different liquids, for a frequency range from 0.4 to 4 kHz. The spring constant k obtained experimentally is also included for convenience. We observe from Table 6.3 that only mercury has close agreement between experimental and theoretical values of c, since it is relatively difficult to cause cavitation effects in such a dense liquid. Otherwise the discrepancy between theory and experiment is exceedingly large indeed.

TABLE 6.3. REDUCTION OF DAMPING CONSTANT DUE TO CAVITATION

Liquid	C_{exp} (KG-s/m)	C_{theor} (KG-s/m)	k_{exp} (KG/m)
Oil, HVI 40	2.3×10^2	7.25×10^3	1.61×10^4
Silicone (MS 200/20)	3.94×10^2	7.68×10^3	1.48×10^5
Apiezon, Oil J	3.98×10^2	1.57×10^6	1.48×10^5
Mercury (Hg)	4.51×10^2	0.68×10^3	3.57×10^5

In hydraulic machinery applications, we define a cavitation number σ as follows:

$$\sigma = \frac{p_\infty - p}{\frac{1}{2} \varrho V^2}, \tag{6.40}$$

where p is the pressure at any particular point of interest, p_∞ the undisturbed free-stream value of pressure, ϱ the mass density of the liquid, and V a reference velocity. We observe that if we increase p_∞ by a prescribed increment, then p is automatically increased by the same amount; thus the value of σ remains constant at a given location. Let us now suppose that p is reduced until $p \to p_v$, where p_v is the vapour pressure of the liquid at the given temperature. The cavitation number then acquires a critical value σ_c where

$$\sigma_c = \frac{p_\infty - p_c}{\frac{1}{2} \varrho V^2}. \tag{6.41}$$

The reduction in p to the value p_c can be achieved in any of three ways:

(a) a reduction in the overall pressure acting on the system: this affects p_∞ and p equally;
(b) an increase in sliding speed; or
(c) consideration of other locations within the system where p has lower values than originally specified.

Equation (6.41) proposes a critical cavitation number σ_c which identifies the start of cavitation. We may state the determining criterion briefly as follows:

$$\begin{cases} \sigma_c > \sigma & \text{or} \quad p_c < p \quad \text{No cavitation.} \\ \sigma_c \leqslant \sigma & \text{or} \quad p_c \geqslant p \quad \text{Cavitation.} \end{cases}$$

We thus have a yardstick for predicting and measuring the risk of cavitation at specific locations in a given lubrication system.

Photographic studies of the growth and subsequent implosion of bubbles have been made both in the vicinity of solid boundaries and as a result of inception by ultrasonic waves. The importance of cavitation may be judged from the fact that erosion in metals can be directly attributed to it, and its onset may be triggered by the existence of surface roughness at a neighbouring boundary.

6.8 Miscellaneous

Non-Newtonian Liquids

Table 6.4 shows a broad classification of Newtonian and non-Newtonian lubricants. Indeed, many other forms of non-Newtonian behaviour exist which do not appear in the table, and they obviously modify the original theory of Reynolds. Most gases, non-colloidal liquids, and lubricants follow the simple law of Newton within a limited range of shear stress or strain rate. The simplest type of non-Newtonian behaviour is typified by Bingham fluids, whose characteristic stress versus strain rate dependence appears in Table 6.4 and Fig. 6.24. The principal forms of power law variation are pseudoplastic and dilatant respectively, dependent on the value of the index N in Table 6.4. For $N > 1$, dilatant behaviour is approximated, whereas for $N < 1$ we have pseudoplastic characteristics, normally typified by rubbers and certain plastic materials (Fig. 6.24).

TABLE 6.4. NEWTONIAN AND NON-NEWTONIAN LUBRICANTS

Lubricant classification		Relevant equation
Newtonian		$\dfrac{\partial V}{\partial y} = \dfrac{\tau}{\mu}$
Non-Newtonian	Bingham	$\dfrac{\partial V}{\partial y} = \dfrac{\tau + \tau_0}{\mu}$
	Viscoelastic	$\dfrac{\partial V}{\partial y} = \dfrac{\tau}{\mu} + \dfrac{1}{G}\dfrac{d\tau}{dt}$
	Power law	$\dfrac{\partial V}{\partial y} = \left(\dfrac{\tau}{\mu}\right)^{1/N}$

Notation: y = coordinate perpendicular to film. N = power law index. τ_0 = constant. τ = applied shear stress. G = shear modulus.

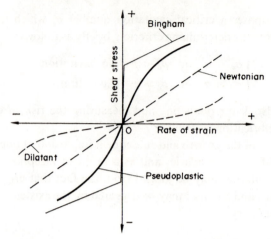

FIG. 6.24. Shear stress vs. rate of strain for various non-Newtonian lubricants.

Only the viscoelastic form of non-Newtonian behaviour in the above classification is time-dependent, and we note that several forms of such behaviour (each having slightly different characteristics) can exist besides the form listed in Table 6.4. We also observe that a lubricant which follows Newton's law of viscosity under static or slow-speed conditions may well exhibit viscoelastic traits under impact or high-speed conditions.

Generalized Sommerfeld Number

The generalized Sommerfeld number in lubrication theory S_0 is a basic design parameter defined as the ratio of viscous to pressure forces in a thin lubricating film thus:

$$S_0 = \frac{\tau L^2}{pL^2} = \frac{\mu U}{pL},$$ (6.42)

where τ and p are shear and pressure stresses and Newton's law has been used to obtain the final form

$$S_0 = \frac{\text{viscosity} \times \text{speed}}{\text{load}}.$$

The Sommerfeld number is of fundamental significance in lubrication problems, and it constitutes the most important parameter in bearing design. This will become apparent in the many types of bearing application in Chapter 15.

Having examined the fundamental principles of hydrodynamic or thick-film lubrication in this chapter, we will next consider the effects of thinning the film to the point where part solid contact may occur. Whereas with bulk lubricants under fully hydrodynamic conditions properties are to a large extent isotropic and homogeneous despite temperature gradients, this is no longer true in boundary lubrication. The properties of thin films in mixed and boundary lubrication are difficult to identify, since chemical and physical interactions occur within the bounding surfaces. Such complexities will occupy our attention in the next chapter.

CHAPTER 7

BOUNDARY LUBRICATION

7.1 Introduction

Fully hydrodynamic lubrication presupposes the presence of a fluid completely separating the rubbing parts. Since the solid parts do not touch, there can be no wear, and the viscosity of the fluid determines the friction. The oiliness characteristics of the lubricant are therefore of no concern in the fully hydrodynamic regime.

When a machine is started from rest, however, it often happens that the bulk of the intervening lubricant has been squeezed out during the idle period, and only an adherent surface film remains. This film may be only a few molecules thick, but it can still prevent seizure of the parts and permit sliding to take place. The adherent film of lubricant is bonded to the metal substrate by very strong molecular adhesion forces, and it has obviously lost its bulk fluid properties. Under very severe conditions even the absorbed film may be scraped or burned off, and a solid lubricant such as graphite (or a chemically formed coating such as metal oxide or sulphide) is required to offset seizure. Severe conditions under which perfect lubrication is impossible are classified as conditions of imperfect or boundary lubrication. Both the properties of the underlying surfaces (usually metals) and the physical and chemical structure of the lubricant become of primary importance.

If conditions are such that it is impossible to sustain complete hydrodynamic separation of two sliding surfaces, then mixed lubrication often accompanies the boundary lubrication phenomenon. Mixed lubrication involves part solid contact at the tips or peaks of major asperities and the presence of bulk lubricant in the intervening void spaces. Boundary lubrication is concerned with complex adsorption and physico-chemical interactions which determine the exact nature of solid contact in mixed lubrication, although in some cases the bulk liquid between asperities is non-existent.

Figure 7.1 shows a plot of the coefficient of friction versus the generalized Sommerfeld number, S_0 (see Section 6.8). We observe that larger values of S_0 correspond to greater film thicknesses and hence to fully hydrodynamic conditions, and there is a distinct proportionality between f and S_0. The asymptote Oa passing through the origin of coordinates can be regarded as the Petroff[†] equation in plotted form, and it coincides with the predictions of classical theory at very large values of the Sommerfeld number. This is particularly rele-

[†] See Section 6.1.

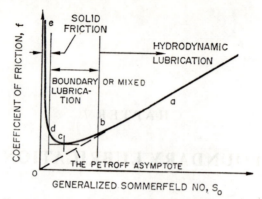

FIG. 7.1. Lubrication regimes for sliding surface pairs.

vant to journal bearing operation.[†] Very small values of S_0 imply solid friction (or seizure, in the case of clean metals) and the transition zone between solid and fully hydrodynamic operation may be classified as mixed or boundary lubrication as shown. Indeed, the phenomenon of boundary lubrication may indicate coefficients of friction which are either in excess of hydrodynamic values or close to the minimum point *c* in the figure. In the former case, the effective shear strength of the boundary film exceeds that of the bulk lubricant, so that larger frictional readings are recorded. The following reasoning is offered to explain the minimum friction point *c*.

Let the boundary lubrication condition be expressed in terms of the individual contributions of liquid, solid, and deformation components to the total frictional coefficient f_{BL} thus:

$$f_{BL} = f_{liq} + f_{solid} + f_{deform}. \qquad (7.1)$$

Here the f_{liq} component is due to shearing of the lubricant film, being identical with τ in Section 6.3 and therefore proportional to the absolute viscosity μ of the bulk lubricant. If we consider Fig. 6.1 as representing the fully hydrodynamic condition of lubrication between two surfaces, it is apparent that the relative speed of sliding ($U_1 - U_2$ or U) is sufficiently great in this case to maintain complete separation of the surfaces. The degeneration from hydrodynamic to mixed and boundary lubrication along the path *abcd* in Fig. 7.1 can be attributed most simply to a reduction in U. The separation between the sliding surfaces is diminished, and the onset of the boundary lubrication regime is characterized by the creation of a solid friction component as rubbing between asperities commences. The rubbing action in turn generates localized heating effects, which reduce the viscosity μ of the surrounding bulk lubricant. *The overall effect may well be that the reduction in* f_{liq} *because of a decrease in* μ *exceeds the value of the new component* f_{solid}: under these conditions the overall frictional coefficient decreases along *bc* to the minimum value *c* in the figure. As speed is further reduced, the fraction of total nominal area participating in rubbing increases, and the solid friction component dominates the boundary interaction. The coefficient of friction therefore rises along *cd* in Fig. 7.1. It is difficult to ascertain exactly what happens

[†] See Chapter 15.

to the deformation component at this time, but it is certain that its rate of change during the transition *abcd* is small compared with changes in f_{liq} and f_{solid}.

The minimum friction condition indicated by the point c would appear to be an ideal objective in the design of machinery, in particular, bearings. Unfortunately, this point is highly unstable, and there is a high probability of total seizure occurring along *de* if the surfaces are clean and free of contaminants. The small additional friction introduced by operating in the fully hydrodynamic region is but a small penalty to pay for stable and long-life performance.

We now examine in detail the nature of boundary friction for metals and elastomers.

7.2 Boundary Lubrication of Metals

Figure 7.2 shows the condition of mixed or boundary lubrication for a metal sliding on metal. If the total or apparent area of contact between the sliding surfaces is denoted by A,

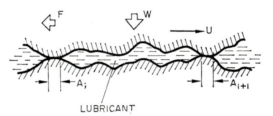

FIG. 7.2. Boundary lubrication for metal-on-metal.

we may write the friction force F as the sum of solid friction at asperity peaks, liquid friction in the voids, and the ploughing contribution F_{plough} (see eqn. 3.18) thus:

$$F = A[\alpha_W s_{\text{solid}} + (1 - \alpha_W) s_{\text{liq}}] + F_{\text{plough}}, \qquad (7.2)$$

where α_W is the fraction of area A at which solid contact occurs and s is the shear strength. Since the load W is supported both by solid contact at asperity peaks and by hydrodynamic pressure generation in the voids, we can write

$$W = A[\alpha_W p^* + (1 - \alpha_W) p_{\text{hydro}}], \qquad (7.3)$$

where p^* is the plastic yield pressure of the softer metal and p_{hydro} is the hydrodynamic pressure generated between asperities. We note that whereas $s_{\text{liq}} \ll s_{\text{solid}}$ in eqn. (7.3), p_{hydro} is only slightly less than p^* in eqn. (7.3). It is therefore convenient to define an average pressure $\bar{p}$ such that $p^* > \bar{p} > p_{\text{hydro}}$ and eqn. (7.3) can then be approximately represented in the form

$$W = A\bar{p}. \qquad (7.4)$$

By dividing F by W, we obtain from eqns. (7.2) and (7.4) the following expression for the coefficient of boundary lubrication f_{BL}:

$$f_{BL} = \frac{F}{W} = \alpha_W \left(\frac{s_{\text{solid}}}{\bar{p}} \right) + (1 - \alpha_W) \frac{s_{\text{liq}}}{\bar{p}} + f_{\text{plough}}, \qquad (7.5)$$

10*

where the ploughing contribution ($f_{\text{plough}} = F_{\text{plough}}/A\bar{p}$) may be henceforth neglected. We observe that the second term on the right-hand side of this equation is also very small compared with the first term.

If we now assume dry contact between the sliding metal surfaces, we can write for the load W and frictional force F

$$W = A\alpha_D p^*,$$
$$F = A\alpha_D s_{\text{solid}} + F_{\text{plough}},$$

so that by putting $p^* \doteq \bar{p}$ and neglecting the ploughing force, the coefficient of friction under dry conditions becomes

$$f_{\text{dry}} = s_{\text{solid}}/\bar{p}. \tag{7.6}$$

From eqns. (7.5) and (7.6)

$$\frac{f_{BL}}{f_{\text{dry}}} = \alpha_W + (1 - \alpha_W)\frac{s_{\text{liq}}}{s_{\text{solid}}}. \tag{7.7}$$

Indeed, the second term on the right-hand side of eqn. (7.7) is negligibly small, so that for metal-on-metal sliding

$$f_{BL} \ll f_{\text{dry}}, \tag{7.8}$$

which is an agreement with experimental results.

No mention has been made in the above analysis of the nature of solid or liquid contact, and the equations in this sense can apply either to mixed or boundary lubrication. It is the identification of the components f_{solid} and f_{liq} in terms of surface properties, adsorption phenomena, and properties of the bulk liquid which determines the boundary lubrication condition in particular instances. These interactions will be discussed in a later section.

7.3 Boundary Lubrication of Elastomers

Consider now the case of an elastic body sliding on a rough, rigid base in the presence of an interfacial liquid (Fig. 7.3). It is difficult to neglect elastohydrodynamic effects for such a combination, but let us assume that the speed of sliding U is sufficiently low to neglect fluid entrainment over asperity peaks (see Chapter 8 later for a detailed account of this important effect), while at the same time it is sufficiently large to create hydrodynamic lift in the void spacing between neighbouring asperities. A condition of boundary lubrication exists under these circumstances, but, as distinct from Fig. 7.2, the fluid film is continuous between the surfaces. The film existing at asperity peaks is extremely thinned and has properties which are quite distinct from the bulk lubricant in the voids. Whereas the metal–liquid interface is more readily defined in Fig. 7.2, this cannot be regarded as valid in Fig. 7.3, particularly near asperity peaks where physico-chemical interactions cause the lubricant to assume some of the properties of the draped elastomer. These latter areas constitute a fraction β_W of the total projected area A between the surfaces, and by analogy with eqn. (7.3)

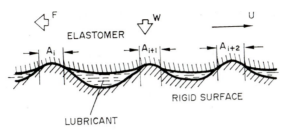

FIG. 7.3. Boundary lubrication for elastomer-on-rigid-base.

for metals we can write for the frictional resistance F,

$$F = A[\beta_W s_{\text{peaks}} + (1 - \beta_W) s_{\text{bulk}}] + F_{\text{hyst}}, \qquad (7.9)$$

where s_{peaks} is the shear strength of the interacting lubricant layer at asperity peaks and s_{bulk} the shear strength of the bulk liquid. Also,

$$W = A[\beta_W p_0 + (1 - \beta_W) p_{\text{hydro}}], \qquad (7.10)$$

where p_0 is the elastic (or viscoelastic) mean pressure near asperity peaks. If we write an average pressure $\bar{p}$ such that $p_0 > \bar{p} > p_{\text{hydro}}$, then eqn. (7.10) reduces to the approximate form

$$W = A\bar{p}, \qquad (7.11)$$

and by combining eqns. (7.9) and (7.11)

$$f_{BL} = \frac{F}{W} = \beta_W \frac{s_{\text{peaks}}}{\bar{p}} + (1 - \beta_W) \frac{s_{\text{bulk}}}{\bar{p}} + f_H^W, \qquad (7.12)$$

where the last component $f_H^W (= F_{\text{hyst}}/A\bar{p})$ is the hysteresis contribution to boundary friction under wet conditions.

For the case of dry contact existing between the sliding surfaces, we can write for the load W and the frictional force F,

$$\left. \begin{array}{l} W = A\beta_D p_{\text{dry}}, \\ F = A\beta_D s_{\text{solid}} + F_{\text{hyst}}^D, \end{array} \right\} \qquad (7.13)$$

where s_{solid} is the shear strength at asperity peaks under dry conditions, β_D is the fraction of total area where dry contact occurs, and F_{hyst}^D is the frictional force due to hysteresis effects under dry conditions. From eqn. (7.13), it follows that the coefficient of friction is given by

$$f_{\text{dry}} = \frac{s_{\text{solid}}}{p_{\text{dry}}} + f_H^D, \qquad (7.14)$$

where f_H^D is the hysteretic coefficient of friction under dry conditions. We note that p_{dry} corresponds very closely to the mean Hertzian value of pressure at asperity tips, and it is reasonable therefore to assume that $p_{\text{dry}} \doteq \bar{p}$. If the hysteresis term in eqn. (7.14) is neglected for the moment, it can be seen that this equation is identical with the simple adhesion theory (see Chapter 3).

By subtracting eqn. (7.14) from eqn. (7.12), we can write the following expression:

$$f_{BL} - f_{dry} = \frac{\beta_W}{p}[s_{peaks} - s_{bulk}] + \frac{1}{p}[s_{bulk} - s_{solid}] + (f_H^W - f_H^D). \qquad (7.15)$$

It should be remembered that $s_{peaks} \ll s_{solid}$ and $s_{bulk} \ll s_{peaks}$, so that the first bracketed expression on the right-hand side of eqn. (7.15) is positive and the second is negative. If we neglect the third hysteresis term for the moment, then we can say that

$$f_{BL} \doteq f_{dry}, \qquad (7.16)$$

whereas for metals the coefficient of boundary friction is considerably less than the dry coefficient, in accordance with eqn. (7.8). The following data for nylon shows that the addition of a lubricant may increase the coefficient of sliding friction compared with dry conditions:

Lubricant	f
None (dry)	0.46
Water	0.52
Ethylene glycol	0.58
Perfluorolube oil	0.58

It is not at all clear what the net effect of boundary lubrication is in determining the relative magnitudes of f_H^W and f_H^D for the same forward speed of travel V. If we assume reasonably large values of V, then we conclude that:

(a) both hysteresis terms f_H^W and f_H^D are appreciable (see Section 4.3 and Fig. 4.5); and
(b) the dry condition produces a slightly higher operating temperature.

There is therefore a tendency to reduce f_H^D in comparison with f_H^W. On the other hand, however, the wet condition attempts to separate the sliding surfaces by hydrodynamic pressure generation (see Chapter 8), and this tends to diminish f_H^W slightly in comparison with f_H^D. We can only say in the general case that

$$f_{BL} \gtrless f_{dry} \qquad (7.17)$$

and conclude that the addition of a lubricant to an elastomer–rigid surface sliding system does not necessarily imply a reduction in the resulting coefficient of sliding friction. This behaviour is in marked contrast to metals, as a comparison of eqns. (7.8) and (7.17) will show.

7.4 Molecular Structure of Boundary Lubricants

The majority of lubricants are oils that are compounds of carbon and hydrogen with perhaps a small amount of other elements. There are two main categories: (a) *mineral oils* (derived from petroleum) and (b) *vegetable oils* (derived from animals and vegetables).

The molecules of mineral oils are generally in the shape of long chains of carbon atoms with hydrogens attached, or of rings with long side chains, as shown in Fig. 7.4. The relatively high viscosity that oils possess can be pictured physically as due to the entangling and intermeshing of their very long chains. Another result is a tendency for the chains under certain conditions to align themselves parallel to one another or perhaps to pack closely together, thus forming a relatively dense and rigid layer on a solid surface. This tendency is greatly accentuated by the addition of small amounts of other types of oil to the pure hydrocarbon.

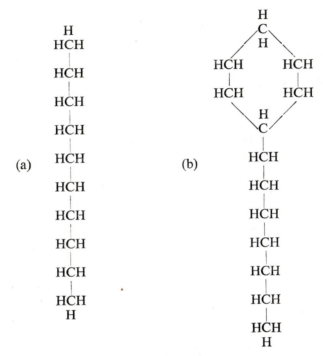

FIG. 7.4. Molecular structure of mineral oils. (a) Straight-chain hydrocarbon, and (b) ring hydrocarbon with side chain.

Vegetable oils are a mixture of a wide variety of substances, including chiefly tri-glycerides and then "fatty" acids and esters (Fig. 7.5). The tri-glycerides have three long carbon chains hooked through a carboxyl linkage to a nucleus of three carbon atoms as shown. Fatty acids and esters are essentially long-chain carbon atoms (again with hydrogens attached), the former having a hydroxyl (—OH) or a carboxyl (—COOH) radical on one end. All of these compounds occur naturally in animal and vegetable oils and can be removed or concentrated by refining. Often some of the carbon atoms in a chain are not bonded to as many hydrogen atoms as possible, and double bonds between two carbons are formed: such molecules are said to be unsaturated. Generally, unsaturation is likely to occur in vegetable oils and saturation in mineral oils. The region of the double bond is chemically active and will absorb other molecules, notably oxygen from the air if the oil is heated (as in an automobile engine). This phenomenon has two results:

(a) it produces acidity of the oil, and leads to severe chemical attack of bearing surfaces,[33] and

(b) oxidation produces gum and varnish deposits which are usually harmful in engines.

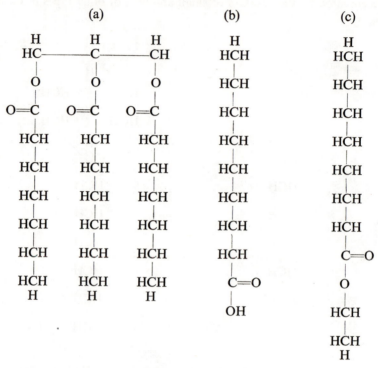

FIG. 7.5. Molecular structure of vegetable oils. (a) A tri-glyceride, (b) a fatty acid, and (c) an ester.

All of the molecular structures shown in **Figs.** 7.4 and 7.5 have a common feature (namely, a long chain). When in the vicinity of a solid surface, the molecular chains tend to orient themselves so as to appear perpendicular to the given surface, with one end attached to it. A long chain is therefore desirable to provide a maximum physical separation between two bearing surfaces. Another feature of these long-chain molecules in contact with a solid surface is their tendency to adhere in clusters. If a sufficient number are present, they will completely coat the surface with a monomolecular film of great lateral strength.

The third structural characteristic of a molecule which is conducive to the formation of a strong adherent film is its possession of an active radical. Whereas the molecules of vegetable oils (and of synthetic additives) have an active radical at one end of the chain, those of mineral oils have none. The possession of an active radical disturbs the symmetry of the molecular chain structure and hence of the electric charge distribution, so that one end becomes predominantly positive and the other end negative. Such a molecule is said to be "polar", and the radical ends of polar molecules have a strong affinity for metal surfaces. These ends attach themselves to such surfaces, while the rest of the molecule tends to stand out normal to the surfaces. The affinity to adhere is termed the *free energy of adhesion*, and although a function of both oil and surface it is believed to depend principally on the nature of the oil.

TABLE 7.1. FRICTION REDUCING PROPERTIES
OF OLEIC ACID

Lubricant	Friction coeff., f
Pure mineral oil	0.360
2% oleic acid + mineral oil	0.249
10% oleic acid + mineral oil	0.198
50% oleic acid + mineral oil	0.198
Pure oleic acid	0.195

The optimum molecular structure for a boundary lubricant consists of a long, straight hydrocarbon chain with an active radical at one end, as we have seen. Compounds having such a structure occur naturally in animal and vegetable oils, but they are noticeably absent in mineral oils. The situation may be remedied in the case of mineral oils by adding a small percentage of an active polar compound (such as oleic acid). These additives, having a great affinity for metal surfaces, migrate to them through the mineral oil, and thus form a strong monolayer or chemical coating which excludes the mineral oil molecules. Only a small

TABLE 7.2. MOLECULAR
LENGTHS OF TYPICAL FATTY
ACIDS

Lubricant	Molecular length (Å)
Castor oil	5.5
Ricinoleic acid	5.2
Oleic acid	10.8
Palmitic acid	21.4
Stearic acid	23.8

fraction of 1% of additive is required to coat the surface in this manner. Table 7.1 shows that a significantly greater percentage (in this case 10%) of oleic acid is required to reduce the coefficient of sliding friction to its minimum or near-minimum value. Other typical fatty acids include castor oil, ricinoleic acid, palmitic acid, and stearic acid, and their molecular lengths are measured in Ångstrom units as Table 7.2 indicates. Since the action of

fatty acids in reducing boundary friction is bonding to a metal surface with sufficient adherence to resist being torn off (when the rubbing surfaces slide over each other) together with
normal orientation of the chains to the surface, it follows that the longer the length of the
molecular chains (and hence the greater number of carbon atoms) the lower the frictional
coefficient. The evidence for this is to be found in Fig. 7.6, which compares the frictional
performance of normal fatty acids and normal paraffin hydrocarbons[34] as a function of

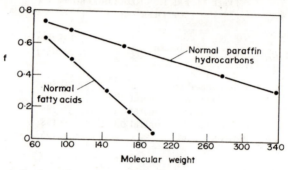

FIG. 7.6. Friction vs. molecular weight of normal fatty acids and normal paraffin hydrocarbons.[34]

molecular weight. The latter, of course, can be regarded as some measure of molecular chain
length. The paraffin hydrocarbons do not have the "oiliness" action of fatty acids in permitting relative slip near a solid boundary, and therefore the same reduction in friction
with increasing chain length cannot be expected.

   The bonding between the active end of the fatty-acid molecule and the solid surface suggests the formation of a soap film as a result of a chemical reaction. Thus, fatty acids are
most effective as friction reducers where the nature of the underlying metal permits a definite
chemical reaction to take place (such metals include copper, cadmium, zinc, magnesium,
and, to a lesser extent, iron and aluminium). For non-reactive surfaces (such as nickel,
chromium, platinum, silver, and glass), the fatty acid shows very little effect.

   We summarize many of the findings in this section by noting that four requirements must
be met by any mineral oil additives (whether fatty acids or metal soaps). These are:

(a) The molecule of the additive should contain an active radical to give strong adhesion
    to the bearing surfaces.
(b) The molecule must also be of the long, straight-chain type, having the active radical
    at one end.
(c) The additive must be present in a concentration of at least a few per cent but not in
    such quantity as to contribute disadvantageous bulk properties (such as acidity and
    gummy deposits) to the lubricant.
(d) Either its own "transition" temperature[†] or that of its reaction product with the bear-
    ing surface should exceed the maximum anticipated operating temperature.

[†] The "transition" temperature defines an upper limit beyond which attraction and orientation in boundary lubricant films ceases entirely as a result of thermal agitation, and molecules that were in the surface
layer disperse through the bulk lubricant.

The most promising alternative to the hydrocarbon family of compounds which have been considered exclusively in this section are the silicones. These are synthetic substances characterized by long chains composed of alternate silicon and oxygen linkages, as shown in Fig. 7.7. A silicone molecule may also contain active radicals, and silicones are superior

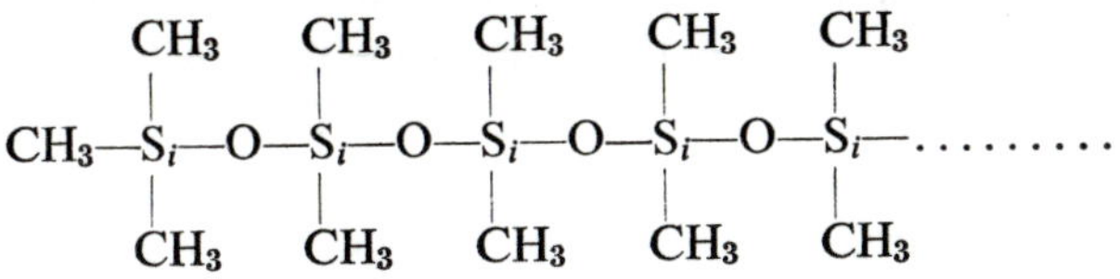

$$CH_3-\underset{\underset{CH_3}{|}}{\overset{\overset{CH_3}{|}}{Si}}-O-\underset{\underset{CH_3}{|}}{\overset{\overset{CH_3}{|}}{Si}}-O-\underset{\underset{CH_3}{|}}{\overset{\overset{CH_3}{|}}{Si}}-O-\underset{\underset{CH_3}{|}}{\overset{\overset{CH_3}{|}}{Si}}-O-\underset{\underset{CH_3}{|}}{\overset{\overset{CH_3}{|}}{Si}}-\cdots\cdots$$

FIG. 7.7. Molecular structure of silicones.

to hydrocarbons in having a higher viscosity index and in being non-inflammable. They find use as hydraulic fluids in air and marine applications.

Water is sometimes used as a boundary lubricant for the rubber bearings of marine propeller shafts. Greases are mixtures of minerals and soaps (with or without a solid filler) usually having the consistency of a semisolid gel at room temperature; they find application at high temperatures, or perhaps in inaccessible locations, or where liquid lubricants may damage a product through leakage.

7.5 General Properties of Metallic Films

The contact of metals and elastomers is illustrated schematically in Figs. 7.2 and 7.3 respectively, but although helpful in deriving the relevant boundary lubrication equations in Sections 7.2 and 7.3, the figures do not indicate a distinction between boundary and bulk lubricants. Figure 7.8 shows a generalized macro-rheodynamic model of boundary lubrication which (in the case of metals) shows a transition from bulk to boundary lubricant to

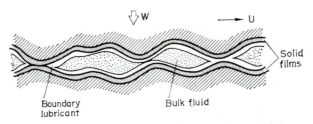

FIG. 7.8. Generalized macro-rheodynamic model.[35]

boundary solid as we approach the contours of the sliding surfaces. The solid films present are usually metallic oxides and have a thickness of approximately 100 Å. Superimposed on the solid films are monolayers or multilayers of boundary lubricant having a thickness of about 30 Å. Sometimes the solid films are discontinuous and incomplete because of chemical and physical effects during their formative period.[35] Even if originally continuous, they may become detached during sliding, thus exposing clean metal surfaces and creating a much greater frictional resistance. Boundary films may range from extremely thin (such as a monolayer of chemisorbed soap), to thick (such as a 1000 Å film of iron sulphide). During their formation they are first physically adsorbed on the surface in question by a relatively weak bonding mechanism. In many cases the physically adsorbed films react chemically to form a new entity–*chemisorbed* films, which are characterized by strong bonding energies. The formation of both adsorbed and reaction films depends critically on temperature. Thus the weak bond of a physically adsorbed film is further reduced by increased temperature, whereas chemically bonded films increase their rate of formation with increased temperature.

Other properties besides adhesion (or tenacity) and rate of formation, are melting point, shear strength and hardness. It appears that thermal softening is one of the major factors leading to the failure of solid films with a consequent high friction and wear between surfaces. A correlation between melting point and failure temperature has been established for a variety of organic films, and in all cases the correspondence of melting points and sudden rise in friction is striking. Soft metal films (such as lead and zinc) fail at their melting points, and soap films and chemisorbed fatty acids fail to lubricate above their melting points.

The shear strength of boundary films varies from a very low value of 0.25 kg/mm for calcium stearate (coefficient of friction = 0.05) to 130 kg/mm for iron. High shear strength is generally indicative of high friction values, thus coefficients of friction as high as 1.8 have been recorded for copper oxide films.[36] In general, friction is a function of the cleanliness factor k of surfaces[†] (where $k = s/\tau$, s = shear strength of solid film or surface contaminant, τ = critical shear strength of bulk metal). This can be shown by putting $p^* = 3\tau$ and $\alpha = 3$ in the plasticity eqn. (3.21), and after rearrangement substituting the simple theory of adhesion relationship in eqn. (3.20) thus:

$$\left.\begin{array}{l} \bar{p}^2+9s^2 = 9\tau^2, \\ \bar{p}^2 = 9s^2[(1/k^2)-1], \\ f = \dfrac{s}{\bar{p}} = \dfrac{1}{3\sqrt{[(1/k^2)-1]}} \,. \end{array}\right\} \tag{7.18}$$

Figure 7.9 shows a plot of this equation, and it has been verified by many experiments using varying contaminating films on metal.[1] Thus MoS_2 film on steel gives low friction and Fe_2O_2 gives high friction. The theory expressed by eqn. (7.18) also suggests that even if different combinations of contaminant and surface are used, the friction remains the same provided that k remains unchanged. This is confirmed by experiment; thus in order to obtain low friction with a copper surface, a low shear strength film such as stearic acid is required, and with this combination $f = 0.09$. The same frictional coefficient may be obtained with materials of a higher level of shear strength, such as lead sulphate on steel. In all boundary lubrication experiments using solid films, the location of the shear plane remains unknown, and a further understanding of boundary lubrication will undoubtedly emerge when this is known.

Optimum properties should include a high melting point and thermal stability and low shear strength. Unfortunately, high melting point is usually associated with high shear

[†] See Section 3.7.

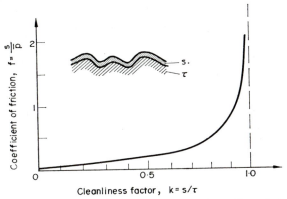

Fig. 7.9. Coefficient of friction as function of cleanliness factor.

strength (e.g. iron sulphide films), and compromises must be made. Molybdenum disulphide is an exception, since it possesses both high thermal stability and low shear strength simultaneously.

7.6 Adsorption of Lubricants

We have considered exclusively in the previous section the behaviour of solid metallic films. In boundary lubrication a thin lubricant film is superimposed on the solid film (as shown in Fig. 7.8) whose shear strength is considerably less than that of the metal. On the other hand, if flexible or polymeric materials are considered, the long-chain molecules of superimposed lubricant films are little different from the molecular structure of the underlying material. Thus even in cases where strongly adsorbed films are present, their shear strength and sliding coefficient of friction cannot be very different from that of the polymer itself. In the latter case, the bulk properties of the polymer may or may not be affected to some extent by the lubricant. Thus the distinction in behaviour after lubrication can be regarded as not between elastomeric and metal surfaces specifically but rather between permeable and impermeable materials. Three distinct effects are discussed in this section as they relate to the friction of polymers or elastomers in a lubricating environment. These are *softening, lubricity*, and *porosity*.

Softening

Laboratory tests have shown a distinct rise in the coefficient of sliding friction when natural rubber and leather specimens are soaked in water for varying periods of time and then caused to slide on a selection of rough and smooth surfaces. Figure 7.10 indicates these trends for a natural rubber slider. The results show that the coefficient of friction is increased in all cases by prolonged soaking up to a maximum soaking period of about 30 min. Further soaking after such a period causes little change in frictional behaviour. The fine sandpaper appeared to give the highest and the teflon surface the lowest coefficient of friction both before and after soaking, but the largest increase in coefficient was obtained on the smooth aluminium surface and the smallest on the fine sandpaper texture. Similar results were obtained using leather sliders.[2]

The results in Fig. 7.10 can only be explained by softening of the surface layers of the slider material during soaking. We note that although both the fine sandpaper and aluminium surfaces have comparable friction coefficients after soaking, only the sandpaper has a hysteresis component of friction (see Chapter 4). The aluminium surface being smooth and nominally dry has a very high adhesion. Since the total coefficient of friction f consists of adhesion and hysteresis components (as seen in Chapter 4) which are pressure dependent as follows:

$$f = K_1 \left[\underbrace{\frac{E'}{p^r}}_{\text{Adhesion}} + \underbrace{K_2 \left(\frac{p}{E'} \right)}_{\text{Hysteresis}} \right] \tan \delta, \tag{7.19}$$

and since E', $\tan \delta$, and $K_1 K_2$ are constant in a given experiment, we conclude that changes in friction must be due to changes in K_1 or p or both. The notation for the variables in eqn. (7.19) may be found in Chapter 4. Shore hardness measurements for the rubber slider taken immediately after each soaking period showed no detectable changes in the pressure p as a consequence of the soaking and hence softening effect. Thus the observed changes in f from Fig. 7.10 must be due to surface rather than bulk effects, and this can be interpreted

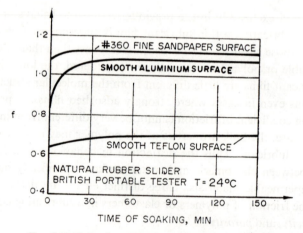

FIG. 7.10. Softening effects on coefficient of friction.

as a softening of the surface layers of the slider. The constant K_1 in eqn. (7.19) and hence the area of intimate contact increases markedly with soaking time. This produces a dramatic increase in the coefficient of sliding friction f for the aluminium surface as a result of increased flexibility of the surface layers. The percentage increase in total coefficient due to soaking and softening is necessarily smaller in the case of the sandpaper texture because a substantial part of f in this case is due to hysteresis and therefore invariant.

The teflon surface undoubtedly exhibits a lubricity condition at the sliding interface as discussed in the next paragraph. This reduces the effective shear strength and ultimately reduces the increase in K_1 which would otherwise occur. It is possible to produce a bulk softening effect by selecting a lubricant other than water for the soaking tests; in this case, it is anticipated that by reducing p in eqn. (7.19) the same trends will be evident as those shown in Fig. 7.10. It is important to stress, however, that no bulk changes in slider properties were observed in the tests already conducted, and the softening effect has therefore been confined to the surface layers.

Lubricity or "Oiliness"

Experience has indicated a marked superiority at moderate temperatures in the ability of certain animal and vegetable oils to keep smooth surfaces slippery as compared with mineral oils of the same viscosity. This ability has been called "lubricity" or "oiliness". The term is useful in denoting the quality of a lubricant which makes it effective under conditions of boundary lubrication where viscosity is not important.

Another definition of oiliness adopted by the Society of Automotive Engineers indicates the uncertainty which exists as to its actual origin. This definition describes oiliness as a term "signifying differences in friction greater than can be accounted for on the basis of viscosity when comparing different lubricants under identical test conditions". No physical property of a fluid has yet been found to correlate with the results of oiliness tests, but it appears certain that lubricity effects reflect the adsorption of lubricants on particular surfaces. Adsorption and lubricity characteristics are much more significant in the case of elastomers and polymers compared with metals. For elastomers the long-chain molecular structure is not unlike that of a superimposed lubricant, so that adsorption may readily occur and the properties of the adsorbed film are similar to those of the elastomer. According as we increase the chain length of the lubricant (by progression through a homologous series), it becomes increasingly difficult for the lubricant molecules to penetrate the polymer, and the lubricity or slippery effect decreases.

Oiliness is therefore a boundary friction characteristic of a lubricant which is independent of viscosity. There are many practical examples of the presence or absence of the oiliness phenomenon. Water, for example, has appreciable viscosity but practically no oiliness. A wet finger rubbed on the edge of a wine glass produces enough friction to cause a stick–slip excitation which may eventually create ringing in the form of a musical note. With glycerine (having considerable viscosity) the edge of the glass can still be felt as a harsh, rough surface. On the other hand, a little grease or lubricating oil (having the oiliness characteristic) causes the finger to slip freely along the edge of the glass without vibration. Dry, smooth teflon gives a much lower coefficient of sliding friction than other surfaces having the same approximate contact area, and we may attribute this to a lubricity effect at the sliding interface, as suggested by Fig. 7.10.

Surface Porosity

The absorption of lubricants must be physically dependent to some extent on the surface and bulk porosity of the underlying solids. We distinguish carefully between static and dynamic porosity effects in bodies dependent on the application in mind. Thus the provision of pores within the surface layers of bituminous and asphaltic pavements enables excess rainfall to permeate through the road structure, thereby minimizing the risk of flooding. While this is a static or low-speed phenomenon, dynamic effects may occur on the same road texture due to the passage of a rapidly rolling automobile wheel. Under such conditions the tread elements rapidly attempt to force water through the pores of the road texture and often create a condition of choked flow. It is highly probable that the choking effect which accompanies dynamic loading reduces the volume flow rate through the pores which would occur under "static" conditions. Surface porosity effects may play a vital role in the lubrication mechanism for animal joints by permitting synovial fluid to exude through the cartilage and thereby preventing contact between joints. In bearing design, the effects of porosity are often included to augment the lubrication effect. On a microscopic scale the existence of minute surface pores on certain materials provides a basic adsorption mechanism which tends to remove lubricant films and create a condition of boundary friction.

7.7 Wettability and Contact Angle

Since the addition of a lubricant at a sliding interface greatly alters frictional performance, studies have been carried out on the degree to which various surfaces can be wetted. This is normally done by measuring the contact angle of a liquid (usually water) on the surface in question. A surface is said to be hydrophobic if it tends to repel the wetting agent so that the angle of contact θ is large, whereas hydrophilic surfaces have an affinity for water, giving rise to a small contact angle θ (see Fig. 7.11). If γ_L, γ_{SL}, and γ_S denote the free energies per square centimetre of the liquid, solid–liquid, and solid interfaces (also called their surface tensions), we can write the following relationship:

$$\gamma_S = \gamma_{SL} + \gamma_L \cos \theta, \qquad (7.20)$$

which has been called the Young–Dupré equation. At first it appears that this relation is a trivial balance of horizontal forces in accordance with Fig. 7.11. However, it can also be derived in terms of the conformation of minimum free energy and it has a sound thermo-

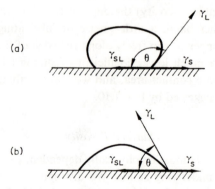

FIG. 7.11. Sessile liquid drop on (a) hydrophobic, and (b) hydrophilic surfaces.

dynamic basis. It is generally agreed that the derivation of eqn. (7.20) as a force balance is merely intuitional, and it leaves as a consequence the quantities γ_{SL} and γ_S undefined physically. Not only are these latter terms impossible to measure, but it is difficult to imagine tensile stresses existing in the surface of a solid which would not penetrate to some extent beneath the surface layer. We can regard eqn. (7.20) as valid for a sessile drop on a surface which is not too deformable.

Calcium carbonate ($CaCO_3$) and glass (SiO_2) can be regarded as reference surfaces which are extremely hydrophobic and hydrophilic respectively. Figure 7.12 shows that whereas the dry adhesion coefficients are comparable, the addition of water at the interface provides

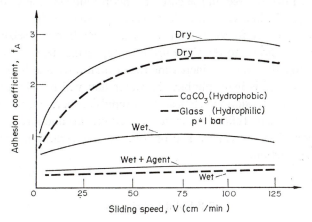

FIG. 7.12. The adhesion of hydrophobic and hydrophilic surfaces.

a distinct difference in adhesion readings. The $CaCO_3$ tends to reject the water, so that complete wettability is not attained, and a relatively high adhesion coefficient of about unity is recorded. On the other hand, the hydrophilic glass surface gives almost complete wettability and $f_A \doteq 0.25$. We note that the addition of a detergent or wetting agent (about 1 part per 1000) reduces the surface tension of the water and permits the normally hydrophobic surface to be readily wetted.

The contact angle θ in Fig. 7.11 varies from $0°$ for complete wettability to some value $< 180°$ for non-wettable surfaces. Table 7.3 lists the values of θ for some common plastics

TABLE 7.3. CONTACT ANGLE
ON PLASTICS

Plastic material	Contact angle θ
Nylon	0°
Perspex	0°
Terylene	41°
Polythene	89°
Teflon	126°

and hence the wettability. There is a direct relationship between the adhesion coefficient measured on surfaces and the value of contact angle θ. When the solid surface exhibits a roughness, there is evidence that the adhesion capability is increased above the value for smooth surfaces. Finally, the wettability of a surface is time-dependent with a tendency to progress from the hydrophobic to the hydrophilic state. This is especially true for terylene (Table 7.3) and for most waxed surfaces.

A knowledge of the contact angle which water makes with a surface is of interest not only because an indication is given of the adhesion developed between the surface and solid ice[1], but also because it will influence the physical behaviour of the water layer formed during the sliding process. For example, in the case of a hydrophilic ski the water formed on the summit of an ice crystal will become attached to and spread over the surface of the ski. The formation and subsequent re-freezing of this water film may lead to a seizure of the surfaces so that the snow sticks firmly to the ski. Furthermore, surface tension effects tend to pull the surfaces closer together, thereby increasing the normal force and hence the friction. On the other hand, if the ski is hydrophobic, the water film will remain attached to the ice crystal (and also surface tension effects lead to a repulsion between the surfaces), so that a lower friction is obtained. Teflon is commonly used on skis and aircraft skids for operation on ice and snow since it creates a hydrophobic surface condition and gives the lowest coefficient of sliding friction.

7.8 Solid Lubricants

A solid lubricant can generally be defined as a material that provides lubrication between two surfaces in relative motion under essentially dry conditions or at asperity peaks in the case of boundary lubrication. The most common of the present dry, solid lubricants are graphite and molybdenum disulphide. Although the use of graphite as a lubricant probably dates back to the middle ages, its use as a bonded solid lubricant is relatively recent.[37] The use of molybdenum disulphide as a lubricating solid† started in the 1940s, and it is currently the most widely used material. Indeed, a strong trend has developed since this time towards the use of higher temperatures and pressures in moving parts, and the development and use of solid lubricants is due to the need for lubrication under these extreme conditions. Other solid lubricants in use include soft metals (such as lead) and metallic oxides, sulphides, selenides, and tellurides.

Solid lubricants have the advantage of good stability at extreme temperatures and in chemically reactive environments. They are generally of lighter weight, they require fewer seals, and the need for a recirculating oil system (with pump and other components) is eliminated. Solid lubricants can be used on components that are difficult to lubricate with conventional liquids.

On the other hand, the coefficient of friction is generally higher than in the case of hydrodynamic lubrication. Another disadvantage is that dry film coatings have finite wear lives,

---

† See Chapter 5 for further details.

and some wear is unavoidable because of solid sliding contact. Solid lubricants also have no cooling capacity.

Bonded and Powdered Solid Lubricants

There are, in general, three distinct ways in which lamellar solids can be applied as solid-film lubricants to metal surfaces:[1]

(a) incorporation of lamellar solid into a suitable resin or binder and glueing to the surface in question;
(b) depositing solid as fine powder and rubbing it into the surface; and
(c) formation of material by chemical reaction at the surface itself.

With regard to the first method, the use of a binder ensures stronger adhesion, and it is widely used in practice especially with molybdenum disulphide (MoS_2). Resin-bonding solid lubricants are generally applied in thin films to metal surfaces, usually by spraying to a thickness of about 12.5 μ. The resins used as the adhesive material or binder (whether phenolic, epoxy, or polyimide) are organic in nature, and the pigment or lubricating solid is usually graphite or molybdenum disulphide. The proper pigment-to-binder ratio is critical to the expected life of the solid lubricant. For high-temperature applications, inorganic bonding using ceramic or salt-based binders is substituted for resin bonding, and for greater thermal and oxidative stability the graphite or MoS_2 is replaced by lead oxide or sulphide, calcium fluoride, etc.

Depositing the solid lubricant as a fine powder is a convenient method of applying the material to a metal surface. The three most common powders are graphite, MoS_2, and teflon, but these have definite limitations. Thus MoS_2 oxidizes at about 400°C, which limits its long-term use in air at very high temperatures. Graphite requires the presence of adsorbed moisture to help its lubricating properties, and it is therefore unsuitable for high-vacuum conditions. Teflon is limited to light loads and temperatures <315°C. These limitations, however, have stimulated the synthesis of other superior lubricating solids, including oxides, sulphides, selenides, and tellurides.[38] Having been deposited as a powder, subsequent rubbing of the material causes most of the powder to disappear, although microscopic studies have indicated that some of the dry lubricant remains trapped in the abrasion grooves which are formed in the underlying surface.

Lubricating solids are added to oils and greases to improve their lubricity characteristics. This tends to delay or prevent seizure in places where the film is disrupted. The choice of how much additive is required undoubtedly depends on the application, but for general use a grease containing about 5% of the lubricating solid is the most effective.

Lubricating Plastics

Non-lubricated plastic bearings fulfil a need that cannot be satisfied by metal bearings, chiefly by reducing or eliminating the need for lubrication. They also help eliminate problems arising from high wear rates, lubricant failure, chemical corrosion, stick–slip motion,

11*

and extreme temperatures. Common materials used in such bearings are nylon, acetals, tetrafluoroethylene (TFE), polimides, and phenolics. Indeed, these plastics are often blended, and small additions of MoS_2 and graphite as dry lubricant fillers reduce the friction and wear levels. Glass fibre fillers increase stiffness and reduce thermal expansion especially for nylon. TFE has the lowest friction of any plastic, and its coefficient of friction decreases with load. It has the highest resistance to heat, although a poor thermal conductor, and is commonly used as a blending agent to increase wear life.

Some moulded bushings have three lubricants added: graphite, molybdenum disulphide, and a viscous oil. Fibre fillers arranged radially serve as wicks for the oil,[37] so that when the bearing warms up and the lubricant is fluidized, true hydrodynamic lubrication is obtained.

Lubricating plastics, in general, are best suited to lightly loaded conditions and moderate speeds, and under these conditions they give extremely long service. Additional load capacity can be obtained, however, by the addition of metal and fibre fillers. The applications of solid lubricants are diverse, ranging from military and aerospace erqipment to the automotive, heavy electrical and metal working industries.

7.9 Stick–Slip Phenomena

We have seen in Chapter 4 that adhesion can be described in fundamental terms as a molecular–kinetic, thermally activated exchange mechanism at a sliding interface. Molecular bonds formed at any instant are stretched, ruptured, and relaxed an instant later, thus giving rise to a dissipative stick–slip process on a molecular level. Although the scale at which these events occur is too microscopic to observe this phenomenon directly, experimental evidence appears to confirm overwhelmingly the nature of the process.

On a macroscopic level, a similar mechanism commonly referred to as "stick–slip" can be observed experimentally. A pre-condition for macroscopic stick–slip to occur is the exis-

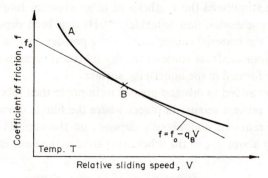

FIG. 7.13. Decreasing friction vs. velocity characteristic.

tence of a decreasing friction velocity characteristic, as shown in Fig. 7.13. Consider the motion of a block of weight W sliding relative to a rough, horizontal base, as shown in Fig. 7.14(a). The block is restrained by a spring k and a viscous damper η attached to a

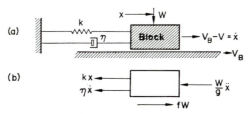

fixed support, the velocity of the base to the right is V_B, and V is the relative velocity (in the negative x-direction) of block and base. At the instant considered, the block is moving to the right with a relative velocity V corresponding to the point B in Fig. 7.13, so that the corresponding coefficient of sliding friction is given by

$$f = f_0 - q_B V, \tag{7.21}$$

where q_B is the slope of the friction–velocity curve at B and f_0 is a constant. Figure 7.14(b) shows the instantaneous forces acting on the block, from which the following equation may be written

$$(W/g)\ddot{x} + \eta\dot{x} + kx = fW. \tag{7.22}$$

Substituting for f from eqn. (7.21) into eqn. (7.22) and putting $V = V_B - \ddot{x}$ results in

$$\ddot{x} + K_3\dot{x} + K_4 x = K_5, \tag{7.23}$$

where $K_3 = -([q_B W - \eta]/W)g$, $K_4 = (kg/W)$, $K_5 = (f_0 - q_B V_B)g$, and q_B is considered constant for the motion. Assuming that the slope of the friction curve q_B is such that $q_B > \eta/W$, the negative damping coefficient K_3 in eqn. (7.23) can be shown to give an exponentially increasing amplitude of vibration.[46] This self-excited "chatter" has many applications in mechanical engineering such as may occur with the driving wheels of a locomotive or the screech of chalk on blackboard, or perhaps the ringing of wine glasses described earlier.

The sinusoidal form of the solution to eqn. (7.23) is not, of course, characteristic of stick–slip oscillations, but it is important to realize that the negative damping coefficient feeds energy into the system and makes the stick–slip phenomenon possible. If we consider an increase in the slope of the friction–velocity curve as speed is reduced (i.e. q_A → see point A in Fig. 7.13), then K_3 becomes even more negative. This causes a violent and dangerous increase in the amplitude of the chatter. A typical example is the behaviour of cast-iron brake blocks on railway vehicles.[46] Since cast-iron displays a rapidly increasing slope of the friction versus velocity curve as speed is reduced, the chatter becomes extremely violent as the train approaches a standstill. At higher speeds, the slope is virtually zero, so that the onset of braking is very smooth. Non-metallic brake blocks, while subject to considerably less variation in friction with speed than cast-iron, have, nevertheless, a falling characteristic which results in some chatter at all speeds.

CHAPTER 8

ELASTOHYDRODYNAMIC LUBRICATION

8.1 Introduction

Elastohydrodynamic lubrication can be briefly described as "the study of situations in which elastic deformation of the surrounding solids play a significant role in the hydrodynamic lubrication process".[39] In most machine applications, forces are transmitted from one component to another by means of large effective bearing areas, but it is not uncommon to find in addition nominal line or point contacts. Typical examples of the latter are gears and rolling contact bearings, and these perhaps constitute the most commonly encountered applications of the elastohydrodynamic phenomenon occurring between metallic surfaces.[40] It has been recognized for years that many loaded contacts of low geometrical conformity behave as though they are hydrodynamically lubricated, yet in the absence of elastohydrodynamic theory, the Reynolds theory of hydrodynamic lubrication fails to predict why adequate lubrication should exist at all under what appears to be the most severe and limiting stress conditions. For example, the line contact of meshing involute gears suggests extremely high pressures (since the area across which forces are transmitted appears to approach zero), and in the absence of elastic distortion of the gear teeth it is difficult to imagine a lubricant capable of resisting such pressures. By allowing for changes in lubricant viscosity with pressure and elastic deformation of the contacting solids, however, it can be shown that (in agreement with experience) adequate lubrication will persist under such conditions.

The two significant effects which occur in elastohydrodynamic situations and are not accounted for in the classical theory, are (a) the influence of high pressure on the viscosity of liquid lubricants, and (b) substantial local deformation of the elastic solids.

These effects drastically change the geometry of the lubricating film which in turn alters the pressure distribution at the contacts. In essence, the hydrodynamic pressure generation must be matched with the elastic pressures in the contacting solids, and a solution to the combined lubrication and elastic equations gives the final elastohydrodynamic condition at the contact spots.

Although the phenomenon of elastohydrodynamic lubrication was first discovered as a result of the need for understanding the lubrication of gears and roller bearings, there has been an increasing interest during the last decade in the lubrication of soft, flexible surfaces made from elastomeric or polymeric materials.[2] Typical applications such as the lubrication of windshield wipers, reciprocating and rotary lip seals, flexible-pad thrust bearings, and automobile tyres sliding and rolling on wet roads, are presented in detail in Part II. Both the rigid–rigid and rigid–flexible types of elastohydrodynamic lubrication are treated fundamentally in this chapter, and it will be seen that distinct differences arise between the two mechanisms. Before proceeding in this direction, it is first necessary to establish the general nature of the elastohydrodynamic problem and the method of solution.

8.2 General Iterative Procedure

The elastohydrodynamic problem involves an iterative procedure to establish a compatibility between the hydrodynamic pressure generated in the lubricating film which separates two elastic bodies in relative motion and the elastic pressures which are developed between the bodies as a consequence of their virtual contact, In simple terms, we assume some initial film thickness which is inserted in the Reynolds equation to obtain a pressure distribution.

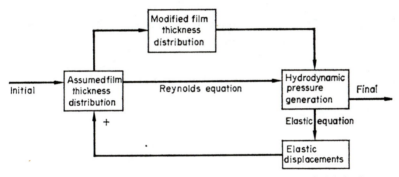

FIG. 8.1. The general iterative procedure in elastohydrodynamic lubrication.

If the latter is now inserted in the appropriate elastic equation, an initial estimate of the elastic displacements is obtained which is then used to modify the original assumed film-thickness distribution. The iteration continues until the modified film-thickness distribution is little different from the assumed distribution in any one iterative step. Figure 8.1 depicts the iterative sequence in schematic form.

Consider as a specific example of the procedure outlined in Fig. 8.1 the case of a rigid cylinder approaching a flexible or elastic plane in the presence of a lubricant. (In point of fact the distortion of the elastic plane 2 in Fig. 8.2 can be considered as the sum of the distortions produced in both cylinder and plane if the cylinder is not assumed to be rigid; thus the example has general validity, and the rigidity of body 1 in the figure does not limit the generality of the model.) Let h_{0_i} represent the centre-line film thickness at any instant measured from the base of the cylinder to the undeformed profile of the elastic plane. The actual film thickness h at any location x below the cylinder and within the area of contact is given by

$$h = h_{0_i} + \frac{x^2}{2R} + w, \tag{8.1}$$

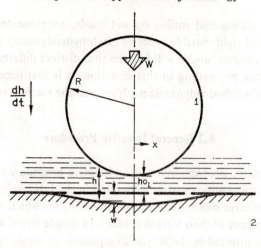

FIG. 8.2. Example of general elastohydrodynamic problem: rigid cylinder on flexible plane.

where R is the radius of the cylinder and w the local elastic deflection of the lower surface. From elasticity considerations[40] we can write for w

$$w = -\frac{2}{\pi E'} \int_{s_1}^{s_2} p(s) \ln (x-s)^2 \, ds + \text{const}, \tag{8.2}$$

where the equivalent Young's modulus E' is defined as

$$\frac{1}{E'} = \frac{1}{2} \left(\frac{1-v_1^2}{E_1} + \frac{1-v_2^2}{E_2} \right), \tag{8.3}$$

and the suffixes 1 and 2 refer to cylinder and plane respectively, as shown. The pressure $p(s)$ in eqn. (8.2) is variable between the limits s_1 and s_2, but the logarithm in the same integrand gives difficulties in numerical integration (since at $x = s$, the logarithmic function approaches the value minus infinity). One means of avoiding these difficulties is to write $p(s)$ in the form of a polynomial thus:

$$p(s) = p_0 \sum_{1}^{N} A_i s^i. \tag{8.4}$$

If we now substitute for $p(s)$ in eqn. (8.2), we obtain[40]

$$\int_{s_1}^{s_2} s^i \ln (x-s)^2 \, ds = f_i = \frac{2}{i+1} \left\{ (s_2^{i+1}-x^{i+1}) \ln |x-s_2| - (s_1^{i+1}-x^{i+1}) \ln |x-s_1| \right.$$
$$\left. - \frac{1}{i+1} (s_2^{i+1}-s_1^{i+1}) - \frac{1}{i} x(s_2^i - s_1^i) \ldots \ldots - \frac{1}{2} x^{i-1}(s_2^2 - s_1^2) - x^i(s_2 - s_1) \right\},$$

which can be readily evaluated. The final expression for displacement w becomes:

$$\frac{\pi w E'}{2 p_0} + \text{const} = \sum_{i=1}^{N} A_i f_i, \tag{8.5}$$

which can be computed. Usually, three terms of the series expression for f_i are sufficient to estimate w accurately.[40]

During normal approach of the cylinder in Fig. 8.2, the following form of the Reynolds equation is appropriate:

$$\frac{\partial}{\partial x}\left(\frac{h^3}{12\mu}\frac{\partial p}{\partial x}\right) = -\frac{\partial h}{\partial t}, \tag{8.6}$$

and the boundary conditions for the pressure p may be taken as

$$\lim_{x \to \infty}\left(\frac{\partial p}{\partial x}\right) = 0; \quad \lim_{x \to \infty}(p) = 0. \tag{8.7}$$

The iteration necessary to establish compatibility between the elastic and hydrodynamic pressure distributions at the interface in Fig. 8.2 may now be described following the general procedure in Fig. 8.1. If we initially neglect w and use eqn. (8.1) in eqn. (8.6) to compute the hydrodynamic pressure p subject to the boundary conditions prescribed by eqn. (8.7), the resulting pressure distribution may be inserted in eqns. (8.2) or (8.5) to compute w. The latter, of course, modifies our original film-thickness distribution. The iteration is usually allowed to proceed until the change in w obtained in any one iterative cycle is sufficiently small to indicate that a compatibility exists between the hydrodynamic and elastic effects.

It is interesting to note that the initial estimate of the pressure distribution under the cylinder is the hydrodynamic effect, since hydrodynamic conditions predominate during the initial stages of the approach. However, the final approach is characterized by the predominance of the elastic pressure distribution, and the final equilibrium of the cylinder on the elastic plane is a function only of the elastic properties of the base. Thus the hydrodynamic pressure distribution converges in the case of normal approach to the Hertzian[†] or elastic distribution.

We must again emphasize the generality of the approach problem in Fig. 8.2. It is a well-known fact that the contact between many machine elements can be represented by two geometrically and kinematically equivalent cylinders. This in turn can be reduced[‡] either to a single elastic cylinder near a plane rigid boundary or a single rigid cylinder near a plane flexible boundary, as shown in Fig. 8.2. In this manner, the elastic, hydrodynamic and kinematic conditions of the original contact are adequately simulated.

Finally, the method of iteration described in this section may or may not be directly applicable in its existing form to particular lubrication problems. Thus in metal-on-metal applications the elastohydrodynamic pressures generated change the viscosity of the lubricant by many orders of magnitude, and in sliding applications thermal effects occur which must be accounted for. These effects modify the iterative procedure suggested by Fig. 8.2, as indicated in the next section.

[†] Hertz, 1896, see ref. 13.
[‡] In the case of lubricated contact.

8.3 Fundamental Parameters

Three significant design parameters in elastohydrodynamic theory are as follows:

$$
\begin{aligned}
\text{Load parameter} \qquad & W' = \frac{W}{E'R}, \\
\text{Speed parameter} \qquad & U' = \frac{\mu_0 U}{E'R}, \\
\text{Materials parameter} \qquad & G = mE',
\end{aligned}
\tag{8.8}
$$

where W' and U' are dimensionless, W is the load per unit width of cylinder (see Fig. 8.2), R the effective radius of the roller pair, E' an effective modulus defined by eqn. (8.3), U the relative sliding speed, μ_0 a constant value of lubricant viscosity, and m the pressure exponent of viscosity according to the following relationship:[†]

$$
\mu = \mu_0 \exp(mp). \tag{8.9}
$$

In the case of metal-on-metal elastohydrodynamic film formation, the most influential parameter above is the speed parameter U' having a range of values from 10^{-8} to 10^{-13}. Since high pressures affect the absolute viscosity μ of the lubricant according to eqn. (8.9), it is convenient to replace p as an independent variable by a reduced pressure q, where

$$
q = \frac{1}{m}[1 - \exp(-mp)] \quad \text{or} \quad \frac{dq}{dx} = \exp(-mp)\frac{dp}{dx}. \tag{8.10}
$$

We note that q is therefore the pressure which would be generated in a fluid having constant viscosity μ_0, all other factors remaining unaltered.

By substituting q for p in the appropriate Reynolds equation[‡] the iterative procedure described earlier can be carried out. However, full analytical solutions to the combined elasticity and hydrodynamic equations have not been found, the difficulty being that the iteration process does not always converge. Thermal effects can be included by making use of the heat conduction equation for the solids[*] and an energy equation for the lubricant.[(39)]

Grubin[(40)] was the first to combine the load, speed, and materials parameters in a design formula for calculating mean film thickness. He assumed that the surfaces of the bounding solids outside the high pressure zone adopt the same profile during wet contact as they would in dry conditions. By further assuming that the pressure at the inlet to the high pressure zone attains very high values, Grubin was able to calculate the mean film thickness from a solution of the Reynolds equation thus:

$$
H = \frac{1.95G^{8/11}\,U'^{8/11}}{W'^{1/11}}, \tag{8.11}
$$

[†] Also eqn. (6.8).
[‡] This can take the form of eqn. (8.6) in normal approach, or eqn. (8.21) in relative sliding.
[*] See eqn. (3.47).

where $H = h/R$, h is the thickness of the parallel part of the film (see Section 8.5), R is the equivalent radius $R_1R_2/(R_1 + R_2)$ in the case of two cylinders (or simply the cylinder radius for a single cylinder contacting a flat surface), and the load, speed, and materials parameters W', U', and G have been defined in eqn. (8.8) previously. Equation (8.11) can also be written

$$\frac{h}{R} = 1.95 \left(\frac{W}{E'R}\right)^{-0.091} (mE')^{0.73} \left(\frac{\mu_0 U}{E'R}\right)^{0.73}. \tag{8.12}$$

A similar expression was obtained by Dowson and Higginson[40] as follows:

$$\frac{h_{min}}{R} = 1.6 \left(\frac{W}{E'R}\right)^{-0.13} (mE')^{0.6} \left(\frac{\mu_0 U}{E'R}\right)^{0.7}. \tag{8.13}$$

Since the power to which E' is raised is low in either expression, and since the range of variation of E' is small for heavily loaded contacts (such as steel and bronze), it follows that h_{min} is virtually independent of E'. Furthermore, it will be shown later[†] that h_{min} is almost independent of load W, so that to a first approximation we can assume that the load parameter W' and h_{min} are independent of each other. Equation (8.13) can therefore be written in the following abbreviated form:

$$h_{min} = \text{const} \, (\mu_0 U)^{0.7}. \tag{8.14}$$

Practical applications of the elastohydrodynamic film phenomenon usually require that the parallel film thickness (h or h_{min}) be expressed in terms of the other variables, as in eqns. (8.11)–(8.14).

We will now consider the normal and tangential approach of selected surface profiles (usually cylinders, spheres or flats) to a plane lubricated base.

8.4 Normal Approach

Cylinders

The case of a rigid cylinder in non-inertial descent on to an elastic, lubricated plane has been shown in Fig. 8.2, and the local deformation w of the plane defined by eqn. (8.2) is used to modify the lubricant film thickness given by eqn. (8.1). The same equations apply if the cylinder is assumed to be elastic and the plane surface rigid. Figure 8.3 shows the relative change in film thickness and pressure distribution for a flexible cylinder approaching a lubricated and plane, rigid base. These results were obtained[41] following the general iterative procedure in Section 8.2. In both Fig. 8.2 and Fig. 8.3 we assume that the pressures attained are not sufficient to alter the lubricant viscosity μ, so that eqn. (8.9) is inapplicable. We observe two distinct effects in Fig. 8.3:

[†] See Fig. 8.10.

(a) the change in pressure distribution from the familiar "bell" shape which typifies the initial stages of the approach to the final Hertzian or elastic distribution is relatively slight, whereas:

(b) the entire shape of the corresponding film thickness changes since the elastic surface of the cylinder inverts during the squeeze motion from a convex to a concave curvature.

The physical explanation for the radical change in film thickness is that the pressure distribution always has a maximum value along the centre line of the cylinder, and this promotes a maximum elastic displacement which eventually produces the concavity in the cylinder surface, as shown in the inset of Fig. 8.3. It is important to realize that Fig. 8.3(a) describes only a relative change in film thickness (h/h_0) as a function of position (x/a) in the area of "contact". During the squeeze action, of course, the centre-line film thickness $h_0(t)$ decreases with time. In the last stages of the approach the edges of the deformed elastic cylinder attempt to seal and trap the lubricant in the cavity. Subsequently, and over a long time period, the lubricant within the cavity exudes laterally and will completely disappear after an infinite time.

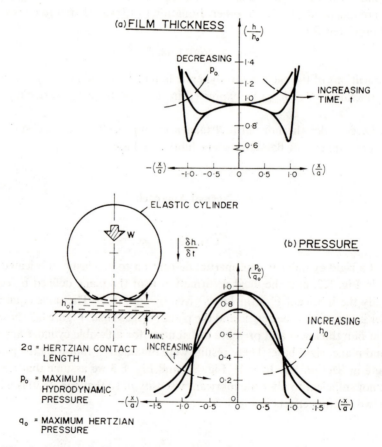

FIG. 8.3. Elastohydrodynamic film thicknesses (a) and pressures (b) for elastic cylinder approaching rigid plane surface.

Spheres

The analysis for a flexible or elastic sphere approaching a lubricated and plane rigid boundary is similar to that for a cylinder except that eqns. (8.1) and (8.2) are replaced by the following:[13]

$$h = h_{0_i} + r^2/2R + w \tag{8.15}$$

and

$$w = \frac{1-v^2}{\pi E} \int\int p \, ds \, d\psi, \tag{8.16}$$

where r is any radius within the contact circle and ds, $d\psi$ are defined as shown in Fig. 8.4. It can be shown simply that eqn. (8.16) may also be written as

$$w = \frac{4(1-v^2)}{\pi E} a \int_0^{\pi/2} p(r) \sqrt{\left(1 - \frac{r^2}{a^2} \sin^2 \psi\right)} d\psi. \tag{8.17}$$

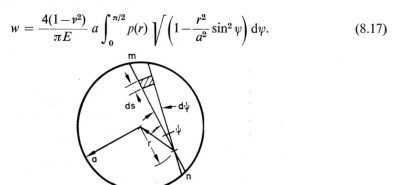

FIG. 8.4. Notation for elastic deflection in contact area of sphere on plane surface.[13]

This equation permits us to calculate readily the elastic displacement w at each location r. Also, the Reynolds equation (8.6) must be written in polar coordinates thus:

$$\frac{\partial}{\partial r}\left(\frac{rh^3}{12\mu}\frac{\partial p}{\partial r}\right) = -r\frac{\partial h}{\partial t}. \tag{8.18}$$

The iterative procedure is then carried out as described earlier. Figure 8.5 shows a comparison of film thicknesses obtained by pressing a rubber sphere on to a flat glass plate in the presence of a silicone fluid[42] using a normal load of 10 g. The experimental curves of film thickness (shown in Fig. 8.5) were constructed from interferometric photographs of the contact patch after (a) 5 s and (b) 5 min time duration. The solid curves in these figures have been obtained by substituting the Hertzian pressure distribution for a sphere into the Reynolds equation and finding the corresponding film thickness. We observe clearly from Fig. 8.5 that whereas the calculated and experimental values of the centre-line film thickness h_0 agree closely both during the initial and final stages of approach, the distribution of film thickness

(a) is far from the Hertzian equivalent at small time values, and
(b) shows a marked deviation from any calculated value at $r = a$ during the entire squeeze action.

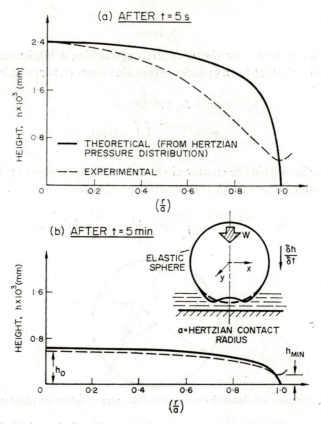

FIG. 8.5. Elastohydrodynamic film thicknesses for elastic sphere approaching rigid plane surface.

There are at least four possible reasons for the discrepancy between theory and experiment in Fig. 8.5(a):

(a) A Hertzian pressure distribution assumes the predominance of elastic effects, whereas during the initial approach the pressure generated is due primarily to the hydrodynamic squeeze contribution.

(b) It is extremely difficult to simulate theoretically the starting conditions of the experiments underlying Fig. 8.5.

(c) The ratio $(h_0/w) \gg 1$ at the start of the squeeze motion, whereas for a near-elastic pressure distribution it is necessary that $(h_0/w) \ll 1$.

(d) The "elastic" body may, in fact, exhibit viscoelastic properties, especially during the initial approach.

A major obstacle in performing the iterative steps for spheres and cylinders is that although the relative height (h/h_0) at any point in the elastohydrodynamic film may be obtained as a function of time during the squeeze motion, no means as yet of calculating the absolute magnitude of h_0 are available.

Flats

Perhaps the most interesting elastohydrodynamic problem in normal approach is the non-inertial descent of a flat, rectangular rigid punch on to a lubricated elastic plane, as shown in Fig. 8.6. Here again, the flexibility of the plane is such that very high pressures do not occur in the contact zone, and there is therefore no change in lubricant viscosity with increasing pressure. During the initial approach, the distribution of pressure is largely

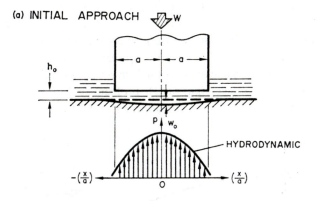

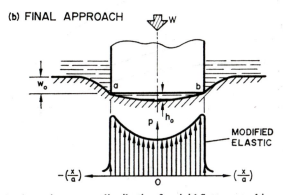

FIG. 8.6. Elastohydrodynamic pressure distribution for rigid flat approaching an infinite elastic plane.

parabolic because of the hydrodynamic effect, and the elastic deflection of the surface of the plane is slight, with a maximum value occurring along the centre line of the punch. After a relatively long period, the pressure distribution inverts to the elastic case as shown in Fig. 8.6(b). In this second position, the flat punch has penetrated into the elastic plane, and the edges *a* and *b* effectively seal off the lubricant beneath the punch. The elastic pressure distribution per unit width perpendicular to the paper and for sharp edges takes the form[13]

$$p(x) = \frac{W}{\pi a \sqrt{(1 - x^2/a^2)}},$$

(8.19)

and the corresponding deflections from elastic theory are as follows:

$$w = w_0 \qquad\qquad\qquad\qquad \text{for } |x/a| < 1,$$
$$w = w_0 \left[1 - \frac{\ln (2x^2/a^2 - 1)}{2 \ln 2} \right] \qquad \text{for } |x/a| > 1. \qquad (8.20)$$

These deflections neglect the additional contribution from the squeeze film trapped beneath the punch. We observe that the pressure distribution given by eqn. (8.19) reaches infinity at the edges a and b in Fig. 8.6(b). In practice, some degree of rounding must exist along these edges, and the pressure can therefore be assumed to attain high but finite values at these locations. *We note in particular that whereas for the elastic sphere or cylinder the distribution of pressure changes slightly during the squeeze motion and the elastohydrodynamic film thickness inverts in shape, the exact opposite is true for the rigid punch example.* No exact theoretical analysis exists for the rigid-punch problem in Fig. 8.6, but an approximate analysis has been attempted.[2] The results appear as a plot of the dimensionless variable

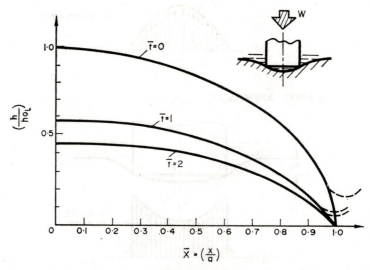

FIG. 8.7. Dimensionless film thickness for flat rigid punch on elastic plane.[9]

(h_0/h_{0_i}) vs. location (x/a) within the film for different time values $\bar{t}$, as shown in Fig. 8.7. The reference height h_{0_i} is the centre-film thickness under the punch at time $t = 0$, and $\bar{t}$ denotes a time ratio t/t_0, where t_0 is a constant. We observe from Fig. 8.7 that the elastohydrodynamic film preserves its relative shape during squeezing. In fact, the solid curves in the figures as $\bar{x} \to 1$ suggest a contradiction, since the existence of zero film thickness at these locations precludes the possibility of any escape for the trapped lubricant, and the squeeze motion cannot therefore take place. For this reason, the more realistic profile with rounded edges is shown dashed in the figure, thus permitting a ready escape path for the lubricant.

8.5 Relative Sliding

The steady-state sliding of an elastic body relative to a rigid plane surface (or of a rigid body relative to an elastic base) in the presence of a lubricant follows the broad iteration procedure dealt with in Section 8.2, except that the Reynolds equation now takes the form

$$\frac{d}{dx}\left[\frac{h^3}{6\mu}\frac{dp}{dx}\right] = U\frac{dh}{dx},\qquad(8.21)$$

where U is the relative velocity of sliding. Figure 8.8 shows the measured film thickness[42] and corresponding pressure distribution[40] for an elastic sphere sliding on glass in the

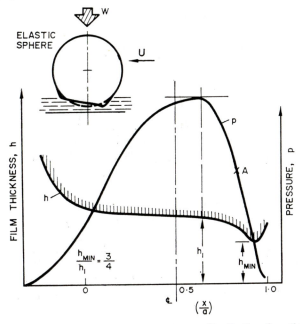

FIG. 8.8. Elastohydrodynamic film thickness and pressure distribution for sliding elastic sphere.

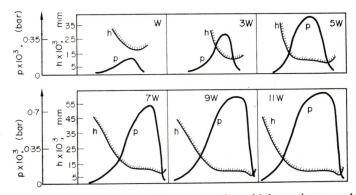

FIG. 8.9. Changes in film-thickness and pressure profiles with increasing normal load.

12 M: PAT: 2

presence of a viscous oil. We observe that a constriction appears in the film towards the rear of the contact length, and this is best explained from continuity conditions. Thus within the front part of the contact zone the pressure build-up is due to a narrowing film thickness, whereas the relatively sudden drop in the pressure curve towards the rear part requires a sharp constriction to maintain continuity of the flow without cavitation. The gradual appearance of the constriction with increasing load is more apparent from Fig. 8.9, which also indicates the accompanying steepening of the outlet pressure curve in the case of a rubber block sliding on a lubricated, bronze drum surface.[40] The pressure curves in Fig. 8.9 appear to approach a near-Hertzian distribution as load is increased.

At higher pressures than those considered in the previous two figures, the effects of pressure on viscosity must be included. Figure 8.10 shows pressure distribution and corresponding

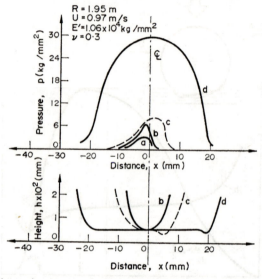

FIG. 8.10. Effects of change in viscosity and elastic deformation on pressure and film thickness.[40]

film thickness variation for a cylinder sliding on a lubricated plane surface. Four distinct cases appear in Fig. 8.10 thus:

(a) constant viscosity lubricant, rigid cylinder;
(b) variable[†] viscosity, rigid cylinder;
(c) constant viscosity, elastic cylinder;
(d) variable[†] viscosity, elastic cylinder.

We observe in particular from this figure:

(a) The vast combined influence of two effects (viscosity change and elasticity) compared with the relatively small influence of each separately.

----
† Pressure-dependent, see eqn. (8.9)

(b) The movement towards a Hertzian distribution of pressure (see *d*).
(c) The existence of the characteristic constriction in the film thickness towards the outlet for an elastic cylinder (*c* and *d*).
(d) The very small variation of film thickness with load.

All the characteristic features of an elastohydrodynamic film are present in Fig. 8.10 with one notable exception—the pressure spike corresponding to the film constriction (this would appear at point *A* in Fig. 8.8). In fact, the values of the material parameter *G* and the sliding speed *U* are sufficiently low in Figs. 8.8–8.10 to eliminate the secondary pressure peak. This condition usually applies in the case of elastomeric materials such as rubber or plastics. For metals, however, we cannot ignore the pressure spike, as discussed in the next section.

8.6 The Pressure Spike

Consider the case of a cylinder sliding relative to a lubricated base surface under conditions of high normal load and low speed as shown in Fig. 8.11. The pressure and film-thickness distributions have been presented as dimensionless plots, where $P = p/E'$, $H = h/R$, and $X = x/a$. Here *a* denotes the Hertzian semi-contact width, and the usual notation applies as used previously. We observe a distinct pressure spike in the vicinity of the film-thickness constriction for two values (a) and (b) of the materials parameter *G*, corresponding respectively to steel and bronze cylinders in mineral oil. We note that the height of the pressure spike increases with the value of *G*, as also does the parallel value of film thickness. For an incompressible lubricant as in the case of Fig. 8.11, *H* can also be shown

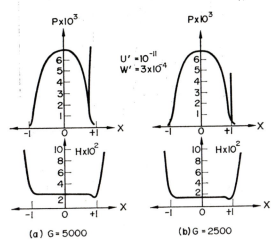

(a) G = 5000   (b) G = 2500

FIG. 8.11. The pressure spike at low speed and high loading.[40]

to be independent of *W*. The figure indicates a near-Hertzian pressure distribution which is characteristic of high loading, and the highest quality surface finish would be required for satisfactory operation under the conditions indicated.

12*

The pressure spike is, of course, a theoretical concept. Because of its relatively sudden rise and fall, it presents generally a difficult problem in measurement and detection. Compressibility of the lubricant effectively moves the spike downstream and reduces its overall magnitude.

Consider now the effects of sliding velocity on the general features of an elastohydrodynamic film confined between metal boundaries. It is generally agreed that changes in sliding speed produce the most dramatic and spectacular effects in both the pressure distribution and film-thickness variation at an elastohydrodynamic contact. Figure 8.12 shows the influence of speed change on the distribution of film thickness for a cylinder in mineral oil.[40] We observe that:

(a) with increasing speed, the constriction occupies an increasing portion of the Hertzian contact zone, *ab*; and

(b) in all cases, the minimum film thickness is about three-quarters that at the maximum pressure point.

The theoretical pressure curve can also be shown to depart radically from the Hertzian distribution as speed increases, and the characteristic pressure spike moves upstream and

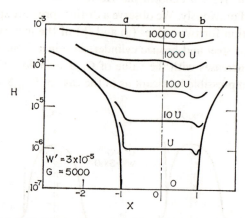

FIG. 8.12. Effects of speed on film thickness variation.

increases in magnitude. We conclude that the compressibility of the lubricant and increase in sliding velocity have opposite effects on the magnitude and location of the secondary pressure peak.

8.7 The Elastohydrodynamic Number

The most common condition of lubricated sliding is where the slider attempts to squeeze out the supporting lubricant during tangential motion. This situation corresponds both to the case of unsteady-state sliding, and to the transient behaviour of the slider after application of the load in cases where a steady tangential speed is preserved. Indeed, Section 8.5 deals with the particular and common example where the speed of sliding and the applied load are constant and steady-state conditions prevail in every respect.

It is helpful, however, to consider the steady sliding condition as being composed of an attempted squeeze motion which would occur if the tangential motion were absent. To illustrate this concept, consider the case of a rigid sphere of radius R squeezing through a lubricant (in the absence of tangential motion) on to a highly elastic plane. For very small displacements of the plane, the time of approach can be expressed as[43]

$$t = \frac{6\pi\mu R^2}{W} \ln\left(\frac{h_{0_i}}{h_{0_f}}\right), \tag{8.22}$$

where the absolute viscosity μ of the lubricant is not affected by the relatively low contact pressures. The applied load is W, and h_{0_i} and h_{0_f} are the initial and final film thicknesses beneath the centre line of the sphere. If instead of the squeeze motion the sphere is assumed to move tangentially with speed V such that the time taken to travel a distance equal to the Hertzian diameter of contact $2a$ is identical with t in eqn. (8.22), then

$$t = 2a/V. \tag{8.23}$$

By eliminating t from eqns. (8.22) and (8.23) and assuming that the ratio (h_{0_i}/h_{0_f}) is constant in a given application, we can write

$$2a/V = \text{const}\,(\mu R^2/W). \tag{8.24}$$

Now, the Hertzian radius of contact a from eqn. (3.3) is as follows:

$$a = \left[\frac{3}{4} WR \left(\frac{1-v^2}{E}\right)\right]^{1/3}, \tag{3.3}$$

where E and v are the Young's modulus and Poisson's ratio respectively for the elastic plane. Substitution of eqn. (3.3) into eqn. (8.24) gives the following result after rearrangement:

$$\left(\frac{E}{q_0}\right)\left(\frac{\mu VR}{W}\right) = \text{const}, \tag{8.25}$$

where the Hertzian pressure[†] q_0 is obtained from eqns. (3.2) and (3.3) as follows:

$$q_0 = W^{1/3} E^{2/3} R^{-2/3}. \tag{8.26}$$

The dimensionless groups defined by eqn. (8.25) describe the lubrication mechanism during the transition from elastohydrodynamic to fully hydrodynamic performance, and the constant can be shown to be proportional to the film thickness h_0 existing below the centre line of the sphere.[2] Thus eqn. (8.25) describes clearly in terms of lubricant and elastic

---

[†] Symbols q_0 and $\bar{q}$ used in this chapter to denote Hertzian pressures, whereas p and q denote hydrodynamic pressures without and with pressure–viscosity variations respectively.

170   *Principles and Applications of Tribology*

properties what value of sliding velocity V is required to just support the load with a given film thickness h_0. The analysis is confined to relatively light loads.

We observe that the ratio $(\mu VR/W)$ is identical with the Sommerfeld number in hydrodynamic lubrication theory.[†] The ratio (E/q_0) may be termed an elastic number, and the product $(E/q_0)(\mu VR/W)$ is called the elastohydrodynamic number EH. In the special instance where the Hertzian and hydrodynamic pressures are identical (i.e. $q_0 = p_0$), it has been shown[2] that $EH = 1$. The elastohydrodynamic number extends the Sommerfeld number concept to include elastic distortion of one or both of the bounding surfaces of the lubricant film during relative sliding.

We have seen in Chapter 4 that the generalized friction coefficient $(f/\tan \delta)$ is a function of the pressure–modulus ratio (p/E), and that both $(f/\tan \delta)$ and (p/E) remain virtually constant as the sliding speed V increases. The constancy of the first ratio stems from the observation that the two components of sliding friction (namely adhesion and hysteresis) exhibit viscoelastic characteristics which closely match the viscoelastic features of $\tan \delta$. From eqns. (4.25) and (4.38) we can write

$$\frac{f}{\tan \delta} = \frac{f_A + f_H}{\tan \delta} = B\varphi'\left(\frac{E}{p^r}\right) + \zeta A\varphi\left(\frac{p}{E}\right) \tag{8.27}$$

with the notation given in Chapter 4. The constancy of the ratio (p/E) as speed is raised follows from eqn. (8.27) since the left-hand side of this equation is constant under dry conditions. The same qualitative explanation is valid under lubricated conditions for sliding speeds ranging from zero to a critical limit which depends on the nature of the sliding system and experimental conditions. Figure 8.13 shows a plot of the generalized coefficient $(f/\tan \delta)$ versus the elastohydrodynamic number EH for smooth and rough spheres sliding on different rubber surfaces covered with a water-plus-detergent solution. The experiments under-

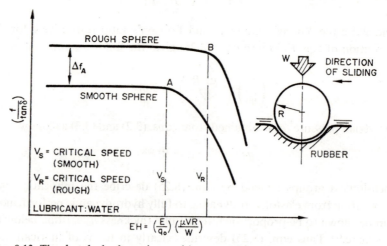

FIG. 8.13. The elastohydrodynamic transition speed for spheres on lubricated rubber.

[†] See Section 6.8.

lying Fig. 8.13 were conducted under different load conditions and over a wide speed range. The constancy of the ratio (f/tan δ) is evident up to a critical value of EH (defined by V_S for the smooth sphere, and V_R for the rough sphere), beyond which the generalized coefficient falls rapidly to a value determined by viscous shear and drag in the lubricant film together with a hysteresis loss.

The drop in generalized coefficient which begins at the critical limit of sliding velocity is best explained from squeeze-film theory. Thus at the lower sliding speeds the time taken for squeezing out the lubricant from the contact area is less than the time taken to traverse a distance equal to the Hertzian contact diameter, and physical contact is established between the sphere and the rubber. The point A in Fig. 8.13 for a smooth sphere corresponds to an exact equality between the times of squeezing and traversal. At still higher speeds, no physical contact is established between sphere and rubber, since there is not sufficient time for the squeeze action to be completed, and the coefficient drops rapidly. The upper curve in the figure corresponds to the case where a micro-roughness has been produced on the surface of the sphere by random abrasion with emery paper. We observe that not only does the surface roughness produce a distinct increment in friction in the low-speed range (this being largely due to adhesion), but more importantly the critical limit or transition speed V_R is greater than for the smooth sphere. This can also be explained from squeeze-film theory, since the time of squeezing is very much reduced when one or both of the approaching surfaces is rough. Thus a greater sliding speed V is necessary than in the smooth case to produce identical times for squeezing and traversal. A similar reasoning explains the steeper drop in the generalized friction coefficient beyond the critical sliding limit V_R for the rough sphere. The results presented in Fig. 8.13 are quite general and far-reaching, and they represent master plots of all variables except surface texture. In general, surface roughness radically alters the performance of ball-bearings and reduces fatigue life.

8.8 Macro-elastohydrodynamics

The previous sections in this chapter have been concerned with the normal approach or tangential motion of individual asperity elements (such as spheres, cylinders, and flats) relative to a lubricated elastic or rigid base. If these elements are now considered to form part of the macro-texture of a rigid surface over which an elastomer slides in the presence of a lubricant, the resulting hydrodynamic–elastic effects occurring on each asperity have been given the name *macro-elastohydrodynamics*. Consider, as an example, the case of an elastomer sliding under load on a lubricated two-dimensional surface having sinusoidal asperities. Figure 8.14 shows in detail the draping effect of the elastomer about a single asperity. In the absence of tangential motion, the normal load W would effectively squeeze out the lubricant at the tip of the asperity. However, the effect of the tangential velocity V is to create a hydrodynamic pressure wedge in the converging part of the lubricant film (the inlet region). This effect is resisted by the elasticity of the elastomer, and a final state of vertical, elastohydrodynamic equilibrium is attained which produces a steady-state distortion in the surface of the elastic body during sliding. The centre part of the film can be assumed to have a virtually constant thickness in the manner of a foil bearing,[44] and in the diverging or outlet region the pressure falls rapidly to perhaps negative values, as shown in Fig. 8.14.

The film thickness in the foil-bearing region is considerably less than the elastic deformation of the sliding member, and it is therefore assumed that the pressure in this region is

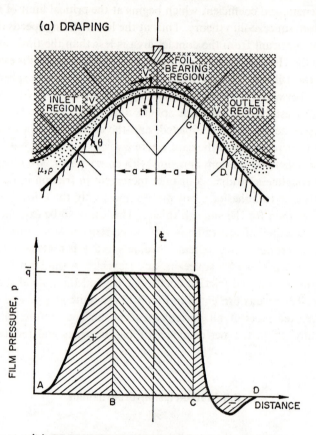

(a) DRAPING

(b) PRESSURE GENERATION

FIG. 8.14. The macro-elastohydrodynamic effect.

constant, with a value equal to the mean Hertzian pressure $\bar{q}$. If we assume initially that the central region supports the entire load per asperity W, we can, therefore, write

$$\bar{q} = W/2aL, \qquad (8.28)$$

where a is the semi-width of the foil-bearing region and L a length perpendicular to the paper. From Hertzian theory we can write for the semi-contact width a between a rigid cylinder and a flexible plane,[2]

$$a = \frac{2}{\sqrt{\pi}} \left[WR \left(\frac{1-v^2}{EL} \right) \right]^{1/2}. \qquad (8.29)$$

By substituting for *a* from eqn. (8.29) into eqn. (8.28),

$$\bar{q} = K\sqrt{W},\qquad(8.30)$$

where $K = \sqrt{(\pi E/LR(1-v^2))}/4$ is a constant proportional to the stiffness of the elastomer. Thus eqn. (8.30) permits a first estimate of the mean pressure $\bar{q}$ in the central region of the asperity.

We also observe that $\bar{q}$ is the maximum hydrodynamic pressure attained in the inlet region. If it be assumed that the inlet film thickness at the point A in Fig. 8.14 is h_0, decreasing to the value h^* at the point B, then the application of the Reynolds equation permits an evaluation of $\bar{q}$ in terms of h^{*2}. The exact variation of the film thickness between the value h_0 and h^* along the length AB is of no importance in performing this calculation since the pressure generated in a converging wedge depends to a first approximation only on the initial and final values of the film thickness and the length of the wedge. It is convenient, therefore, to represent the variation in film thickness along AB either as a straight-line or exponential decay. The final relationship between $\bar{q}$ and h^{*2} takes the following form for an exponential variation in film thickness:

$$h^{*2} = \alpha\left[\frac{\mu V}{\bar{q}}\right],\qquad(8.31)$$

where α is a complex numerical factor expressed in terms of the variation in film shape.[45] By substituting for $\bar{q}$ from eqn. (8.30) in eqn. (8.31), a first estimate of h^* is obtained. Also,

$$h_0 = K'h^*,\qquad(8.32)$$

where the constant $K' > 1$ and is known from the initial geometry of the draped elastomer. The load-carrying capacity of the inlet region is obtained from the equation

$$W_1 = L\cos\theta\int_A^B p\,dx,\qquad(8.33)$$

where p is the hydrodynamic pressure at any location within the pressure wedge (reaching a maximum value of $\bar{q}$ at the point B), and θ is the mean slope of the wedge relative to the horizontal position. In establishing eqn. (8.31) earlier, p has been expressed in terms of h^{*2} and α, and by substituting this relationship in eqn. (8.33) the load capacity W_1 of the inlet region is calculated. In a similar way, the load capacity W_3 of the outlet region is obtained. The summation of the three load support regions should then equal the original applied load W thus:

$$W_1 + W_2 + W_3 = W,\qquad(8.34)$$

where W_2 represents the load support of the foil-bearing region. However, since the initial estimate of W_2 is W, the calculated values of $\bar{q}$ and h^* are incorrect, and an iteration technique must be established so that the final selection of values complies with eqn. (8.34). In performing the iteration, it is assumed that the initial film thickness h_0 at the point A remains unchanged. This is equivalent to the reasonable assumption that conditions at the surface of the sinusoidal asperity are constant with respect to time. We assume that the film thickness h^* changes to the value $h^* + \Delta h^*$, so that a lesser slope is assumed in the inlet region. This corresponds to a reduced value of the constant K' in eqn. (8.32). Two or three iterations may suffice in establishing the correct value of Δh^*, since the solution converges rapidly.

The above elastohydrodynamic problem is not concerned with the change of pressure from a hydrodynamic to a Hertzian distribution over a time interval, as in the case of normal approach shown earlier. Such examples assume a relatively thick lubricant film which is progressively squeezed out from the contact area. *The macro-elastohydrodynamic problem, however, is concerned with a very thin viscous layer of lubricant on the surface of the asperities, where a Hertzian distribution is a reasonably valid estimate of the pressure effect at low sliding speeds. It is the change in film thickness Δh^* at asperity peaks due to hydrodynamic support which is of paramount importance in practical applications of this phenomenon.* It can be seen clearly from eqn. (8.31) that h^* (or Δh^*) increases with sliding speed, and the additional void created by Δh^* is filled by forced feeding from the inlet region.

If the results of the above theory are applied to a random surface, the generation of positive pressure increments on the forward slopes of individual asperities in the texture and

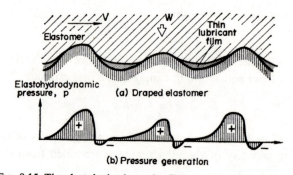

(a) Draped elastomer

(b) Pressure generation

FIG. 8.15. The elastohydrodynamic effect on a random surface.

negative increments on the backward slopes affects the normal "flow" pattern of the elastic body. The positive pressure increments far exceed the negative contribution, as shown in Fig. 8.15, so that there is a net lifting force tending to separate the elastomer from the surface. It is, in fact, this force which produces the increment Δh^* in the film thickness at asperity peaks.

Although the texture can be regarded as random, the asperities are noticeably rounded as shown, and this is a necessary condition for the validity of our theory. However, most engineering surfaces subjected to repeated traversal and wear (such as roads and pavements) acquire a rounded profile at asperity tips, so that this condition is not necessarily limiting. In applications where a high frictional coefficient is required between the elastomer and the surface, we can specify the peak-to-trough dimension of micro-roughness, ε_{MR}, which must exist at asperity tips to counteract the separating effect, thus:

$$\varepsilon_{MR} \geqslant h^*. \tag{8.35}$$

Thus the desired micro-roughness must exceed or equal the film thickness which would otherwise exist at asperity peaks, and under these conditions physical contact is maintained. We note that $h*$ is critically dependent on speed, and the micro-roughness is therefore effective only below a certain slip-speed value V. The iterative approach described in this section represents an inverse form of the general procedure outlined in Section 8.2.

8.9 Elastohydrodynamics and Hysteresis

We have shown in an earlier chapter that hysteresis produces an asymmetry of contact in sliding since the elastomer tends to accumulate ahead of the sliding body and to break contact sooner in the rear part of the contact area. Consider as an example the case of a rigid cylinder on a semi-infinite elastomeric surface as shown in Fig. 8.16. Figure 8.16(a)

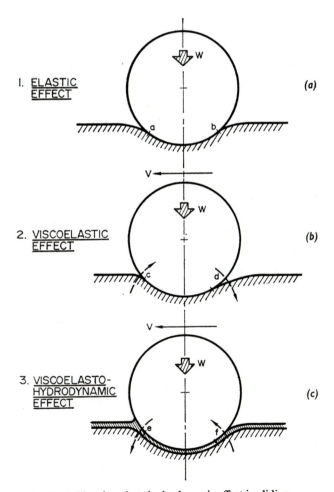

FIG. 8.16. The viscoelastohydrodynamic effect in sliding.

shows a symmetrical contact arc *ab* at zero sliding speed. At a finite forward sliding speed V the contact arc under dry conditions moves forward to the unsymmetrical position *cd*, as shown in Fig. 8.16(b). If a liquid is now introduced between the cylinder and the elastomer while still maintaining the same forward speed V, a hydrodynamic wedge effect is created just ahead of the contact arc tending to separate the surfaces, and a corresponding negative load effect is created behind the contact arc which tends to draw the surfaces together. *Thus Fig. 8.16(c) shows clearly that the elastohydrodynamic effect opposes the hysteresis effect at finite sliding speeds, and tends to restore contact symmetry.* The final position of the contact arc in Fig. 8.16 is *ef*, and it is certain that the point *e* lies somewhere between *a* and *c*. Further comparisons are difficult to make, because both the hysteresis and elastohydrodynamic effects tend to shorten the contact arc, so that $|ef| < |cd| < |ab|$ for the same load W. No attempt has been made theoretically to find the critical speed V at which the elastohydrodynamic effect exactly cancels the hysteresis effect, so that a symmetrical contact arc *ab* is restored.

The above discussion assumes the existence of one asperity which discretely indents an elastomeric plane during sliding. When several asperities exist, the hysteresis and elastohydrodynamic interaction effects are very complex. As shown in Chapter 4, the hysteresis effect itself will tend to restore contact symmetry at very high sliding speeds, and the elastohydrodynamic effect must be superimposed on the contact geometry.

8.10 Conclusion

The subject of elastohydrodynamic lubrication is still in its infancy today, and there is a need for further experimental work on contact geometry for a range of conditions and operating variables. It is certain that the elastohydrodynamic phenomenon is more likely to occur in the case of elastomers rather than metals, since the elastic distortion of one or both members takes place more readily, and in such cases lubricant viscosity is unlikely to be affected by the relatively low contact pressures. The relevance of elastohydrodynamics to a proper understanding of lubricated contact problems is now broadly accepted, and this applies equally to elastic surfaces of all types, whether elastomeric as in the case of tyres and flexible seals, or metallic as in the case of gear teeth and rolling contact bearings.

CHAPTER 9

WEAR AND ABRASION

HAVING considered the mechanisms of friction and lubrication in depth, we must now concern ourselves with the principles of wear and abrasion in both metals and elastomers. Several mechanisms may be involved in wear, either separately or in unison, and this is particularly true of metals. However, despite the fact that the individual mechanisms are well understood, wear as a general phenomenon remains somewhat unpredictable in quantitative terms. Metallic wear is a complex event and includes such complicating factors as work hardening, oxidation of exposed metal, metal transfer, and phase changes in metallurgical composition. The wear and abrasion of rubbers and polymers are no less complex, being dependent on a combination of mechanical, thermal, and chemical processes. More and more attention is being devoted to the question of what exactly constitutes wear, how it should be measured, what interacting factors determine its magnitude, and how it can be minimized and controlled. In this chapter we endeavour to classify and illustrate the various wear mechanisms, so that a qualitative understanding of the phenomenon emerges. The insight and information gained in this manner can then be applied to the design of suitable experiments from which quantitative data can be obtained.

9.1 Wear Mechanisms in Metals

The following classification suggests seven distinct mechanisms for the removal of material from sliding metal surfaces:[46]

(a) Continuous wear.

(b) Scuffing, or galling.

(c) Pitting.

(d) Abrasion.

(e) Chemical corrosion.

(f) Fretting corrosion.

(g) Surface flow.

We will briefly consider each mechanism in turn:

(a) *Continuous Wear*

The main feature of this type of wear is that material is removed as small particles, and these may be carried away in the lubricant without resulting in any gross surface damage.

177

The surface may become smoother rather than rougher. The following four processes may contribute to the removal of the wear particles:

    (i) Mechanical interlocking of asperities which are pressed together (see Chapter 3).

    (ii) Localized adhesion at contact spots following the welding theory (see Chapter 3).

    (iii) Abrasion by hard occasional particles.

    (iv) Erosion caused by cavitation or the corroding effect of the lubricant, leading to the dislodgement of oxidized particles.

At light loads, a fraction of the metallic junctions welded and sheared according to the welding–shearing–ploughing theory[1] becomes detached to form wear particles. If the load is increased to a value such that the average pressure exceeds about one-third of the hardness

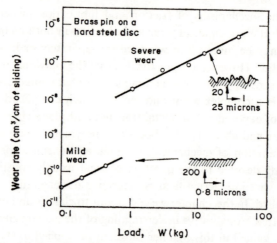

Fig. 9.1. Transition from mild to severe wear by load increase.

of the softer metal, a large increase in the volume of wear particles occurs. This is due to the fact that at high loads the true contact area approaches the apparent contact area in size, and a loose wear particle once formed is unable to escape without producing further particles in a self-accelerating process. The accumulation of loose debris therefore leads to a rapid acceleration of the wear process, and the wear particles thus formed are relatively fine.

Figure 9.1 shows the abrupt transition from mild to severe wear for a brass pin rubbing on a hard, steel disc. Below about 1 kg load, the sliding surfaces become polished (with a microfinish of about 0.8 μ) and the wear debris appears as a fine, dark powder. In the severe wear region, there is considerable subsurface damage; the wear debris consists of metallic flakes and the surface roughness is about 25 μ. Characteristics very similar to those observed in Fig. 9.1 are known to occur in the presence of lubricant films.

(b) *Scuffing*

Certain forms of surface damage do not occur slowly and continuously, but may suddenly cause gross disruption of the mating surfaces and render them incapable of relative motion. Scuffing, or "galling", typifies such a mechanism, the distinguishing feature being the numerous torn patches and blobs of transferred material. The cause of scuffing is basically the high temperatures produced at points of intimate contact as a result of high load and sliding speed. Since frictional heat cannot be dissipated rapidly, an unstable phenomenon develops that leads quickly to localized welding of the opposing surfaces and their subsequent rupture. Hypoid gears are prone to this type of failure. In addition to temperature, it is certain that other factors (such as the conditions of lubrication if a lubricant is used) play a significant role in determining the extent of the scuffing action. Metal transfer in large, discrete blobs occurs even in mixed lubrication or thin film conditions. Micro-hardness tests on scuffed material indicate a very hard surface and a fine grain condition.

(c) *Pitting*

A common type of failure of rolling elements is characterized by the formation of small pits in the surfaces. This is believed to be due to fatigue of the material in the region of maximum shear stress which normally occurs at some depth below the surface. It is not certain, however, whether minute cracks originate in this region and spread towards the surface, or whether the cracks commence at the surface and spread inwards. Pitting can also be produced in polymethyl methacrylate by high velocity liquid impingement which causes minute fractures or roughening of the surface and then pits. In the case of metals, an experimental study of pitting can be performed in a rolling four-ball machine.[46] This consists of three balls which are allowed to rotate in a ballrace, and a fourth larger ball is then loaded on to the three and caused to rotate so as to correspond with the inner race of an angular contact ball-bearing.[†] Experiments show that the number of load cycles to cause pitting failure is given by the expression

$$N \sim 1/W^3 \sim 1/\bar{q}^9, \qquad (9.1)$$

where W is the applied normal load and $\bar{q}$ the mean Hertzian contact stress. It requires only a very small increase in stress to produce a marked diminution in fatigue life. Furthermore, the presence of any surface imperfections (which may act as stress raisers) accelerates the formation of pits and drastically reduces fatigue life.

(d) *Abrasion*

In metal applications, abrasion is defined as a form of erosion which can be attributed to the action of numerous small particles which impinge on a surface. Shot blasting or sand blasting can be envisaged from this viewpoint as an extreme case of abrasion. The impingement mechanism may be less severe than that which is known to cause localized damage such as pitting, and the type of wear which characterizes the abrasive process is relatively uniform and continuous. There are several distinct abrasive mechanisms in metals (such as gouging, grinding, and erosion) as treated in Section 9.7, dependent largely on the size of the wear

---

† This arrangement differs from conventional four-ball tests where the three lower balls are held rigidly and sliding contact occurs. See Chapter 11.

debris. Resistance to abrasion is defined as the ability to withstand non-elastic deformation as a result of groove formation. Abrasion may also occur in hydrodynamic bearings when the dimensions of an abrasive particle exceed the minimum film thickness. In such cases, the particle is driven across the rubbing surfaces, and surface damage appears as scoring.

(e) *Chemical Corrosion*

The bearings of internal combustion engines may suffer because of attack by oxidation products which result either from the decomposition of lubricating oil or from contamination from certain products of combustion. Such chemical corrosion may be severe. However, chemical reactions may offer an advantage by the formation of protective coatings on surfaces, and a film of relatively low shear strength may result which facilitates sliding. Environment plays a decisive role in determining the rate of oxidation. Thus in the presence of water, hydroxides may result, whereas oxides are produced in its absence. The relative proportions of oxide and hydroxide determine the mechanical properties of the film formed, and therefore the degree of protection afforded the underlying metal. At comparatively low temperatures, metal soaps are formed which give protection to metal surfaces, whereas at higher temperatures inorganic metal salts (these may be chlorides, sulphides, phosphides, or phosphates depending on the nature of the additive) are produced by the interaction of additive agents with the underlying bare metal. In the latter case, such reactions require a certain time in which to be effective, and fortunately the rate of reaction is sufficiently high in most practical examples (such as running gears) to enable effective protection to exist. The oxidation of metal surfaces may play an important role in promoting chemical changes in the lubricant, the latter often leading to the formation of a film of polymerized material on the metal surfaces themselves. Such films can be detrimental to performance, being responsible (for example) for piston-ring "sticking" in internal combustion engines. On the other hand, the polymerized film may afford a protection against abrasive wear.

(f) *Fretting Corrosion*

When two mating surfaces nominally at rest with respect to each other are subjected to slight vibrational slip, a particularly serious form of wear may be encountered. The oscillatory motion may break down any natural protective film covering the surface so that metal is broken away at each oscillation. The unprotected metal then oxidizes and the oxide debris acts as an abrasive causing serious damage. The presence of a lubricant restricts the access of oxygen and thus drastically reduces the scale of the damage caused. The effect of moisture is complex: generally, both very dry and very wet conditions increase the magnitude of fretting corrosion.

(g) *Surface Flow*

This final form of metallic wear may occur at sliding speeds which, although not as high as would be required to initiate scuffing, nevertheless cause some heating of bearing surfaces under conditions of perhaps mixed lubrication (see Chapter 7). The heating effect is accompanied by softening and then plastic flow which may appear as a rippling pattern across the surface. Rippling-type failure has been observed in hypoid gears[46] under conditions of low speed and high torque. Prior to failure, the tooth surfaces are deformed in a series of raised and polished ridges running diagonally across the surfaces and resulting in noisy operation. This type of failure does not occur suddenly as in the case of scuffing or pitting, and its onset is characterized by noisy and rough running. Although continuous over a period of time, the surface flow phenomenon differs fundamentally from continuous wear by rearrangement of the macro-geometry of a surface rather than the removal of fine particles.

9.2 Wear of Clean Metals

The shearing of clean intermetallic junctions generally produces a very severe form of wear, and four distinct possibilities exist:

(a) The interface is slightly weaker than either of the two sliding metals. Shearing will then occur at the interface itself and the wear will be extremely small. This appears to apply to the sliding of tin on steel.

(b) The interface is stronger than one of the sliding metals but weaker than the other. Shearing occurs within the softer metal, and fragments of this are left adhering to the harder surface. This mechanism applies when lead slides on steel.

(c) The interface is stronger than one metal and occasionally stronger than the other. There is a marked transfer of the softer metal to the harder, and occasionally fragments of the harder metal are plucked out. A common example of this phenomenon is the sliding of copper on steel.

(d) The interface is always stronger than both metals, and shearing takes place a short distance from the interface. The relative sliding of similar metals typifies this condition. The interface becomes heavily work-hardened (see Section 3.8), and this results in heavy damage to both surfaces.

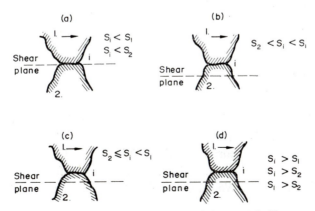

FIG. 9.2. The shearing and wear of clean metals. [1]

The four possibilities are depicted in Fig. 9.2. The extent of the wear may vary considerably between (a) and (d) perhaps by a factor of 1–100, whereas the friction may be substantially the same in all cases. This can be explained by assuming that *although all the junctions formed contribute to the friction, not all contribute to the wear process.*

Consider a simple wear model in which individual metallic junctions are formed and sheared. Let a junction have radius a and yield pressure p^*, and let it support a load δW. Then

$$\delta W = p^* \pi a^2.$$

The volume of a hemispherical wear fragment after shearing of a junction is $2\pi b^3/3$, where b is the radius of the fragment and is much smaller than a. This wear fragment is formed in a sliding distance $2a$, so that the volume worn per unit sliding distance is

$$\Delta v = \frac{2\pi b^3/3}{2a} = k \frac{\pi a^2}{3} = \frac{k \, \delta W}{3p^*}, \qquad (9.2)$$

where $k = (b/a)^3$. Assuming a distribution of junction sizes and geometrically similar wear fragments, the total volume Δv worn per unit sliding distance is given by

$$\Delta v = \frac{k}{3p^*} \Sigma \ W = \frac{kW}{3p^*}, \qquad (9.3)$$

where W is the total applied load. We observe that all junctions involved in the friction process produce a wear fragment of lesser proportions than the dimensions of the junction if this theory is valid. Consequently, the factor k relates the size of a wear particle to the radius of the junction from which it originates. As an alternative interpretation, we might assume that the radii of the wear fragment and junction are identical, but that only a fraction of the total number of friction junctions produce a wear fragment. Here, of course, k denotes such a fraction. In point of fact, it does not matter which interpretation is used, provided we realize that in practice k has a very small value indeed. For clean metals, k varies from 0.01 to 0.1, whereas for surfaces protected by oxide or lubricant films,[†] k may be as low as 10^{-6} or 10^{-7}. The simplified theory given by eqn. (9.3) shows that wear as a volumetric measure is directly proportional to normal load, and inversely to the hardness p^* of the metal. It is necessary for the equation to be valid that the wear process produces discrete particles or lumps rather than the transfer of a complete surface layer.

9.3 Loose Wear Fragments

We have seen that the production of loose wear fragments is an inherent part of the wear mechanism in metals, whether such fragments occur in discrete large-scale "blobs" or in the form of a fine powder. The theory in Section 9.2 indicates the relationship between total worn volume, load, and metal hardness, but no quantitative information is available on the size of loose wear fragments. We cannot project the scale or nature of the wear process

[†] This is regarded as a very mild form of wear.

from eqn. (9.3) in terms of individual particle size, and our interpretation of the factor k is largely intuitional.

The following simple mechanism remedies the situation by permitting the mean size of detached wear fragments to be estimated. We postulate that in order for a particle to become loose, the elastic energy stored within it must equal or exceed the free energy of adhesion,[†] binding it to the substrate.[47] If the hardness of the metal is p^*, then the elastic stored energy per unit volume of the fragment is proportional to $p^{*2}/2E$, where E is the elastic modulus. Since the volume of a fragment is v, the strain energy available upon release of a fragment is thus proportional to $p^{*2}v/E$. Now for pairs of similar metals, the ratio (p/E) can be taken as constant,[‡] and we can, therefore, write

$$\text{Available strain energy} \sim pv \sim pd^3,$$

where d is the mean dimension of a wear particle.

It is assumed, of course, that the loose wear particles evolve from the sliding of one metal surface on another, although our attention so far has been confined to the asperities of one surface from which wear particles originate. For simplicity, we now assume similar sliding metals, and define W_{aa} as the specific surface energy of either metal at the frictional interface. The surface energy of a particle or fragment about to be dislodged becomes

$$\text{Particle surface energy} = W_{aa}(\pi d^2/4), \tag{9.4}$$

assuming for simplicity a hemispherical shape. By relating the right-hand sides of the last two equations, we obtain the criterion

$$d \geqslant \text{const } (W_{aa}/p^*). \tag{9.5}$$

If the surface energy–hardness ratio is measured in centimetres and d in microns, the constant in eqn. (9.5) has the value 60,000 for most metals. Figure 9.3 shows experimental verifi-

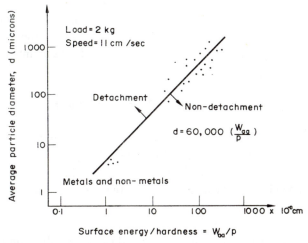

FIG. 9.3. Wear particle size as a function of surface energy/hardness ratio.[47]

[†] Or surface energy.   [‡] $p/E = 3 \times 10^{-3}$ for most metals.

13*

cation of this equation, using a wear tester in which a fixed annular specimen of metal is pressed against a like specimen mounted in a lathe. Wear fragments falling from the sliding interface are collected for later examination. In order to find a typical particle diameter, the wear particles were sieved through different meshes. The average particle was then defined as that which has all heavier particles exactly equal in weight to all lighter particles. The diameter of this particle was assumed to be identical with the opening of a sieve which would just allow the particle to fall through. When a lubricant is used, the fragments are much smaller in size, and it is suggested that the surface energy W_{aa} is correspondingly smaller. Equation (9.5) shows clearly that only particles above a certain size will become detached. Furthermore, metals with a large surface energy relative to their hardness should give large, loose wear particles—and this has been confirmed by experiment.

A distinction must be made between the mechanisms of metal transfer from one surface to another and the formation of wear debris. By using a leaded brass pin rubbing on a tool-steel ring, it has been shown[48] that the *amount of wear debris steadily increases with the duration of sliding, but the amount of metal transferred to the ring and adhering to its surface soon reaches an equilibrium value.* The explanation is that the contact pressure between pin and ring will be increased at those regions where the ring surface has already been raised infinitesimally by previous transfer. These will be the areas where further welding and metal transfer are most likely to take place. The transformed fragments gradually increase in size until the mechanical stresses which they suffer when they pass beneath the pin become sufficient to detach them as loose wear particles. A fresh part of the ring is now re-exposed and the transfer–wear process continues. In this matter, a saturation level for the transfer rate is reached, while the wear itself steadily increases. The results indicate that the increase in wear rate with load is accompanied by an increase in the mean size of the wear fragments rather than an increase in their number.

Apart from the scale of wear debris, the manner in which the wear process occurs varies markedly with the ductility of the metals involved. Thus experiments have shown that with hard, non-ductile materials (such as hard copper), the shearing of junctions occurs directly across the neck of the welded asperities. On the other hand, a ductile metal (such as lead) fails in such a manner that material at the junction tears and rolls up into a whorl.

9.4 Mechanisms of Wear in Elastomers

Although our treatment of metallic wear is by no means complete, it is helpful here to consider the forms which elastomeric wear assume so that a broad comparison is feasible. Three distinct wear mechanisms can be identified when an elastomer slides on a rigid base surface, and they depend to a very large extent on the existence and nature of the surface roughness of the base:

(a) *Abrasive wear.* A sharp texture in the base surface causes abrasion and tearing of the sliding elastomer. Micro-cutting and longitudinal scratches are observed on the abraded elastomer.

(b) *Fatigue wear.* If the base surface has blunt rather than sharp projections, the surface of the elastomer undergoes cyclic deformation, and failure due to fatigue eventually occurs.

(c) *Roll formation.* On smooth surfaces, a new mechanism of wear[49] specific to highly elastic materials causes roll formation at the sliding interface, and eventually tearing of the rolled fragment.

We observe that abrasive and fatigue wear occur on rough surfaces, whereas roll formation is characteristic of smooth surfaces having a high coefficient of friction. Abrasive and roll wear are usually very severe, but fatigue wear is relatively mild. However, although fatigue wear is the least damaging mechanism, it predominates in most sliding operations, and it characteristically requires a relatively low coefficient of friction between the elastomer and the abradant. Experiments have shown that elastomers and rubbers which have a high abrasion resistance have a relatively low coefficient of friction, and this observation confirms the general theory that an increase in friction produces more wear or abrasion. The abrasion resistance of rubbers with a high frictional coefficient can be increased only by reducing the coefficient itself; thus powdering the surface of the rubber with talc or adding exuding anti-friction lubricants (such as silicone oil) during compounding, will considerably increase the abrasion resistance.

The basic parameters which influence wear are pressure, sliding velocity, frictional coefficient, surface texture, elastic modulus, strength, and fatigue resistance of the elastomer. The effect of elastic modulus is illustrated by the following example. On concrete roads, abrasive wear of tyre treads becomes severe[49] at high loads when $E > 60$ kg/cm; it therefore might appear unwise to increase the hardness of tread rubbers. However, soft rubber compounds (having low E values) may also produce severe abrasive wear since the contouring of the concrete asperities results in a relatively large contact area. Optimum values of hardness and Young's modulus obviously exist, but the complexity of the deformation processes makes their determination difficult in a given example.

The nature of the three basic wear mechanisms will be examined later in more detail.

9.5 Wear Measurement

Wear debris can be measured either as a weight loss or as a change in volume or dimension of one or both sliding members. The following six wear criteria have been proposed and used:

1. *Linear wear rate*

$$K_L = \frac{\text{thickness of layer removed}}{\text{sliding distance}} = \frac{h}{L}.$$

2. *Volumetric wear rate*

$$K_V = \frac{\text{volume of layer removed}}{\text{sliding distance} \times \text{apparent area}} = \frac{\Delta v}{L A_a}.$$

3. *Energetic wear rate*[†]

$$K_E = \frac{\text{volume of layer removed}}{\text{work of friction}} = \frac{\Delta v}{FL}.$$

4. *Gravimetric wear rate*

$$K_W = \frac{\text{weight of layer removed}}{\text{sliding distance} \times \text{apparent area}} = \frac{\Delta W}{LA_a}.$$

5. *Abradibility*

$$\gamma = \frac{\text{volume abraded}}{\text{work of friction}} = \frac{\Delta v}{FL} = \frac{\Delta v}{fWL} = \frac{(\Delta v/WL)}{f} = \frac{A'}{f}.$$

6. *Coefficient of abrasion resistance*

$$\beta = \frac{\text{work of friction}}{\text{volume abraded}} = \frac{1}{\gamma} = \frac{1}{K_E} = \frac{f}{A'},$$

where L is the sliding distance, F the friction force, f the coefficient of sliding friction, and A' the abrasion factor. We observe that the energetic wear rate and the abradibility are identical, each being equal to the reciprocal of the coefficient of abrasion resistance. Furthermore, $K_W = \varrho K_V$, where ϱ is the density of the abraded material. A new set of modified wear parameters can be obtained by replacing the sliding distance in the appropriate expressions with the diameter of a contact spot.

These definitions are very useful in establishing theories of wear, and they are particularly applicable to the case of elastomer-on-rigid surface frictional events. Another parameter not listed here is the mean diameter of loose wear particles in metallic wear, according to eqn. (9.5). One practical problem which constantly arises in wear tests is the small size of the total wear debris obtained in a reasonable test time. This difficulty can be overcome by the use of radioactive tracers or isotopes on one of the sliding members, which gives a highly sensitive and accurate measurement of wear debris. The technique involves computation of the difference in radioactive count at one of the sliding surfaces before and after a test, and then relating this difference to the thickness of the abraded layer. In cases where a lubricant is used, an alternative indicator of wear is the increase in radioactivity of the lubricant itself.

9.6 Intrinsic and Pattern Abrasion

When a rubber or elastomer is abraded without a change of direction, sets of parallel ridges are often found on the surface of the sample and at right angles to the direction of motion. These ridges are called "abrasion patterns", and they are illustrated in Fig. 9.4 for the case of natural rubber. When the formation of a pattern is prevented by periodically changing the direction of travel, intrinsic abrasion is said to occur.[50, 51] The existence of

---

[†] Or energy index of abrasion.

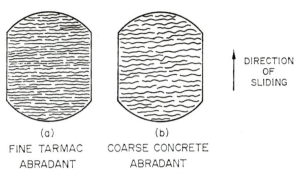

(a)   (b)

FINE TARMAC   COARSE CONCRETE
ABRADANT   ABRADANT

FIG. 9.4. Abrasion patterns on filled natural rubber.

pattern abrasion causes additional wear, which may be considerable under certain conditions. This can be visualized by considering a vertical cross-section through the abrasion pattern, as shown in Fig. 9.5. The sawtooth profile is opposed to the direction of abrasion, and during motion of the abradant the teeth are bent backwards, thus exposing their underside to the action of the abradant. At the same time, part of their surface is protected from abrasion at the rear. The result is an undercutting effect. The teeth wear progressively thinner until the crests are torn away, leaving blunt edges. In the meantime the ridges continue

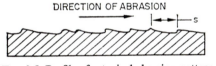

FIG. 9.5. Profile of a typical abrasion pattern.

to grow out from the bulk material, and the pattern is self-perpetuating. There appears to be a large difference in scale between the mean size of abraded particles in intrinsic abrasion and those produced by pattern abrasion in the manner outlined here.

It has been shown[2,51] that the mean spacing s between ridges in the elastomer surface is directly proportional to the volumetric wear rate K_V, and also a function of the mean particle size d, thus

$$s \sim K_V \sim d^{2/3}. \tag{9.6}$$

This relationship has been verified by experiments. In the case of two concrete surfaces (classified as fine and coarse in this text) used as abradants on a single rubber compound, the results shown in Table 9.1 were obtained experimentally:

TABLE 9.1. ABRASION EFFECTS ON ROAD SURFACES

Type of concrete surface	K_V 10^{-3} cm/m cm	Spacing s (cm)	Ratio (K_V/s)
Coarse	1.57	0.832	1.887
Fine	0.41	0.234	1.750

The load in these experiments was 1.65 kg/cm² and the sliding velocity 48 cm/s. The virtual constancy of the ratio (K_V/s) seems to confirm eqn. (9.6) provided the applied load remains constant.

9.7 Abrasive Wear

Perhaps the most severe form of wear in both metal-on-metal and elastomer–rigid pair combinations, is the result of abrasion by sharp projections in the mating surface. We will consider here each abrasion process separately.

Metals

Abrasive wear is very common in running machinery, and research into the mechanism of abrasion takes the form of examining the friction and wear of solids sliding on hard, abrasive papers. One complicating factor is the clogging of the emery paper with abraded and transferred metal fragments. This effect can be appreciable for grit sizes less than about 70 μ in diameter, and it has a strong influence on friction and wear performance. Thus for copper sliding on fine emery paper, $f \doteq 0.3$, which is similar to the value obtained for copper oxide sliding on copper oxide.[1] By contrast, the friction of copper on coarse emery paper is about 0.7; here the abrasive grains are sufficiently large that clogging does not occur.

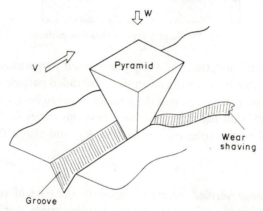

FIG. 9.6. Grooving and formation of a wear shaving–hard pyramidal indenter sliding on metal surface.

Figure 9.6 shows the grooving of a metal surface and the formation of a wear shaving when relatively large† abrasive grains are used.

By comparing the grooving and load support areas in Fig. 9.6 for a given depth of penetration of the indenter, it is possible to estimate the total volume of grooves formed per unit

---

† Greater than approximately 70 μ.

sliding distance. The solution is identical to that expressed by eqn. (9.3). If we now define the abrasion resistance R' as the reciprocal of the worn volume, it is found that

$$R' = 20\,p^*/W. \tag{9.7}$$

This relationship has been broadly confirmed from experiments. We note that provided the particles have sharp cutting edges, the exact shape of the abrasive particles is insignificant. *The main criterion is that a fixed fraction of the volume of formed grooves appears as wear shavings.* Furthermore, for a fixed hardness p^*, the abrasive wear of metals is less than the abrasive wear of brittle solids and greater than that for elastomeric materials such as rubber.

We can conveniently classify abrasive wear in metals into three distinct categories: (a) *gouging abrasion*, (b) *grinding abrasion*, and (c) *erosion abrasion*. All three may occur simultaneously on a wearing component, but usually the type which predominates can be recognized.

TABLE 9.2. TYPICAL WEAR RATES FOR DIFFERENT ABRASIVE PROCESSES

Type of abrasive wear	Type of metal	Typical examples	Wear rate (mils/hr)
Gouging abrasion	Austenitic manganese steel	Hammers in impact pulverizers Shovel dipper teeth Wearing blades on coarse ore scrapers Crusher liners[a] Chute liners[a]	5–1000 5–500 5–100 2–20 0.1–10
Grinding abrasion	Low-alloy, high-carbon steel	Grinding balls in wet siliceous ores Grinding balls in wet cement slurries Grinding balls in dry cement clinker	0.15–0.45 0.05–0.15 0.005–0.015
Erosion abrasion	Pearlitic white iron	Sandblast nozzles Pump runner vanes[b] Agitator impellers[b]	100–1000 0.1–5.0 0.05–1.00

[a] For siliceous ores.   [b] Abrasive mineral slurries.

(a) *Gouging abrasion*: This type of wear is characterized by high stresses and results in macro-deformation of the surface. Abrasive wear rates are usually expressed in mils per hour, this being the rate of metal removal normal to the surface material. Gouging abrasion has wear rates that are higher than for grinding or erosion abrasion (except for such techniques as sandblasting which cause high-velocity erosion). Table 9.2 gives typical wear rates for all three types of wear.[52]

(b) *Grinding abrasion*: Grinding abrasion occurs when two wearing surfaces rub together in a gritty environment with sufficient force to produce a crushing action in the mineral particles or other abrasive trapped between the surfaces. The broken abrasive grains are sharp and capable of scoring the hardest steel. Deterioration then occurs from scratching, local plastic flow, and micro-cracking. Although Table 9.2 indicates that the wear rates for grinding abrasion are relatively low, the wear rate in kilograms per ton of material ground is very high—since large surface areas are involved in the grinding action.

(c) *Erosion abrasion*: The primary factors causing erosion wear are velocity and low impact, with motion taking place parallel to the surface. Erosion abrasion requires hard and sharp particles to be effective, but as distinct from grinding wear their original smoothness or angularity is important, since there is little change with use. The abrasive particles may be suspended and carried in a fluid such as air or water.

For low impact speeds, the best criterion for wear resistance is to use the hardest steel possible with a high carbon content. The latter factor gives a higher proportion of hard carbides in the structure. At higher impact speeds, tougher steels are recommended to withstand the impact. However, although increasing the hardness of a low-cost steel by using heat treatment may be the most significant method in improving abrasion resistance, changing the nature of the surface is another important possibility. Table 9.3 lists the more common techniques for different materials and their effects.

A final word must be said about the mechanism of polishing. The abrasion process in such cases produces such fine dust particles that the macro- or micro-texture of the surface becomes smoother in time. The polishing action has been described as continuous wear in the region of low load application (see Section 9.1 and Fig. 9.1).

TABLE 9.3. SURFACE CHANGING TECHNIQUE FOR ABRASION RESISTANCE

Technique	Applicable to;	Effect

Electroplating
  Chromium, nickel
	Most ferrous and non-ferrous
	  materials
		Hard, smooth, thin surface
		  layer

Anodizing	Aluminium, magnesium, zinc	Hard, thin, oxide layer

Impregnation
  Carburizing, cyaniding,
  carbonitriding
	Low carbon steels	Surface hardness

Spraying
  Metallizing, flame plating,
  ceramic spraying
	Metallic and non-metallic
	  base materials
		Layers of interlocked, partly
		  oxidized particles

Hard facing Gas and arc welding	Ferrous and non-ferrous metals	Heavy alloy or carbide layer
Chill casting	Grey iron, some steels	White iron surface
Flame hardening	Iron and steel	Local surface hardness

Elastomers

The most comprehensive tests for abrasive wear in elastomers appear to have used different rubbers and a needle to scratch the surface under controlled conditions.[53] Later, a small hemisphere of 1 mm diameter was used.[51] These shapes represent extremes of sharpness and roundedness, and therefore cover the broadest possible range of shape of asperities in a rough track. Although such asperities vary considerably in contour, it can be stated in broad terms that they must exhibit either a sharp or a rounded peak, and no other possibilities exist.

Consider the case of a cylindrical slider moving with relatively low velocity on transparent rubber. Figure 9.7 shows the resulting photoelastic stress distribution in terms of the spacing of isochromatic fringes in the neighbourhood of the area of contact. A stress concentration at the rear of the contact area is indicated by the closeness of the fringe pattern, and the

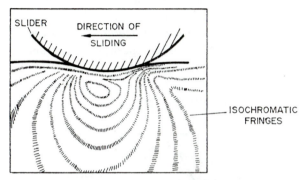

FIG. 9.7. Photoelastic stress distribution in transparent rubber.

rubber is in tension at this point. Now traces left on the rubber track after passage of the cylinder with large normal loading indicate a discontinuous series of tears across the track surface.

The periodic nature of such damage indicates the existence of a macroscopic stick–slip mechanism during sliding; the rubber adheres locally to the cylinder and is stretched in the direction of travel until the elastic restoring force exceeds the limiting friction, whereupon the rubber snaps back and the process repeats. Any failure due to repeated sliding would be expected to take the form of the opening of tears at right angles to the travel direction and originating at the rear of the contact area where the maximum stress concentration exists.

Let us now visualize the replacement of the blunt slider in Fig. 9.8 with a sharp needle point. Here tractive forces are larger than in the case of the cylinder because of mechanical

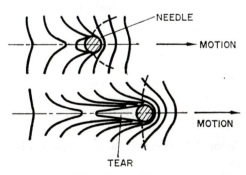

FIG. 9.8. Two successive stages in the deformation and tearing of a rubber surface by a needle.[53]

interlocking between rubber and needle. Consequently, the surface damage is more severe, and only one passage of the needle over the rubber track may be necessary to produce tearing and subsequent detachment of rubber particles. Figure 9.8 shows the distortion of reference lines equidistant from each other, and this in turn indicates the stress concentration in the vicinity of the needle point. Although the stress concentration is highest in front of the needle, frictional adhesion between rubber and needle at this location prevents tearing. The rubber therefore tears instead at the point where it first loses contact with the needle, and the tears develop laterally as indicated by the dotted lines in Fig. 9.8.

For very sharp asperities, we conclude that the cause of abrasion is local stress concentration produced by frictional adhesion and mechanical interlocking between the asperity peaks and the rubber or elastomer track. The chemical composition of the rubber itself (whether filled or unfilled) appears to have no effect on the mechanism of abrasion for sharp asperities.

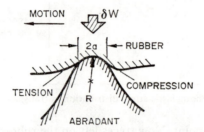

FIG. 9.9. Interaction of elastomer and abradant during sliding.

We now propose a simplified theory of abrasion for elastomeric materials. Consider an asperity in the abradant of mean tip radius R which presses into a sliding elastomer, as shown in Fig. 9.9. The applied load δW creates a contact width of $2a$, and we can write

$$a = a(R, \delta W, G),\qquad(9.8)$$

where G is some elastic constant which characterizes the elastomeric material. Two dimensionless parameters (a/R) and $(\delta W/GR^2)$ can be formed from these variables and they are related by an equation of the form

$$\frac{a}{R} = C_1 \left(\frac{\delta W}{GR^2}\right)^{\alpha},\qquad(9.9)$$

where the constant C_1 and exponent α must be evaluated from experiments. Let there be n^2 abrasive particles per unit area, so that the total load W is given by

$$W = \Sigma\, \delta W = n^2\, \delta W.\qquad(9.10)$$

Now, since the abrasion factor A' is proportional to the total volume Δv of detached elastomer particles, we can write

$$A' \sim n^2 a^3,\qquad(9.11)$$

where the volume of a single detached particle is assumed proportional to the third power of the contact width.[2] By substituting for n^2 from eqn. (9.10) and for a from eqn. (9.9) into eqn. (9.11)[†]

$$A' = C_2 \left(\frac{WR}{G} \right), \tag{9.12}$$

where C_2 = constant. From the earlier definitions in Section 9.5, we may write for the coefficient of abrasion resistance β,

$$\beta = \frac{f}{A'} = C_3 \left(\frac{Gf}{WR} \right), \tag{9.13}$$

where C_2 = constant and f the coefficient of sliding friction. We observe at once the similarity between β in eqn. (9.13) for elastomers and the abrasion resistance R' in eqn. (9.7) for metals. Both parameters are inversely proportional to applied load and directly proportional to some measure of the particular contact conditions. These measures are, of course, the elastic constant G in the case of elastomeric materials or the plastic yield pressure p^* for metals.

9.8 Fatigue Wear

The subject of fatigue failure in materials has such wide implications that one hesitates to introduce the topic. At the same time, however, our treatment of wear would be incomplete without listing the essential characteristics of the fatigue mechanism. In the case of metals, the fracture surfaces of engineering components which have failed as a result of fatigue generally exhibit two typical regions:

(a) A relatively smooth region which results from the nature of crack propagation, being generally accentuated by the rubbing together of the two faces of the crack as it opens and closes and progresses slowly across the section. This region is frequently discoloured by the ingress of oil, dirt, and corrosion deposits.
(b) A more coarse, fibrous, jagged, or "porous" region, similar to that which occurs in a static tensile failure. This represents the area of sudden failure or rupture of the component.

The word fatigue itself, of course, suggests failure under repeated loads, and it is a well-known fact that the number of stress cycles to failure is a function of the amplitude of the reversing or alternating stress as shown in Fig. 9.10. Stress concentration in components due to sudden change in section or the presence of notches will drastically reduce the "fatigue life" or number of cycles to probable failure indicated in Fig. 9.10, and methods exist[54] for estimating these effects.

[†] It is found by experiments that $\alpha \doteq \frac{1}{3}$.

Fatigue in the manner described may indeed cause internal failure of a machine component subjected to alternating stress or strain, but as tribologists we are, of course, more concerned with fatigue failure as an external mechanism at a frictional interface. Fretting corrosion is an example of a surface fatigue mechanism which arises from localized stick-slip events, as described earlier in Section 9.1. Experience indicates quite clearly, however, that the mechanism of surface fatigue in elastomeric materials is far more significant than

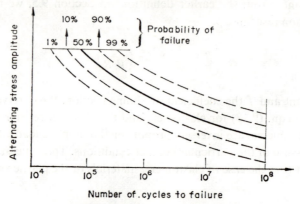

FIG. 9.10. Typical fatigue failure curve for metals.

in the case of metals. This arises from the flexibility of elastomeric solids which enable them to conform to the roughness of a rigid base during sliding, so that cyclic loading always exists. When the asperities of a rough track are rounded rather than sharp, failure may occur within the surface layers of a sliding elastomer following repeated compression, expansion, and reversed shearing stresses. This fatigue mechanism produces a relatively mild form of wear, and it becomes significant when cyclic stresses are present over long periods of time and where the adhesion is relatively small. The wear itself takes the form of mild abrasion and shredding of the surface layers.

9.9 Wear by Roll Formation

The previous mechanisms of abrasive and fatigue wear require for their very existence that the rigid base surface (in the case of elastomeric friction) exhibits a distinct macro-roughness. We have seen, too, that sharp asperities produce the abrasive action, and rounded asperities the less severe but continuous fatigue mechanism. On smooth surfaces we now introduce a new type of wear specific to highly elastic materials, and this is known as wear through roll formation. This occurs when a relatively high coefficient of friction exists between the elastomer and the smooth surface and when the tear strength of the elastic material is low. The elastomer then simply yields or fails internally because of severe deformation before slippage at the interface takes place. The failure occurs when the surface layers of the material are in a state of maximum strain, and a cut or small crack may appear perpendicular to the direction of attempted sliding (i.e. in transverse direction). The local direction of a cut, of course, depends in complex fashion on the nature of the local stress condition and perhaps also on molecular heterogeneities in the structure of the elastomer. The growth of these cuts is not likely to separate wear particles from the surface of the elastomer: a much more likely event is the gradual internal tearing of the elastic material, so that relative motion in the contact area is possible without complete slippage. The latter movement is possible if the layer of rubber separated off during tearing winds into a roll, as shown in Fig. 9.11. With such a mechanism, subsequent relative motion takes place under conditions of rolling friction, accompanied by continuous internal tearing of the elastic

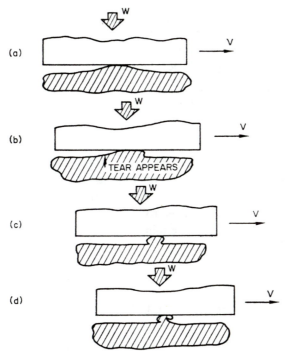

FIG. 9.11. Successive stages in the formation of a roll.

material near the interface, and the accumulation of the separated shred in the form of a roll.

The rolled shred is in a stressed condition during its formation, and its elongation depends on its cross-sectional dimensions which are variable. The force which produces such elongation depends on the resistance to tearing of the material at the place where the shred separates off. Failure of the shred occurs when the elongation reaches a critical maximum value, and the resulting detachment of the rolled shred completes the elementary act of frictional wear.

It is clear that wear by roll formation can occur only when there exists a certain combination of external conditions and elastic properties of the elastomer under test. As indicated earlier, it is necessary that a high coefficient of friction exist at the interface and that the elastomer has a low tear strength. The latter depends critically on the temperature generated as a result of sliding friction.

We now consider briefly the total frictional power P_f involved during the elongation of a shred and its subsequent tearing from the surface layer. Thus

$$P_f \;=\; P_t \;+\; P_e \;+\; P_H. \tag{9.14}$$

| Power used to tear shred from surface layer | Power used to elongate shred | Power used in hysteresis losses which accompany roll formation |

The components P_t, P_e, and P_H can be expressed in terms of other variables, and since from Section 9.5 we can write for the coefficient of abrasion resistance

$$\beta = \frac{\text{work of friction per unit time}}{\text{volume abraded per unit time}} = \frac{P_f}{\Delta v / \Delta t},$$

it follows that β takes the very general form

$$\beta = \beta(T', E, R_d, t, b, r), \tag{9.15}$$

where T' is the characteristic tear energy of the elastomer, E the elastic modulus, R_d the dynamic resilience (equal to the ratio of energy returned to a system during a half-cycle of vibration to the energy expended), and t, b, and r are the average values of the thickness, width, and radius respectively of a shredded roll.

The mathematical condition which determines the probability of this type of wear occurring in a given example is as follows:

$$P_f \leqslant fWV, \tag{9.16}$$

where f is the coefficient of sliding friction at the interface. This states simply that the power required to cause internal failure of the material must be less than that which would be dissipated as a result of sliding friction at the interface.

The wear pattern which occurs on rubber when the abrading surface is smooth is remarkably similar to the characteristic ridge formation obtained on rough surfaces (see Fig. 9.4). The following graph of gravimetric wear rate versus load for natural rubber sliding on abrasive paper and on hard rubber shows the contributions of the various mechanisms of abrasion to the overall wear rate. In the case of the abrasive paper, wear is due primarily to fatigue at low loads, with the contribution of abrasive wear becoming more dominant according as W is increased. For the natural rubber–rubber combination on the other hand, the mechanism of wear is probably due to roll formation and shredding. It is seen that at low loads the abrasive paper contributes substantially more wear than the rubber–rubber combination. At high loads and contact pressures, however, the conditions for wear by roll deformation become very favourable in accordance with eqn. (9.16), and the total wear due to this mechanism may even exceed that occurring on abrasive paper, as shown in Fig. 9.12. Experiments indicate[49] that when the coefficient of friction exceeds the value of about 1.15, there is a rapid and drastic increase in wear rate, and this phenomenon is satisfactorily explained by the roll-formation concept.

In practice it is certain that a combination of all three forms of wear occurs, and it is difficult to isolate the separate contributions of each mechanism to the overall wear effect. The subject of wear is not fully understood, and it is possible at present only to indicate qualitatively the contributing causes. As the elastomer becomes harder and stiffer, the mechanism of abrasive wear dominates. Thus on ebonite and plastics, scratching and split-

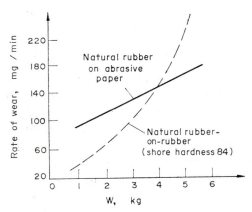

FIG. 9.12. Load dependence of wear rate for natural rubber.

ting are the common indications of surface degradation. There is a parallel between the physical understanding of wear effects and frictional events, which is quite helpful in qualitatively determining the causes of wear in special applications. According as the elastomer hardens and stiffens, its frictional behaviour approaches that of a rigid body, and this can be interpreted as a lessening of the hysteresis component of friction. In a similar manner, the fatigue mechanism of wear (which is a bulk effect) becomes insignificant, and surface phenomena predominate according as we proceed from viscoelastic to harder bodies. It is therefore understandable from this analogy alone that the wear of resins, plastics, and plexiglass is almost completely abrasive, as indicated by surface scoring and longitudinal scratches.

9.10 Effects of Speed and Temperature

Metals

Section 3.14 has dealt at length with the effects of sliding speed and temperature on metallic friction, and the changing wear mechanism as speed increases has been described in detail. The general trend, as seen in Fig. 3.17, is that the friction falls to remarkably low values even under vacuum conditions in the high-speed range.

The development of frictional hot spots cannot itself account for such low friction values. At moderate sliding speeds a series of temperature flashes with peaks reaching the melting point of one of the metals can be observed,[1] yet we consider that conditions in this range are isothermal. The reason is that those material properties which play a major role in the sliding mechanism are not appreciably affected by such hot spots. At higher speeds, however, we can describe conditions as adiabatic. Here the small contact regions are sheared so rapidly that heat is generated at a rate which is much faster than the rate at which it can be conducted away. This weakens the metal where the deformation begins, so that the shearing is confined to a thin zone at the interface. Indeed, some junctions may melt even before they are disrupted and while the underlying metal is relatively hard. Such intense localized heating will substantially reduce friction forces and yet have little effect on the ability of the surface irregularities to support the load. At very high sliding speeds a metal behaves as though a thin molten layer were superimposed on a hard substrate, and shearing of the liquid layer gives rise to the very low friction values. For brittle metals such as bismuth and antimony, the metal surface at a critical rubbing velocity begins to break up in a brittle manner producing a stream of tiny fragments. In this case, the explanation is that plastic strains are propagated within the metal at much slower speeds than elastic strains, so that junctions are sheared before plastic yielding occurs.

Figure 9.13 gives a simplified summary of the effects of speed on friction and wear for five metals. The results were obtained for a steel ball sliding at two predetermined speeds (100 and 600 m/s) on the various surfaces. We observe that all the metals except bismuth show a decrease in both friction and wear with increasing speed, and that molybdenum and tungsten give extremely low rates of wear.

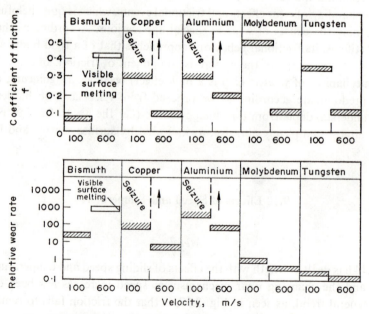

FIG. 9.13. Friction and wear data for steel ball on metal surfaces. [1]

The friction and wear of metals at very high sliding speeds are generally determined by the shearing of a thin molten film of metal, and extremely high rates of strain may be reached within the film itself. If softening and plastic flow of asperities occur at lower speeds, the friction and wear decrease with increasing speed. Since bismuth fractures in brittle fashion, it is an exception to the rule, as shown by Fig. 9.13.

Elastomers

In contrast with the wear of metals it can be shown that the wear and abrasion of elastomers are essentially viscoelastic phenomena. This follows from the fact that both adhesion and hysteresis can be thought of as viscoelastic occurrences under almost any operating condition, as we have established in Chapter 4.

Figure 9.14 shows the abradibility γ plotted versus sliding velocity for three unfilled rubbers.[55] We observe that all of the curves indicate a minimum value of γ. A qualitative

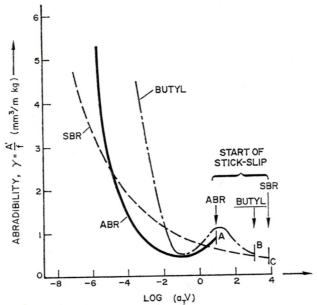

FIG. 9.14. Master curves obtained experimentally for the abrasion of three unfilled rubbers.[55]

explanation for the existence of one or two minima can be obtained from Fig. 9.15, which graphically depicts the variation of f, A', and γ with the logarithm of V at a given operating temperature. Thus Fig. 9.15(a) shows that the coefficient of friction f as a function of sliding velocity exhibits two characteristic peaks due to adhesion and hysteresis respectively. This curve can be obtained by superimposing the separate contributions of adhesion and hysteresis to the frictional coefficient in accordance with eqn. (4.6). Figure 9.15(b) shows the variation of the abrasion factor A' with log V, also at constant temperature, according to eqn. (9.12). This latter equation indicates that A' is inversely proportional to the elastic

14*

constant G, and the variation of G or G' with log V can be obtained from Fig. 4.10. If we now use the relationship $\gamma = A'/f$ to compute the abradibility at each sliding speed V, the curve in Fig. 9.15(c) is obtained, and two distinct minima are observed. In fact, each minimum corresponds approximately to a maximum in the friction–velocity curve, so that we may speak of an adhesion minimum C and a hysteresis minimum D. Now it is possible[56] that the adhesion peak can be minimized or eliminated by placing magnesia powder between the elastomer and the track. Furthermore, the addition of carbon black to rubbers

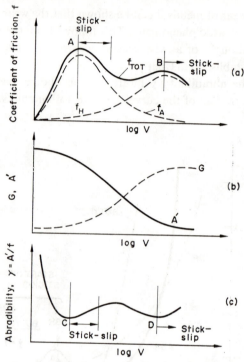

FIG. 9.15. The viscoelastic nature of abrasion.

flattens the adhesion peak into a plateau, with the exception of butyl which still preserves a distinct, sharp peak.[56] Thus acrylonitrile–butadiene and styrene–butadiene rubbers can generally be expected to have not only flattened adhesion peaks but also a reasonably flat abrasion minimum, as shown in Fig. 9.14. The points A, B, and C in the same figure indicate the start of stick–slip, and they correspond very approximately to the onset of a positive slope on the abradibility curves. Figure 9.15(a) shows clearly that these points also qualitatively suggest the start of a negative slope of the friction curves, so that the analogy between friction and abrasion appears to be well established. It must be emphasized, however, that abrasion is an extremely complex and sometimes unpredictable phenomenon, and only the broadest correlation is possible.

Figure 9.16 shows the temperature dependence of f, A', and γ at a given sliding velocity for the case of four different, unfilled rubbers.[57] We observe that both the abrasion factor A'

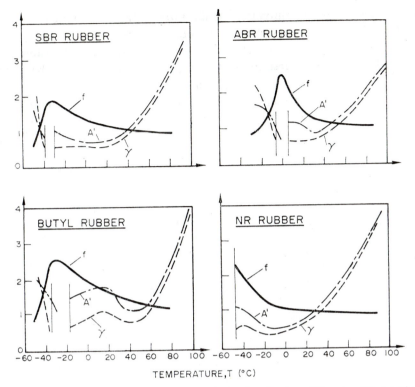

FIG. 9.16. Temperature dependence of friction and abrasion for four different unfilled rubbers.[57]

and the abradibility γ increase rapidly in value at higher temperatures. Now a mirror-image of this characteristic should indicate the velocity dependence of A' and γ, and we see at once that this generally agrees with the trends in Figs. 9.14 and 9.15(b). We conclude that the abrasion of rubber has distinct viscoelastic properties, and it can be predicted in broad terms from frictional data according to the relationship $\gamma = A'/f$.

9.11 Summary

We have examined in some detail the characteristic wear mechanisms in metals and elastomers, and the phenomena appear to be quite distinct. In general, high wear accompanies high friction coefficients, but otherwise there appears to be little correlation between friction and wear except perhaps in the case of certain highly elastic materials (such as rubber as seen in the last section). Only a very limited number of materials are intrinsically able to give both low friction and wear, the outstanding examples being carbons, graphite, and some polymers.

The wear resistance of the surface layers of metals is often markedly improved by diffusion treatments, the most widely used examples being carburizing, nitriding, and sulphiding.

Ferrous surfaces can also be chromized or siliconized so that galling becomes virtually impossible. Chemical techniques (such as the introduction of intermediate polymers like neoprene or nitrile rubber) may be used to modify the surface properties of elastomers, with the objective of increasing wear resistance. Both metals and polymers may be reinforced by impregnation with carbon-fibre reinforced resins to give virtually any desired elasticity.

We conclude that although wear accompanies friction in almost every conceivable application, it can be controlled only by first understanding the basic laws which determine its magnitude in a given application and then applying the most recent techniques available to minimize its effects.

CHAPTER 10

INTERNAL FRICTION

10.1 Molecular Structure of Matter

We are well aware from early exposure to physics that matter is composed of molecules which vibrate and move with thermal energy. Such molecular activity gives rise to properties of matter which can be measured in a laboratory and therefore quantified. It also facilitates an understanding of the essential distinction between solids, liquids, and gases. Consider the curve of attraction and repulsion between neighbouring molecules of any material as a function of their centre spacing λ, as shown in Fig. 10.1. The point O corresponds to the equilibrium position of a pair of molecules at a temperature of absolute zero. For finite temperatures T the molecules vibrate about this point as a result of their thermal energy

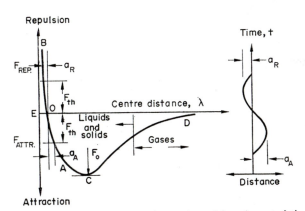

FIG. 10.1. Typical molecular attraction and repulsion characteristic.

in such a manner that the centre distance varies cyclically between the limits $(\lambda - a_R)$ and $(\lambda + a_A)$, where a_A and a_R are the amplitudes of the vibration during the attractive and repulsion stages respectively. In any single cycle of vibration, the maximum repulsive and attractive forces must be identical $(= F_{th})$ and this implies that $a_A > a_R$ as seen from Fig. 10.1. Let F_0 denote the force of attraction which a molecule at zero temperature (i.e. at the point O in the figure) must surmount to "escape" from its neighbour at C. At temperature T this barrier force is numerically equal to $(F_0 - F_{th})$, since the F_{th} component is provided by the vibration of the molecules due to thermal energy. As T is increased, the value of F_{th} and the amplitude of vibration increase correspondingly, until at the instant when $F_0 = F_{th}$ the molecules have sufficient energy to surmount the barrier at C and escape along the path CD.

TABLE 10.1. EFFECTS OF PHASE CHANGE ON MOLECULAR ACTIVITY

Process	Phase	Intermolecular forces	Average spacing (Å)	Ratio (Δ/d_0)	Molecular arrangement	Types of statistics needed
Heat Addition	Solid	Strong	$\doteq 3$	$\ll 1$	Ordered	Quantum
	Liquid	Medium		$\doteq 1$	Partially disordered	Quantum and classical
	Gas	Weak	$\doteq 30$	$\gg 1$	Disordered	Classical

Δ = amplitude of random thermal movement: d_0 = average molecular spacing.

The increase in vibrational activity due to the addition of heat along the path $OACD$ in Fig. 10.1 corresponds to a change of phase from solid to liquid to gas. We observe that in the solid state the relatively closely packed molecules have a prohibitive barrier F_0 to surmount, and this decreases with heat addition until escape in the gaseous state occurs at C. The average spacing of molecules in the solid and liquid phases is given by $d_0 \doteq 3$ Å, where $d_0 = OE$ in Fig. 10.1. The corresponding average spacing of molecules in the gaseous state is about $10d_0$, or 30 Å. Table 10.1 shows the progressive decrease in intermolecular forces which accompanies the change of phase from a solid to a liquid and gas. We also note the great increase in the amplitude of random thermal movement of the molecules and the accompanying transition from an ordered to disordered molecular arrangement. This implies simply that whereas in the solid phase the spatial orientation of the molecular motion is controlled by the basic structure of the material (being greatest along planes of minimal interatomic attraction), no such limitation applies to the gaseous phase, and the thermal motion is completely random in direction. The progression from an ordered to a disordered state of molecular motion can be regarded as a natural law more commonly referred to as the second law of thermodynamics.

The motion of molecules within solid, liquid, or gaseous matter creates an internal energy dissipation as a consequence of impact, stretching, twisting, vibration, and sliding. This mechanism is often called internal friction, and it takes different forms dependent chiefly on the phase in question. For a solid body internal friction may appear as a hysteresis or energy loss loop, and several parameters exist to measure its magnitude in a given application. For liquids and gases, internal friction appears as viscosity. These aspects will occupy our attention in succeeding sections.

We observe finally that whereas virtually the whole of the subject matter of this book excluding this chapter is devoted to the principles and applications of surface frictional phenomena, we concern ourselves here with the internal or bulk mechanism. This is necessary to unify and synthesize the subject of tribology, and it also helps our understanding of surface effects. Thus hysteresis friction, as we have seen in Chapter 4, can be defined as a bulk effect which ultimately appears at an interface although it may have no relevance at very large depths within a dynamic body.

10.2 Internal Friction in Solids

Investigations of internal friction in solids date back to the end of the eighteenth century.[58] Early workers used the decay in the amplitude of motion of a torsion pendulum suspended in a vacuum to study the phenomenon. Even when all external sources of energy absorption are removed, the periodic oscillation decays and eventually disappears because of the action of an internal dissipative mechanism within the system. Many methods have been developed for investigating internal friction in solids associated with bending, compression, torsion, and pure shear stresses. In general terms, the stresses are developed by giving a member an initial deflection and then releasing it—or perhaps by an initial impulse. As an alternative to observing the amplitude decay in free vibration, the motion of the member can be sustained as a steady low-frequency vibration by driving it continuously from an external oscillator. Such drivers may take the form of mechanical or electromechanical transducers of various types (e.g. electromagnetic, magnetostrictive, piezoelectric, etc.).

Several types and sources of internal friction have been investigated in the past. Thermoelastic internal friction is associated with stress inhomogeneity. Materials subjected to a quickly applied normal stress experience a temperature change. If there is a variation of stress throughout the material, there will be a corresponding temperature variation accompanied by irreversible heat flow. The ultimate result of this effect is the conversion of mechanical into thermal energy. There are a variety of conditions which produce stress inhomogeneity in a solid (e.g. variation in size and orientation of crystallites in polycrystalline materials, or external loading conditions, or heat transfer within a travelling wave, etc.). Another type of internal friction, as we shall see, is associated with plastic flow. As the frequency of vibration is raised, internal friction losses occur due to thermal relaxation on a molecular scale, and at very high frequencies due to wave propagation effects.

From the viewpoint of engineering applications and practical significance, the existence of internal friction in solids offers three distinct advantages:

(a) The limitation of stress magnitudes at near-resonance vibrations.
(b) An increase in the endurance limit of materials which exhibit a high damping capacity (or large internal friction).
(c) A reduction in noise levels. Thus plastic materials with an average damping capacity which is 10 times that of steel are used in machine parts, and are considerably less noisy than steel parts when subjected to the same impact and vibration forces.

The chief disadvantage of internal friction is, of course, the generation of internal heat. With plastics, this not only means poor conduction away from "hot spots" (since plastics are essentially insulators), but also a reduction in strength due to temperature rise.

10.3 Methods for Determining Internal Friction

There are about five accepted methods for determining the magnitude of internal friction. These are:

(a) Determination of the amplitude decay in free vibration.
(b) Determination of the hysteresis loop in the stress–strain curve during forced vibration.
(c) Determination of the resonance curve in forced vibration.
(d) Measurement of energy absorption during forced vibration.
(e) Determination of sound wave propagation constants.

All of these methods except the first are concerned with forced or externally imposed vibration which supplies energy dissipated through internal friction and enables the system to operate in the steady state. We will consider each method in more detail as follows.

(a) *Determination of Amplitude Decay in Free Vibration*

Consider the natural decay of amplitude with time for a simple vibrating spring–mass system as shown in Fig. 10.2. The damper shown in the inset of the figure is symbolic, and represents the internal dissipation of energy within the spring because of repeated compres-

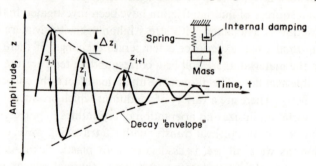

FIG. 10.2. Amplitude decay for spring–mass system in free vibration.

sion and extension. The logarithmic decrement β is defined as the logarithm of the ratio of successive amplitudes, thus:

$$\beta = \ln \left(\frac{Z_{i-1}}{Z_i} \right) = \ln \left(1 + \frac{\Delta Z_i}{Z_i} \right) \doteq \frac{\Delta Z_i}{Z_i} \qquad (10.1)$$

where Z_{i-1} and Z_i are successive amplitudes and $\Delta Z_i = Z_{i-1} - Z_i$ as shown. In practice it is difficult to read Z_{i-1} and Z_i even at very low frequencies because they are so close together, and in any event their difference is small. For greatly improved accuracy it is common to read Z_{i-1} and then count the number of cycles n of the vibration until the amplitude Z_{i-n} is exactly one-half of Z_{i-1} thus:

$$\beta = \frac{1}{n} \ln 2 = \frac{0.693}{n} . \qquad (10.2)$$

When the vibrating system is operated in a vacuum and all external sources of damping are removed, β gives a measure of internal friction or damping capacity within the system. Furthermore, β is related to the more usual measure of internal friction[†] tan δ by the following equation:

$$\beta = \pi \tan \delta. \qquad (10.3)$$

Sometimes we use the concept of dynamic resilience R_d to compare the energy dissipated in any one cycle of free vibration with that in the succeeding cycle thus:

$$R_d = \frac{\text{energy expended in one cycle}}{\text{energy expended in next cycle}} = \frac{Z_{i-1}^2}{Z_i^2}$$

$$= 1 + 2\left(\frac{\Delta Z_i}{Z_i}\right) = 1 + 2\beta \quad [\text{from eqn. (10.1)}]$$

$$= 1 + 2\pi \tan \delta \quad [\text{from eqn. (10.3)}] \qquad (10.4)$$

$$= \exp\left[\log\left(Z_{i-1}/Z_i\right)^2\right] = e^{2\beta} \quad [\text{from eqn. (10.1)}]$$

for $\Delta Z_i \ll Z_i$.

The above method was the first (and at one time the only) method of measuring internal friction. The original torsional pendulum apparatus of Coulomb consisted of a specimen to be tested (usually a wire) attached to a rigid support at one end and carrying an inertia disc at the other end. The wire was about 0.23 mm in diameter and 115 mm long, and the period of inertia of the disc was about 6 s. The system had the advantage of simplicity, but there were inherent errors such as the dissipation of energy at the support (never quite rigid). Furthermore, the frequency range was very limited, and the apparatus was confined to low-stress levels. Later investigations used modifications of this equipment, some using laterally vibrating bars and forks.

(b) *Determination of Hysteresis Loop in the Stress–Strain Curve during Forced Vibration*

Let us assume that as a result of the application of a sinusoidally varying force the corresponding deformation in a specimen is given by $\delta = \delta_0 \cos \omega t$. If η is the viscosity of the dashpot which simulates energy dissipation due to internal friction according to a simple Voigt model of viscoelastic behaviour,[‡] then the energy dissipated per quarter-cycle is given by

$$E_d = L \int \eta \delta' \, d\delta = L\eta\omega\delta_0^2 \int_0^{\pi/2} \sin^2 \omega t \, d(\omega t) = (\pi/4) \, L\eta\omega\delta_0^2, \qquad (10.5)$$

where L is some characteristic length dimension. Consider now the well-known hysteresis loop obtained by plotting applied stress σ versus resultant strain ε during the cycle, as shown in Fig. 10.3. If the enclosed area is given by $4a$, then it follows that during a quarter-cycle of vibration

$$a = \frac{E_d}{L^3} = \frac{\pi}{4} \eta\omega \left(\frac{\delta_0}{L}\right)^2, \qquad (10.6)$$

[†] See Section 4.4.
[‡] See Fig. 4.7 (a).

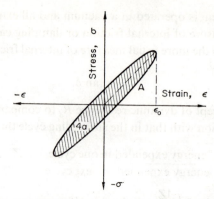

FIG. 10.3. Hysteresis loop during forced vibration.

where a is the energy dissipated per unit volume of material during the quarter-cycle. The strain energy E_s or energy acquired by the specimen during the same quarter-cycle is represented by the area A in Fig. 10.3, where

$$E_s = A = \int_0^\varepsilon (E'\varepsilon) \, d\varepsilon = \tfrac{1}{2} E' \varepsilon_0^2,$$
$$= \tfrac{1}{2} E' (\delta_0/L)^2. \tag{10.7}$$

Now, the dynamic stiffness D is defined by the relationship

$$D = 100(a/A), \tag{10.8}$$

being proportional to the ratio of energy expended to energy stored per cycle of vibration. From eqns. (10.6)–(10.8) and (4.24A), it follows that

$$D = 50\pi \left(\frac{\eta\omega}{E'}\right) = 50\pi \tan \delta. \tag{10.9}$$

We observe also that $D = 100\alpha$, where α is the energy loss fraction defined in Chapter 4, or, more generally, the specific damping capacity γ.

The dynamic resilience R_d in forced vibration is defined by

$$R_d = e^{4D/100} = e^{4a/A} = e^{2\pi \tan \delta}, \tag{10.10}$$

which has the same final form as eqn. (10.4) for free vibration.

Precision is relatively low with the hysteresis loop method, since the dimensions of the loop are of the order 80 mm long by 2 mm wide. This difficulty may be surmounted in different ways by the use of optical levers, prisms, mirrors, photoelectric cell, and oscilloscope for loop magnification. The method is generally suited, however, to the measurement of relatively large values of internal friction, since the width of the loop increases in such cases compared with its length. Although high stress amplitudes may be reached, the frequency range is limited and the apparatus is relatively complicated.

(c) *Determination of Resonance Curve in Forced Vibration*

With this method, the specimen to be tested is subjected to an alternating force of variable frequency (usually electromagnetically generated). In addition to finding the resonant frequency and amplitude, it is also necessary to find two other frequencies for which the amplitude is $1/\sqrt{2}$ times the maximum or resonant value. The ratio of the bandwidth $\Delta\omega$

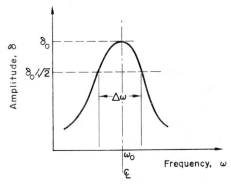

FIG. 10.4. Resonance curve in forced vibration.

between these two frequencies to the resonant frequency ω_0 is regarded as a measure of internal friction (Fig. 10.4). Calling this ratio $1/Q$,

$$\frac{1}{Q} = \frac{\Delta\omega}{\omega_0}. \qquad (10.11)$$

Since the effect of damping is to flatten a resonant peak, it is apparent that the greater the ratio $\Delta\omega/\omega_0$ the greater the internal damping. The symbol Q is a figure-of-merit taken from electrical engineering practice, being a measure of the degree of tuning or sharpness in an electrical circuit. Electromagnetic, electrostatic, piezoelectric or magnetostrictive transducers may be used to vibrate the specimen.

(d) *Measurement of Energy Absorption during Forced Vibration*

We have seen in the second method that the area of the hysteresis loop is a direct measure of internal friction per cycle of vibration provided the system is operated in a vacuum. Another simpler method of measuring energy absorption is to record directly the power input to the oscillator during the motion. Since the function of the oscillator is to supply energy dissipated through internal friction and thereby maintain a steady amplitude of vibration, the logic of the method is apparent. From earlier equations:

$$\gamma = \text{specific damping capacity}$$

$$= \frac{\text{energy absorbed per cycle}}{\text{energy stored per cycle}} = E_d/E_s$$

$$= (\pi/2)\tan\delta = \alpha = D/100.$$

The quantity E_d measures the energy absorbed or dissipated per quarter-cycle of vibration according to eqn. (10.5), and by calculating γ we may compare the relative damping capacities of different materials.

(e) *Determination of Sound Wave Propagation Constants*

Due to the rapid development of ultrasonics, a variety of methods exist in this class today. Essentially, the velocity attenuation of a wave through a solid is a measure of the internal friction characteristic of the material. Two methods will be described briefly here.

(i) *Direct Observation of Amplitude of Travelling Wave in Medium*

By means of a quartz crystal fastened to one end of a cylindrical rod specimen, a short frequency pulse (perhaps 100 cycles of a 5 megacycle per second wave) is sent into the specimen. The pulse is reflected back and forth between the ends of the rod many times, and the crystal is used to detect each reflection (as a pulse viewed on an oscilloscope). The velocity of propagation is found from the transit time between echoes, and the attenuation from the decrease in amplitude of successive pulses.

Consider the following wave equation:

$$\bigtriangledown^2\varphi = \frac{1}{c^2}\left(\frac{\partial^2\varphi}{\partial t^2} + k\frac{\partial\varphi}{\partial t}\right),\qquad\qquad(10.12)$$

where φ is the deviation of a local value from the average disturbance being propagated in a wave, c is the velocity of propagation, and k a dissipative term. In every instance of wave propagation, a local property of the medium is perturbed from a state of equilibrium, and a force arises tending to restore the system to the equilibrium state. The variable oscillates about its equilibrium at a given point, and this in turn perturbs adjoining points and sets them in oscillation. Thus a wave is propagated outwards from the initial disturbance. Solutions to the above wave equation give the value of φ at any instant in time, and the attenuation can be expressed in terms of k. When applied to the quartz crystal method described above, φ can be interpreted as the amplitude of vibration and k as a measure of internal friction. The method applies equally to shear, dilatational,[†] longitudinal, or transverse waves.

(ii) *Determination of Complex Reflection Coefficient of Solid Material*

This method essentially measures the intensity of an ultrasonic wave within a liquid—first when unobstructed, and then when a thin sheet of the material is placed in the path of the sound. The transmission loss is a function of the angle of incidence as the specimen is rotated. From this data, the attenuation constant of the material is found.

Ultrasonic methods have the advantage of being the most suitable at very high frequencies, although the apparatus is complicated and the stress level low.

---

† Pertaining to volumetric expansion or contraction.

10.4 The Fitzgerald Internal Friction Apparatus

In most methods of experimentally determining internal friction in solids, a forced vibration is applied to a sample of the material, appropriately supported, and some measure of
either stress and strain or energy change is made according to Section 10.3. One of the most
highly developed and frequently used devices for making such measurements in elastomers
is the Fitzgerald apparatus shown in Fig. 10.5. Two coils placed in separate radial magnetic
fields are mechanically coupled to each other and to the sample material by virtue of being

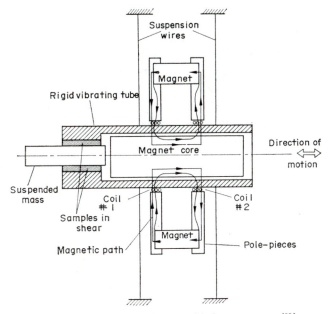

FIG. 10.5. The Fitzgerald internal friction apparatus.[59]

wrapped on a rigid, light-weight metal tube. This tube is suspended by wires so that it can
move only in the axial direction, thereby putting the samples in shear as shown. The axial
motion of the tube is resisted by the inertia of a heavy, suspended mass acting through the
samples. The coils are isolated from each other both electrically and magnetically.

An alternating current I flowing through either coil interacts with the magnetic field to
produce an alternating axial force F on the coil-and-tube arrangement, setting up an oscillation of the tube proportional to the complex shear compliance[†] J_s^*. When the coils vibrate
in the magnetic field as a consequence of F, an induced voltage E is created being proportional to the tube velocity V. The latter is in turn proportional to J_s^*. We note that

$$F \sim I \quad \text{and} \quad V \sim E.$$

[†] Note: $J_s^* = \dfrac{1}{G^*}$, where G^* is the complex shear modulus. See Section 4.4.

By defining electrical and mechanical impedances for the system as follows,

$$Z_{el} = E/I, \quad Z_{me} = F/V,$$

we can therefore establish the relationship

$$Z_{me} = \Gamma/Z_{el}, \tag{10.13}$$

where Γ is some coupling constant. The mechanical impedance Z_{me} is obtained from eqn. (10.13) when the electrical impedance Z_{el} is known or measured, and vice versa. Since $F \approx \sigma$ and $V \approx \varepsilon/\text{time}$, it follows from eqns. (4.13) and (4.14) that

$$Z_{me} \sim G^* \sim G'(1+j \tan \delta). \tag{10.14}$$

Thus the mechanical impedance gives a direct measure of the tangent modulus, and hence of the magnitude of internal friction.

The Fitzgerald apparatus is very versatile, particularly with regard to frequency selection. Besides measuring the damping capacity or internal friction of materials, it can also be used to predict electrical from mechanical properties in accordance with eqn. (10.13). Another important application is the behaviour of non-elastic materials during extrusion and plastic flow.

10.5 Chief Theories of Internal Friction

Having established that an internal energy-dissipation mechanism exists when materials are subjected to changing stress patterns, and having discussed different methods of measuring such a mechanism, we must examine briefly the nature and origin of internal friction. Several theories have been proposed as follows:

(a) Dislocation theory.
(b) Thermoelastic theory.
(c) Relaxation or creep theories.
(d) Viscoelastic theory.

Each theory will now be described briefly.

(a) *Dislocation Theory*[†]

Plastic flow is explained by this theory as the sum of individual displacements occurring at flaws. A flaw is a spot within the bulk of a material where although the average density may be the same as the surroundings, the density is greater at one part of the flaw due to "packing" of the molecules and less at another part. Owing to this density variation, each flaw is the seat of internal stresses, and there is a tendency for two separate flaws of opposite orientation (called a pair of dislocations) to merge. Externally applied shearing stresses

----

[†] For metals only.

may pull them apart in specific, crystallographically determined directions, causing slight accumulation of molecules on two opposite faces of the crystal.

Elasticity theory shows that the external shear stress will overcome the natural tendency for the flaws to merge only if the shear stress τ is greater than a certain minimum value

$$\tau > G \frac{\lambda}{2\pi h},$$

where λ is a molecular distance and h the separation distance between two flaws at right angles to the direction of shear. This value of τ is the yield stress for a pair of dislocations. Let there be n pairs of dislocations per unit area in the two-dimensional case so that

$$h \sim 1/\sqrt{n}.$$

By substituting for h in the previous expression, the yield value of τ becomes

$$\tau = CG\lambda\sqrt{n}, \tag{10.15}$$

where C is a constant.

Now the displacement resulting from the separation of a pair of dislocations has the same order of magnitude and is proportional to $\lambda(=\lambda l$, where l is a distance that can be thought of as a mean free path in kinetic theory). Thus λl is the shearing displacement due to τ. For n pairs of dislocations, the total shearing displacement d is given by

$$d = \lambda ln. \tag{10.16}$$

By eliminating n from eqns. (10.15) and (10.16),

$$\tau = CG \sqrt{\left(\frac{\lambda d}{l}\right)}, \tag{10.17}$$

which indicates a parabolic relationship between shear stress and resulting displacement. This closely approximates the actual relationship due to shear hardening beyond the elastic range. The value of l for copper and iron at room temperature is about 10^{-4} cm. Experiments have shown that eqn. (10.17) is surprisingly accurate in relating stress and shearing displacement, so that the dislocation theory can be regarded as a valid underlying cause of the phenomenon of work hardening[†].

A simple experiment to demonstrate internal friction and work hardening is the continuous back-and-forth bending of a thin piece of aluminium strip using one's fingers. As the bending proceeds, it becomes progressively more difficult to deform the material. This is due to work hardening, which can be thought of as a diminution with time of the number of pairs of dislocations or prospective slip planes within the material. The aluminium may have a large number of dislocations per unit area at the start of the experiment, but these are rapidly reduced making the work of bending more difficult.

---

† See Chapter 3.

(b) *Thermoelastic Theory*

This theory differs fundamentally from the dislocation theory by giving prominence to irreversible heat flow which accompanies stress variations within a material. When subjected to stress, heat will flow between a material and its surroundings, or the temperature will be changed, or both. For example, when a thin wire of uniform cross-section is subjected to an adiabatic increase in tensile force ΔF, the resulting change in temperature ΔT is given by

$$\Delta T = T\alpha L\Delta F/C_p, \tag{10.18}$$

where T is the absolute temperature in K, α the linear coefficient of thermal expansion (1/K), L the length of the wire (cm), and C_p the heat capacity of the system (in ergs/K).

When an isolated wire is stretched uniformly, there will be a uniformly distributed change in temperature of the wire. As long as the state of stress is uniform at any instant, no internal heat flow will take place. If the uniform stretching occurs at infinitely slow speed, there will be zero temperature change in the wire. If, however, there are finite variations of stress within the body, there will be finite temperature gradients, and internal heat flow may occur. Since heat flow through a finite temperature difference is essentially an irreversible process, some of the mechanical energy stored in the body during the period of stressing will be converted into thermal energy. If the stressing is periodic, the heat flow will tend to be periodic likewise.

There is one characteristic frequency ω_0 of stress variation, at which the percentage of mechanical energy converted into heat is a maximum. For $\omega \ll \omega_0$, the temperature gradient is small, since the temperature equalization process is much more rapid than the rate of temperature rise. This is very nearly an isothermal process. For $\omega \gg \omega_0$, the variation in applied stress is so rapid that there is not sufficient time for a finite quantity of heat to flow. This is close to an adiabatic process. As $\omega \to \omega_0$, $\Delta Q \to$ maximum, where ΔQ is the amount of heat given off in an isothermal process by an element dv. The logarithmic decrement β can be obtained[58] from the following integral:

$$\beta = \frac{\pi(\omega/\omega_0)^2}{2TU[1+(\omega/\omega_0)^2]} \int_V \Delta Q\, \Delta T\, dv, \tag{10.19}$$

where U is the total vibrational energy of volume v, and ΔT is the temperature rise in an element dv during an adiabatic process.

We observe that the logarithmic decrement β in eqn. (10.19) is directly related to thermal effects (whether ΔQ or ΔT). The latter arises from stress variations within a body, either because the body material is non-homogeneous or because there exists a finite stress distribution throughout the specimen.

(c) *Remaining Theories*

The remaining relaxation, creep, and viscoelastic theories apply to elastomeric materials such as rubber or plastics. The relaxation theory states that when a given stress is applied to a viscoelastic specimen, the stress required to maintain a fixed strain decreases in time (see *AB* in Fig. 10.6(a)). If now the load causing the initial stress is removed, there is an immediate elastic recovery to *C* (which may be below zero as indicated in the figure). The residual stress then causes further relaxation to the point *D*. The creep theory states that if a material is instantaneously loaded to some stress σ which is then maintained constant, the specimen deforms or "creeps" viscoelastically (see *EF* in Fig. 10.6(b)). If the load is again removed instantaneously, there is an immediate elastic recovery from *F* to *G* followed by a gradual viscoelastic recovery to *H*. There will generally be a permanent deformation *HJ* as

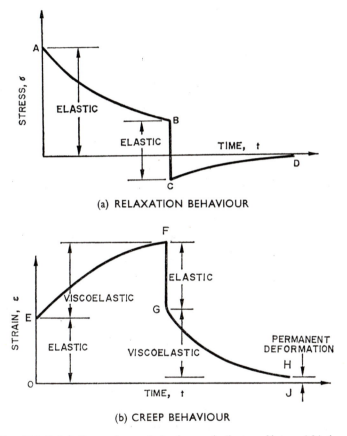

FIG. 10.6. Relaxation and creep behaviour as indicator of internal friction.

shown. Both the relaxation and creep effects are a consequence of slow internal molecular movements as the elastomeric specimen adjusts to a new stress or strain environment.

The elements of viscoelastic theory have been presented earlier in Section 4.4. The dynamic mechanical characteristics of a material describe its deformation or strain when subjected to a time-varying stress. A knowledge of the dynamic properties of materials over a wide frequency and temperature range is necessary to properly select materials most suitable for use in many applications, including the suppression of noise and vibration. The tangent modulus is a measure of internal damping in solids as we have seen, and values may

15*

TABLE 10.2. VALUES OF TAN δ FOR VARIOUS MATERIALS

Material	Minimum tan δ	Maximum tan δ
Metals	0.0001	0.001
Glass	0.001	0.005
Concrete, bricks	0.001	0.01
Sand, shale	0.01	0.05
Wood, cork, plywood	0.01	0.20
Rubbers, plastics	0.001	10.0

range from as low as 0.0001 for metals to as high as 10 for rubbers and plastics, as shown in Table 10.2. It is remarkable that for almost all materials another measure of internal friction in solids (namely, the logarithmic decrement β) is related to the Young's modulus E as follows:

$$\beta E = \text{const.} \tag{10.20}$$

Thus highly elastic materials such as steel have very little internal damping, and noise propagates readily throughout metallic structural members. On the other hand, elastomeric materials such as rubber have very high damping, but their extremely small values of Young's modulus and low rigidity make them virtually useless as structural members. We shall see in Part II that a very effective compromise solution to this problem in the construction industry is layered beam design. This is essentially a sandwich construction of alternating steel and plastic layers—the one giving structural rigidity and the other noise suppression through internal damping.

10.6 Internal Friction in Liquids: Viscosity

We have dealt with the property viscosity μ in earlier chapters, and we think of it as that physical property of a fluid which offers resistance to relative motion or deformation. It is essentially an internal fluid friction produced by the molecules of a lubricant as they flow past one another. The greater the relative motion, the greater is the internal resistance to shearing action that the lubricant offers. Of all the physical and chemical properties that a lubricant may possess, there is general agreement that the characteristic called viscosity is by far the most important. Thus the viscosity of a lubricating oil is considered the best single index of the use for which the oil is to be recommended. In bearings of all kinds—in gear-boxes, hydraulic systems, engines—in fact, wherever a lubricant is used, it is the viscosity which determines the friction loss, heat generation, mechanical efficiency, load-carrying capacity, film thickness, lubricant flow rate, and, in many cases, wear.

Sir Isaac Newton in 1668 conducted the well-known experiment of rotating one concentric cylinder relative to another with a lubricant interposed in the annular spacing, as shown in Fig. 10.7. He observed that a steady tangential force F was required to be applied to the rotating cylinder to shear the liquid at constant speed. Let A denote the swept area of one of the cylinders, V the surface rotational speed, and h the annular thickness of the lubricant. By measuring F for a variety of liquids at different speeds V and film thicknesses h, the following law was postulated:

$$F/A = \text{const } (V/h) \tag{10.21}$$

at any one temperature. Newton termed the constant in eqn. (10.21) the coefficient of viscosity (or absolute viscosity) μ, being a unique property of the liquid used in a particular

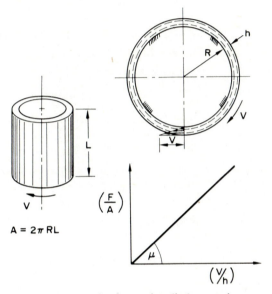

FIG. 10.7. Newton's concentric cylinder experiment.

TABLE 10.3. TYPICAL VISCOSITIES AT 21°C[34]

Fluid	Centipoises	Microreyns
Castor oil	>2000	>290.0
Honey	1500	217.5
SAE 50 oil (heavy)	800	116
Glycerine	500	72.5
SAE 30 oil (medium)	300	43.5
Olive oil	100	14.5
SAE 10 oil (light)	70	10.1
SAE 5 oil (extra light)	32	4.65
Ethylene glycol	20	2.9
Mercury	1.5	0.217
Turpentine	1.45	0.21
Water	1.00	0.145
Octane	0.54	0.078
Air	0.018	0.0026

test. Since (F/A) is the shear stress τ, we define absolute viscosity μ as the ratio of shear stress to velocity gradient. Kinematic viscosity ν is defined as μ/ϱ where ϱ is the density of the liquid.

Table 10.3 shows a list of typical fluid viscosities at 21°C, expressed in centipoise and microreyn units. One centipoise is 10^{-2} poise, where the poise unit is named after the French physician Poiseuille, being measured in dyne-s/cm² units. One microreyn = 10^{-6} reyn, the latter in honour of Sir Osborne Reynolds and expressed in lb s/in² units. The conversion from centipoises to microreyns is achieved by simply dividing by the factor 6.9. We observe from Table 10.3 the extremely broad range of values for μ, differing by a factor of 10^5 from castor oil at one extreme to atmospheric air at the other.

We must bear in mind, of course, that the viscosities listed above apply at room temperature and for normal pressures. The absolute viscosity of a lubricant varies considerably with temperature and to a lesser extent with pressure according to the following relationships:

$$\mu = K/T^\beta, \qquad (1.5 \leqslant \beta \leqslant 3.0),$$
$$\mu = \mu_0 \, e^{mp}, \tag{6.8}$$

where K is a constant and the index $m \doteq 10^{-8}$ m²/kg for most oils and lubricants, being considerably less than the temperature index β. Thus a heavy-duty truck engine using SAE 50 motor oil has a viscosity of about 7 centipoise at its operating temperature of about 120°C, although the viscosity indicated in Table 10.3 at room temperature is more than one hundred times greater. A more precise viscosity–temperature relationship for liquids is given by

$$\log_{10} \log_{10}(\nu + 0.8) = n \log_{10} T + C, \tag{10.22}$$

where ν is the kinematic viscosity in centistokes[†], and n and C are constants for a given oil or lubricant. Figure 10.8 shows schematically the dependence of μ on temperature and pressure in accordance with eqn. (6.8) or (10.22) for the liquid range. The trends as we approach the limits of the liquid phase are indicated by the dashed lines in the figure, and we observe in general that the characteristics of one curve are the inverse of those of the other curve.

As we shall see in Section 10.8, the mechanism of viscosity on a molecular level is entirely different in liquids and gases. The molecules of liquids are relatively closely packed together, and they oscillate randomly in the small space available to them. When a shearing stress is applied to a liquid, the molecules exchange places along the line of shear and the general effect is one of slipping as in the case of lamellar solids (see Chapter 5). The greater the freedom of molecular exchange, the less the viscosity (as opposed to gases), and hence we anticipate correctly that viscosity reduces as temperature is raised. The molecular exchange mechanism in liquids has been ingeniously simulated by Prandtl's Gedankenmodell, following an approach similar to his adhesion model.[2, 60]

[†] 1 centistoke = 1 centipoise/spec. gravity = cm²/s.

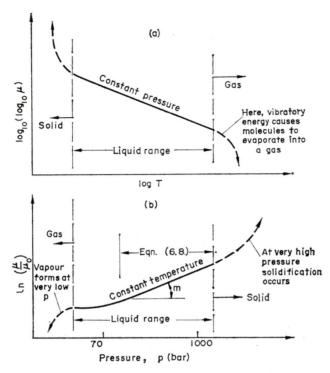

Fig. 10.8. Variation of viscosity with temperature and pressure.

10.7 The Measurement of Viscosity

A multitude of instruments exist for measuring both absolute and kinematic viscosities in liquids, and we will indicate here the principle of operation underlying the more common methods.

Rotational Viscometer

The original experiments of Newton described in the last section and illustrated in Fig. 10.7 were subsequently repeated by Petroff to measure the friction on a lightly loaded journal bearing, assuming that the journal ran concentrically within the bearing lining. Both Newton's and Petroff's experimental equipment can also be described as rotational viscometers, since the apparatus provides a direct measure of absolute viscosity μ as we have seen. It is more common to measure the frictional torque T on the inner cylinder in Fig. 10.7, and the rotational speed N in r.p.m., from which the absolute viscosity of the lubricant is obtained as follows:

$$\mu \sim \frac{T}{R^3 N}\left(\frac{h}{L}\right),\tag{10.23}$$

where L is the length and R the radius of the cylinder.

Falling Sphere Viscometer

Perhaps the simplest and most direct method of measuring the absolute viscosity of a lubricant is to allow a small sphere (usually a ball-bearing) to fall at steady speed through a specified distance in the liquid, while observing the temperature of the lubricant and the time taken for the drop to take place. Figure 10.9 shows the complete fallingsphere appara-

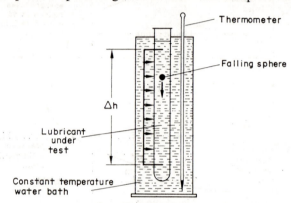

FIG. 10.9. The falling sphere viscometer.

tus. It is assumed that the ball or sphere falls through the liquid at uniform speed, and care must be taken to match the weight or size of the ball with the viscosity of the lubricant. The method should be used in practice only with lubricants of relatively high viscosity, such as motor oils or glycerol. Applying Stokes' law to the steady-state motion of the falling ball,

$$6\pi\mu V R \quad \equiv (4\pi/3)\,(\varrho_B - \varrho_L)\,R^3 \qquad (10.24)$$

$\underset{\text{force}}{\text{Stokes' resistive}}$ $\underset{\text{acting on ball}}{\text{Net downward force}}$

from which (with $V = \Delta h/t$), we derive the following expression for μ:

$$\mu = K't, \qquad (10.25)$$

where $K' = \frac{2}{9}((\varrho_B - \varrho_L)/(\Delta h))R^2$. Here, K' is the instrument constant (for a given diameter and material of ball), t is the time taken for the ball to fall through a distance Δh in the lubricant, R the ball radius, and ϱ_L and ϱ_B the respective densities of lubricant and ball. We observe that the lubricant density has been included in the instrument constant K', since ϱ_L varies little from one lubricant to the next. It is convenient in practice to allow a number of identical spheres to fall in turn through the liquid, taking the time t with a stopwatch in each case, and permitting the balls to accumulate at the bottom of the vessel.

Capillary Tube Viscometer

The measurement of viscosity by the flow of a given lubricant through a capillary tube is perhaps the most common method used today. If the flow is caused by a constant pressure difference Δp across the tube, the instrument usually measures absolute viscosity. On the other hand, it may be more convenient to cause the flow by means of a head of the lubricant—in this case, since density effects are introduced, it is usual to measure kinematic viscosity directly.

Consider an instrument of the latter type as shown in Fig. 10.10. The central chamber holds the lubricant sample and its lower end has a short capillary tube as shown followed by an enlarged opening containing a cork stopper. Surrounding this chamber is a constant-temperature bath as shown. When the bath and oil sample have reached the desired test temperature (usually 38°, 54.5°, or 99°C), the stopper is removed permitting the oil sample

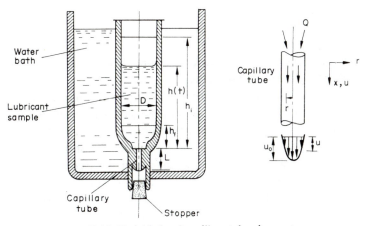

FIG. 10.10. Variable-head capillary tube viscometer.

to flow out through the capillary tube. The time taken for the height of the lubricant in the centre chamber to fall from h_i to h_f is noted, being proportional to the kinematic viscosity of the lubricant.

Consider the flow through the capillary tube as shown on the right-hand side of Fig. 10.10. Writing a balance between the viscous and pressure stresses acting on any fluid element within the tube, we obtain the following:

$$\frac{du}{dr} = -\frac{r}{2\mu}\left(\frac{dp}{dx}\right),$$
(10.26)

where the pressure gradient dp/dx across the capillary tube is given by $\varrho gh/L$. Here h is the instantaneous height of lubricant in the chamber above the capillary entrance, ϱ is the density of the lubricant, and L the length of the capillary tube. By integrating eqn. (10.26) and making use of the boundary conditions,

$$u = 0 \quad \text{at} \quad r = r_0 \quad \text{and} \quad u = u_0 \quad \text{at} \quad r = 0,$$

we obtain for the maximum velocity of flow:

$$u_0 = \frac{\varrho gh}{4\mu L}r_0^2.$$
(10.27)

Since the flow rate $\dot{Q}$ is given by

$$\dot{Q} = \pi r_0^2 \bar{u} = \pi r_0^2 u_0/2 = (\pi D^2/4)(\mathrm{d}h/\mathrm{d}t), \tag{10.28}$$

it follows from eqns. (10.27) and (10.28) that

$$\int_0^t \mathrm{d}t = \frac{2D^2}{r_0^4} \left(\frac{\mu L}{\varrho g}\right) \int_{h_f}^{h_i} \frac{\mathrm{d}h}{h}$$

or
$$v = K''t \tag{10.29}$$

where K'' is an instrument constant defined by

$$K'' = [gr_0^4/2D^2L \ln (h_i/h_f)].$$

Thus, the kinematic viscosity v of the lubricant is directly proportional to the time of efflux t of the lubricant through the capillary tube.

The most widely used viscometer in the United States which uses the above principle for determining the viscosity of lubricating oils is the Saybolt Universal viscometer. The number of seconds may be directly used to measure kinematic viscosity in SUS (Saybolt Universal Seconds) at a given test temperature, or the more usual centistoke units may be obtained using eqn. (10.29). A similar type of instrument called the Redwood viscometer is used widely in England, and on the European continent the corresponding instrument is the Engler viscometer. The essential difference between these three types of instrument is the volume of lubricant discharged through the capillary tube, as indicated in Table 10.4. In place of eqn. (10.29), the following modified form is used:

$$v = At - B/t, \tag{10.30}$$

where the constants A and B have different values for the Saybolt, Redwood, and Engler instruments, as shown in Table 10.4. The second term containing the B constant in eqn.

TABLE 10.4 COMPARISON OF CAPILLARY-TUBE VISCOMETER
CONSTANTS

Viscometer type	A (cm/s)^2	B (cm^2)	Q (cm^3)
Saybolt Universal	0.22	180	60
Redwood	0.26	171	50
Engler	0.147	374	200

(10.30) is a correction factor to allow for the initial acceleration of the lubricant through the capillary when the stopper is removed—and also the final deceleration when the stopper is replaced.

10.8 Turbulent Flow

The mechanism of viscosity in liquids as described in the previous sections can be viewed as the consequence of momentum transfer by molecular motion, and it is the only source of internal energy dissipation under laminar flow conditions. When turbulent flow prevails, however, an additional set of stresses† arises because of the fluctuating nature of the flow. These stresses express in mathematical terms the transfer of momentum by macro-motion of the liquid associated with the fluctuating components of velocity.

Consider an element of fluid moving with velocity components u, v, and w at any instant under laminar flow conditions, as shown in Fig. 10.11(a). In point of fact, we consider that u, v, and w are identical with corresponding mean values $\bar{u}$, $\bar{v}$, and $\bar{w}$ respectively, the latter

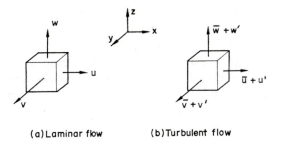

(a) Laminar flow   (b) Turbulent flow

FIG. 10.11. Velocity components in laminar and turbulent flow.

representing average values of velocity over a very short time interval, $\varDelta t$. Figure 10.11(b), by contrast, depicts the instantaneous values of velocity u, v, and w in the case of turbulent flow. In this case,

$$\left.\begin{aligned} u &= \bar{u} + u' && \text{in } x\text{-direction,} \\ v &= \bar{v} + v' && \text{in } y\text{-direction,} \\ w &= \bar{w} + w' && \text{in } z\text{-direction,} \end{aligned}\right\} \tag{10.31}$$

where, as before, $\bar{u}$, $\bar{v}$, and $\bar{w}$ are time averaged quantities, and u', v', and w' the corresponding time-dependent fluctuating components. The nature of u' (or v' or w') is characterized in Fig. 10.12. If eqn. (10.31) is substituted into the well-known Navier–Stokes equations in fluid mechanics,[29] it is found that six additional stress components associated with the fluctuating components appear.

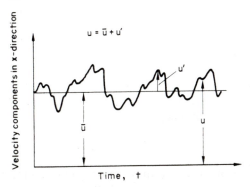

FIG. 10.12. Typical fluctuating velocity components.

† Called Reynolds's or apparent stresses, after Osborne Reynolds.

These are as follows:

$$\overline{\varrho u'^2}, \quad \overline{\varrho v'^2}, \quad \overline{\varrho w'^2}, \quad \overline{\varrho u'v'}, \quad \overline{\varrho u'w'}, \quad \overline{\varrho v'w'},$$

where the stroke or bar over each quantity denotes an average value over a short time-interval. The first three of these are pressure stresses and the remainder shear stresses.

We can gain a physical picture of how these additional stresses appear by considering the motion of small discrete elements of fluid, as shown in Fig. 10.13. The turbulent motion is assumed to consist of a steady, mean flow together with a series of exchanges between regions of different mean velocity. Each exchange is considered to consist of the movement of a fluid element of mass m a short distance Δy in the y-direction (say, y_1 to y_2 for the lower element in the figure)—and the corresponding movement of an element of equal mass m in the reverse direction. After travelling a distance Δy, each element is assumed to acquire the mean velocity of the fluid or liquid then surrounding it.

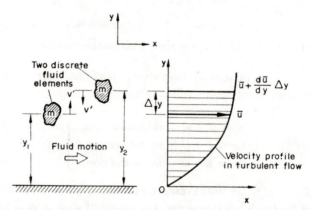

FIG. 10.13. Prandtl momentum transfer or mixing-length theory for turbulent flow.

We observe from Fig. 10.13 that

$$\left.\begin{aligned} u' = \left(\frac{d\bar{u}}{dy}\right)\Delta y \quad \text{for small } \Delta y \\[4pt] u' \sim v' \sim w'. \end{aligned}\right\} \tag{10.32}$$

and, in the general case,

Consider now that the two discrete elements of fluid in Fig. 10.13 form parts of two layers of fluid having a relative velocity $\Delta u = u'$. From Newton's law, the force F exerted between the two layers[†] is given by the following:

$$F = \dot{m}\,\Delta u = (\varrho v' A)\,u',$$

where $\dot{m}$ is the time rate of change of mass across the interface of area A. From this last equation, we can obtain the average value of (F/A) for a small time interval, this being

[†] The force F tends to retard the faster moving layer and to accelerate the slower layer.

identical with the Reynolds stress $\tau_{R, x}$ in the x-direction thus:

$$\tau_{R,x} = \overline{(F/A)} = \overline{\varrho u'v'}. \tag{10.33}$$

Similar expressions may be obtained for the remaining Reynolds stresses.

We now substitute for the fluctuating velocity components from eqn. (10.32) in eqn. (10.33):

$$\tau_{R, x} = \overline{\varrho u'^2} = \overline{\varrho u'v'} = \overline{\varrho u'w'} = \varrho \overline{\left(\frac{d\bar{u}}{dy}\right)^2 (\Delta y)^2} = \varrho \left(\frac{d\bar{u}}{dy}\right)^2 \overline{(\Delta y)^2} = \varrho l^2 (d\bar{u}/dy)^2. \tag{10.34}$$

Here the length l is a characteristic parameter of turbulent flow, called Prandtl's mixing length. It is defined as the root-mean-square value of all lateral fluctuations (such as Δy in Fig. 10.13) which arise from turbulence. Whereas in laminar flow, shearing stresses arise from viscous effects alone, they are due to a combination of viscous and momentum exchange effects in the case of turbulent flow. We may therefore write for turbulent conditions:

$$\tau_{\text{total}} = \tau_x + \tau_{R, x} = \mu \left(\frac{d\bar{u}}{dy}\right) + \varrho l^2 \left(\frac{d\bar{u}}{dy}\right)^2 = (\mu + \mu_T) \frac{d\bar{u}}{dy}, \tag{10.35}$$

where μ_T is a turbulent viscosity coefficient defined by $\varrho l^2 (d\bar{u}/dy)$, by analogy with the absolute viscosity μ of the fluid. *We observe that μ_T is generally larger than μ by several powers of 10, and it varies from point to point in the fluid having zero value at a wall or boundary. In turbulent flow, μ_T contributes virtually all of the internal friction associated with momentum exchange and velocity fluctuations in the fluid. In some texts μ_T is replaced by the "austausch" or exchange coefficient A.* By dividing by ϱ we obtain from μ_T the turbulent eddy diffusivity of momentum, ε thus:

$$\varepsilon = \mu_T/\varrho,$$

which is analogous to the kinematic viscosity ν in laminar flow.

10.9 Internal Friction in Gases

Whereas we readily accept a physical model to represent the action of viscosity in liquids, we remain somewhat unconvinced in the case of gases. Even the definition that "viscosity in gases arises from an interchange of momentum between alternate bulk-velocity layers" is not completely satisfying. Perhaps the difficulty arises from the virtually random motion of molecules within a gas, so that it is difficult to grasp the meaning of a layer in the same sense as for a liquid. We accept more readily the idea that some molecules are inevitably moving faster than others, so that their interaction by momentum exchange tends to reduce the velocity of the faster molecules while increasing that of the slower ones, and that this irreversible process of attempted speed equalization gives rise to internal friction and hence viscosity.

Since we will not attempt to represent the motion of individual molecules, and since differences in velocity are most widespread in a gas, it is inevitable that one region of a gas will exhibit on the whole a mean velocity which differs by an increment from that of another region. Figure 10.14 shows two such regions, having mean velocities $\bar{u}$ and $(\bar{u}+\Delta u)$ in the x-direction. Let it be assumed that there are N molecules per unit volume in the gas and that λ is an approximate measure of the distance between regions taken normal to the mean x-velocities. We further assume that the mean velocity of migration of the gas from region

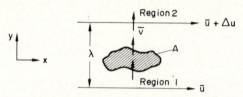

FIG. 10.14. Momentum exchange mechanism in gases.

1 to region 2 is $\bar{v}$ as shown. The volume of gas migrating through a horizontal area A per second is $A\bar{v}$, and the number of molecules per second is $A\bar{v}N$. Thus the mass per second $\dot{m}$ migrating in the y-direction is given by

$$\dot{m} = A\bar{v}(NM) = \varrho A\bar{v},$$

where M is the molecular weight of the gas and ϱ its density.

As in the case of turbulent flow in liquids, we assume that the migrating gas molecules change their x-velocity from $\bar{u}$ to $(\bar{u}+\Delta u)$ over the distance λ. We can then identify a force F exerted between the regions as a result of the migration, where

$$F = \dot{m}\,\Delta u = (\varrho A\bar{v})\,\Delta u$$
$$= (\varrho A\bar{v})\,\lambda(d\bar{u}/dy) \quad \text{[from eqn. (10.32)]}$$

and
$$\tau = F/A = (\varrho\bar{v}\lambda)\,d\bar{u}/dy$$
$$= \mu(d\bar{u}/dy). \tag{10.36}$$

Now λ can be viewed as the mean free path in the gas, or the statistical mean distance traversed by a molecule before collision with another molecule. It is apparent that λ must decrease as the density ϱ of the gas increases, and thus $\lambda \sim 1/\varrho$. Using this relationship in eqn. (10.36), we find that

$$\mu \sim \bar{v}. \tag{10.37}$$

From kinetic theory it can be shown that increasing the temperature of a gas increases the mean kinetic energy of its vibrating molecules, so that

$$T \sim \bar{v}^2, \tag{10.38}$$

where T is the absolute temperature. From eqns. (10.37) and (10.38) we finally obtain the important relationship

$$\mu \sim \sqrt{T}. \tag{10.39}$$

Thus the viscosity of a gas increases with absolute temperature, whereas the viscosity of liquids [see eqns. (6.8) and (10.22)] *decreases radically with temperature rise.* This characteristic of gases is most advantageous in the design of gas bearings, since the likelihood of seizure between the rotating parts decreases with continued usage. Since clearances in gas bearings are much smaller than in the case of liquid bearings, the possibility of seizure between the surface asperities of the moving surfaces must be carefully considered at the design stage.

The magnitude of the viscosity μ in gases is about a tenth of 1% of the viscosity of liquids, and the dissipation of energy through internal friction is relatively small in most practical applications where gas is used as a lubricant.

CHAPTER 11

EXPERIMENTAL METHODS

WE CONCLUDE Part I with a chapter devoted to a summary of experimental methods used in different tribology laboratories. The methods described are not necessarily exhaustive nor even representative of a typical tribology laboratory, but they attempt to indicate the broad range of experimental techniques that are currently used. In the case of elastomer friction, one convenient simplification of any experimental set-up results from applying the Williams–Landel–Ferry transformation method[†] so that frequency and temperature effects are interchangeable. Thus it is necessary only to provide a means of varying the frequency or speed of sliding at constant temperature, or the temperature at constant speed, in order to predict or evaluate the entire frequency and temperature dependence of the frictional force. This conclusion follows from the viscoelastic nature of elastomeric friction as outlined in Chapter 4, and it lessens the cost and complexity of experimental rigs. It is a matter of choice, of course, as to whether we use a variable-speed motor to vary the sliding speed or provide a controlled temperature environment for the experiment.

Ten broad categories of experiments and equipment are distinguished in this chapter, depending either on the application in mind, the ease of establishing an experimental rig, the philosophy underlying a particular experiment, or perhaps the degree of required precision. The list is as follows:

(1) Rotating disc assemblies.
(2) External and internal drum equipment.
(3) Flat-belt apparatus.
(4) Wear and abrasion machine.
(5) Squeeze-film apparatus.
(6) Impact and dynamic testers.
(7) Cross-cylinder apparatus.
(8) Interferometric equipment.
(9) Electron diffraction and microscopy.
(10) Miscellaneous.

These methods are dealt with in detail in the following sections.

† See Section 4.12.

11.1 Rotating Disc Assemblies

The simplest and most convenient rig is the rotating disc assembly, as shown in Fig. 11.1. It consists of a horizontal rotating disc with a superimposed circular sample usually of metallic or elastomeric composition. The mating surface is usually a cone, sphere, or cylindrical pin[†] fabricated from a hard material (typically a metal), and it is mounted as shown on a balanced lever arm and loaded against the rotating surface. For a given disc rotational speed, the relative sliding velocity between the pin or sphere and the disc can be

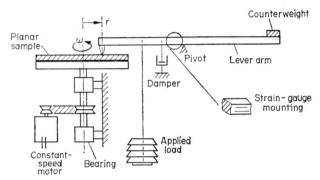

FIG. 11.1. A simple rotating disc assembly for friction measurement.

varied over a wide range by adjusting the radial distance r between the centre of the disc and the contact point. Frictional force is commonly measured by strain gauges mounted on the vertical sides of the lever arm in such a manner as to record the horizontal bending stresses induced in the latter. Alternatively, the strain gauges may be mounted on the shaft between disc and motor to measure the torque produced by the frictional force. The normal load may be applied as a dead weight as shown or by a hydraulic cylinder device. Contact temperatures during sliding are approximated by embedding a thermocouple with suitable insulation in the slider or (in the case of an elastomeric rotating pad sample) by embedding a series of thermo-couples below the surface of the elastomer in the path of the slider.

Measurements obtained with rotating disc arrangements as described above cannot be regarded as precise, although the readings obtained are reliable and broadly consistent. The most serious difficulties arise from vibrations of the lever arm which applies the normal load to the sample and grooving of the rotating disc after a relatively small number of revolutions. Mechanical dampers and the application of the normal load by a hydraulic cylinder (as opposed to dead weights) will greatly reduce the unwanted vibrations, and an electrical filter network will further diminish the spread of the dynamic trace from the strain gauges. The wear or grooving effect can be minimized by either limiting the number of disc revolutions in any one test or by programming the path of the slider to give a constantly decreasing radial distance from the disc centre to the contact spot. Slider wear is normally small or negligible compared with the grooving effect when the disc surface is an elastomer.

[†] In the case of a cylindrical pin, the rotating disc assembly is commonly called a pin-and-disc machine.

16 M: PAT: 2

One drawback of the rotating disc method is the curved path of points on the surface of the disc as they come into contact with the mating frictional slider. To avoid the consequent distortion of the contact area itself, the area of the latter must be quite small. There is the further disadvantage that it is virtually impossible to control lubricant film thickness on a rotating disc because of centrifugal acceleration effects. When contact areas increase and lubrication is required, we commonly consider external or internal drum equipment as a substitute for the rotating disc.

11.2 External and Internal Drum Equipment

The use of either external or internal drum equipment for measuring frictional force represents a more precise and somewhat more complex experimental approach. External drums are more accessible for dismantling or assembling the slider head, but at higher speeds it is impossible to sustain a lubricant film on the external drum surface because of centrifugal effects. By contrast, internal drums provide a stable lubricant film, and the degree of uniformity of film thickness increases with rotational speed. Both drum types

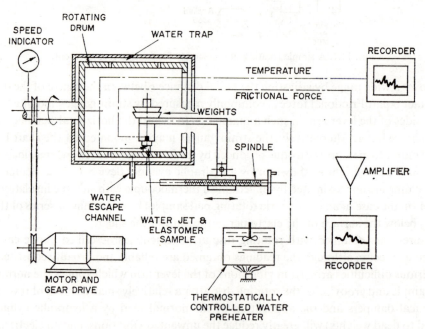

FIG. 11.2. Schematic of internal drum apparatus.[2]

require a variable-speed drive to provide the possibility of altering sliding velocity between the test sample and drum surface.

Figure 11.2 shows a schematic arrangement of an internal drum apparatus and associated equipment. In this example the inside surface of the drum exhibits a distinct surface texture obtained by mounting hemispherical, cylindrical, or conical asperities in different sizes and patterns. Figure 11.3 shows a close-up of the textured internal surface of the drum, in this case for conical asperities having different apex angles and densities. The tips of all conical asperities used must be rounded to prevent excessive wear or rupture of the mating frictional surface when the latter is an elastomer or plastic material. The asperities are grouped in circumferential bands as shown clearly in Fig. 11.3, and if an elastomer sample is used in a particular experiment it can be located directly above a particular band of asperities by the adjustable spindle illustrated in Fig. 11.2. The load is applied to the elastomer sample in

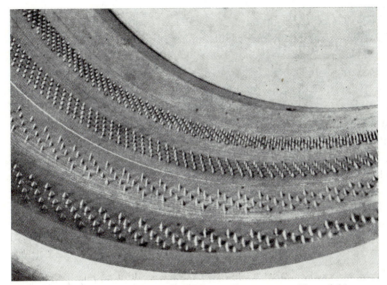

Fig. 11.3. Details of asperity layout on internal drum surface: cones with variable apex angle and spacing density.

this case by means of weights, and the drum is driven by an electric motor and special gearing to permit sliding speeds in the very broad range of 0.05 cm/s to approximately 180 km/hr.

For wet conditions, a water jet supplied from a thermostatically controlled pre-heater is placed just in front of the elastomer sample. Water temperature is variable from room temperature to 80°C. As stated earlier, centrifugal effects during drum rotation produce a stable and uniform water-film thickness, and in the case of Fig. 11.2 the radial escape of excess water is permitted by holes drilled in the drum surface as shown. A stationary cylindrical cage surrounds the entire drum and collects the water which is then returned through a water escape channel.

The frictional force at the elastomer surface produces very small deflections in a steel support block, and these in turn are measured by four temperature-compensated strain gauges. The output signal is then amplified and recorded. Temperatures in the elastomer sample are measured during the experiments by thermocouples mounted at different depths within the elastomer.

Precision machining of the drum segments and careful balancing of the rotor are required to permit such a broad speed range, and this has been accomplished in the model shown in

Fig. 11.2 (see ref. 2). The higher speed range yields valuable data on hysteresis and wear, as well as heating effects, and at lower speeds adhesion may predominate. It is possible to mount the elastomer surface inside the drum and to mount a single asperity in place of the elastomer block, so that the friction of individual asperities may be studied. Likewise, it is possible to study metal-on-metal friction with this apparatus, or indeed any combination of materials. By surrounding the equipment with a controlled environment (this includes temperature and humidity control, with or without vacuum or reduced pressure conditions), the range of possible experiments is exceedingly large.

One distinct disadvantage of internal and external drum equipment is the curvature of the drum itself. Although not a serious problem in applications where the contact area is small (such as the elastomer block–rigid combination described above), corrections must be made for drum curvature in instances where contact takes place over an appreciable length of drum circumference. Thus one of the most common applications of the external drum principle is the tyre-on-drum tester which is a routine and widely used piece of equipment in the tyre manufacturing industry. Here the concave flexure of tread rubber within the contact patch because of drum curvature is a particularly severe test of fatigue and wear resistance. For a comparison of performance in a drum test and on the road, we commonly reduce the length of contact in the former case (by using a lesser contact load) to compensate for increased flexural stresses in the tyre.

The flat-belt apparatus is used in cases where we wish to eliminate entirely the effects of drum curvature.

11.3 Flat-belt Apparatus

Figure 11.4 shows the general layout of a flat-belt apparatus, consisting essentially of a flexible belt passing over two large cylinders. One of these cylinders is driven either by a variable-speed motor drive as shown, or alternatively by a synchronous motor with an intermediate gear reduction unit. Belt slack is accommodated by a simple tensioning wheel

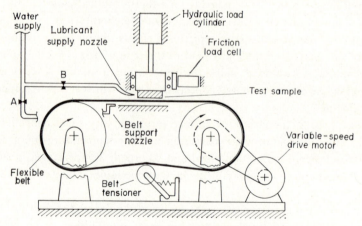

FIG. 11.4. General layout of flat-belt apparatus.

as indicated. The belt itself may be of fabric, rubber, steel strip, or some reinforced composite material, and a surface texture may be incorporated into it either during manufacture or by glueing sandpaper or abrasive sheeting on to one side after assembly of the apparatus. The flat portion of belt between the drive and driven cylinders is normally supported on an air or water cushion to resist distortion by the applied load. In the case of Fig. 11.4, a belt-support nozzle supplies water for the cushion with an exit velocity which is approximately equal to belt speed. The valve *A* which supplies this nozzle may be either manually adjustable or controlled by the speed of the driven cylinder. In the latter case, the proportionality of valve opening and belt speed produces a uniform layer of supporting water of which the thickness is independent of speed.

The normal load on the test sample[†] in Fig. 11.4 is supplied by a hydraulic cylinder, and the side deformation of the sample support structure produced by friction between sample and belt is effectively resisted by frictionless guides as shown. A friction load cell mounted alongside one of the guides is calibrated to record frictional force. If temperature measurements are desired, thermocouples may be embedded at appropriate depths within the sample block or specimen (these are not shown in the figure). A lubricant supply nozzle provides a uniform layer of water on the surface of the belt ahead of the test sample by operating valve *B*.

The apparatus in Fig. 11.4 is capable of sustaining very high speeds of relative sliding between belt and test sample. One limitation is the bursting strength of the belt itself, and this can be alleviated by using a high-strength steel band and by maximizing the cylinder radii in the design. A more practical limitation is rapid wear of the sample particularly if the belt has a rough or emery-type texture and if an interfacial lubricant is not used between sample and belt. The great advantage of the system is the absence of curvature in the contact patch. However, difficulties may arise if the vertical distance between the line-of-action of the friction load cell and the plane of friction is not kept to a minimum value. This is a common dilemma in friction measurement, since the existence of any such distance produces a couple which effectively modifies the applied load *W*. In the case of the flat-belt apparatus there is a particular difficulty that uneven wear of the sample may occur as a consequence of this effect, and special attention should be given to the details of the load and friction measuring systems during the design stage.

11.4 Wear and Abrasion Machine

We have described so far equipment which records friction force on a continual basis, and we have seen that by progressively eliminating the effects of curvature in the contact patch the system design has become more complex. Furthermore, each of the systems described so far is readily adaptable to the measurement of wear, since the latter is an inevitable consequence of any frictional interaction.

---

† This can be a block of fibrous, rubber, metal, or plastic material as shown in Fig. 11.4, or alternatively a cone, sphere, or cylinder of non-deformable material.

We now describe a reciprocating method of measuring wear and abrasion using the machine illustrated in Fig. 11.5. In the simplest terms, the apparatus consists of a travelling platform which is driven at constant speed by a motor and variable-speed transmission, the latter forming an integral part of the platform itself. The base track on which the platform slides is mounted on a pedestal and contains a flat horizontal plate resting on frictionless rollers. This plate in turn is restricted in the longitudinal direction, and it supports a rigid or flexible track. The platform contains a vertical air cylinder capable of applying a range

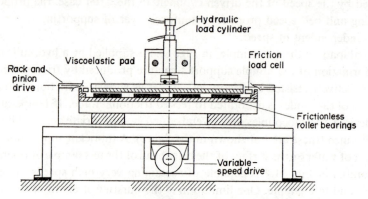

FIG. 11.5. The wear and abrasion machine.

of normal loads from 0 to 50 kgf to a plunger. Various asperity shapes (hemispheres, cones, etc.) of different sizes may be mounted at the end of the plunger. The pneumatic load unit may be displaced laterally as desired to vary the path of the slider on the track. The wear and abrasion machine is capable of performing either accurate low-speed adhesion tests in one pass of the platform, or fatigue–wear tests over a period of days if necessary. Reversing micro-switches are positioned at either end of the track to permit reversal of the direction of travel of the platform, and the spacing between these switches can be adjusted so that the length of travel is variable. The speed of travel can be varied from 0 to 10 cm/s and is indicated by a tachometer. The friction force between the slider and the rubber track is transmitted directly to the flat base plate and recorded by force transducers which restrict the longitudinal motion of the plate.

The number of variables is quite large for such a simple apparatus. Thus the speed, path, length of travel, load, interface contaminant, and surface roughness can be selected in a given adhesion test, or the cyclic frequency, stroke, and number of cycles in a given fatigue or wear test. The most sensitive and accurate method of estimating the degree of wear is the radioactive tracer method. It consists of placing a radioactive tracer on the surface of the track before a test run and observing the decrease in radioactive count during the experiment. The accuracy of this method permits the measurement of wear debris in fractions of a gram, and thus the test time is greatly reduced.

The interface contaminant selected can be water, water with detergent, mineral and vegetable oils with different viscosities, silicone fluids, etc. Surface roughness patterns can be incorporated in both the slider and the track as desired. We may replace the track itself if desired by abrasive or sandpaper strips, and mount an elastomeric or soft metal sample at the base of the load plunger in place of the asperities. This permits a ready evaluation of the effects of a random texture on the abrasive wear produced in the sample. If the sandpaper texture is replaced by transverse rounded ridges, the effects of fatigue wear can be measured. Temperature effects can be recorded as before by embedding thermocouples in one or both of the friction surfaces.

11.5 Squeeze-film Apparatus

Whereas all of the equipment described hitherto in this chapter considers relative tangential motion between surfaces, we deal briefly here with the case of relative normal motion or squeezing. Figure 11.6 shows a simple squeeze-film apparatus used extensively by the author to investigate the effects of surface roughness and angle of inclination on the time of approach

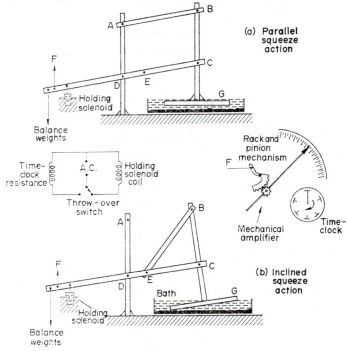

FIG. 11.6. Squeeze-film apparatus.

of plane surfaces having an interposed lubricant.[27, 28] The equipment consists of a beam *CD* pivoted on roller bearings at its centre *D* and carrying a flat plate *G* at one end and a system of balance weights at the other end as shown in the figure. The linkage arrangement provided to ensure that the squeeze action is parallel† appears in the upper diagram. By

---

† Parallel squeezing is such that the plane of the approaching upper plate is always parallel to the plane of the lower surface at all instants during the action of squeezing.

removing the top linkage member *BA* to the new position *BE*, we can convert from parallel to inclined squeezing, as indicated in the lower diagram. The lubricant between the approach surfaces with either arrangement is contained in a transparent or perspex reservoir as shown.

The height of the plate at any instant during squeezing is recorded by a mechanical amplification linkage of the rack-and-pinion type as indicated in the inset of Fig. 11.6. This in turn is transmitted to a pointer which moves over a dial reading in units of as little as $10-20\,\mu$. Time is measured by a precision time-clock reading to hundredths of a second. Both height and time are recorded simultaneously by a fixed camera having an automatic motor-driven film advance mechanism. When the plate has been balanced in the lubricant by a suitable selection of weights, a solenoid mechanism is activated which holds the plate in its initial position while the load to be applied to the plate is, in fact, removed from the weighted end. By operating a throw-over switch, the solenoid is released, thus initiating squeezing of the lubricant and simultaneously causing the time clock to record.

With suitable calibration, this apparatus has been found to be extremely accurate and consistent in the results obtained. By definition, the squeeze action is the application of pressure against the resisting viscous shearing of the interfacial lubricant, and inertia effects within the system are small or negligible. In cases where the latter predominate, a different form of experimental rig is required as shown in the next section.

11.6 Impact or Dynamic Testers

Figure 11.7 shows the general features of an impact or dynamic rig designed to investigate the behaviour of squeeze films under rapid loading conditions. A vertical spindle is supported by a holding solenoid at its upper extremity, and the lower end embodies a square and flat horizontal plate with an elastomeric pad cemented to the underside. The plate, pad, and lower test surface[†] are contained in a transparent bath which may or may not be filled with a lubricant. Dead weights may be applied to the spindle as shown, while the holding solenoid is operative, and a screw adjustment above the weights regulates the initial height of the pad underside above the mean plane of the base. Upon release of the solenoid, the weighted spindle impacts the base surface, being guided in its downward motion by a longitudinal sleeve bearing which is supported by the test stand. An accelerometer mounted on the descending plate records through suitable instrumentation[‡] the history of the approach event.

The impact or dynamic tester illustrated in Fig. 11.7 is extremely simple, yet three significant sets of experiments can be conducted with a view to understanding the influence of several governing parameters:

(a) For a given elastomeric pad material and surrounding lubricant, the effects of surface texture (either regular or random asperities) on squeeze-film performance can be determined.

---

† This may be an assembly of cones, spheres, cylinders, or a random texture.
‡ Instrumentation consists of a charge amplifier and a chart recorder (or oscilloscope).

(b) For a given pad material and surface texture, the damping effects of the lubricant can be isolated and measured.

(c) For a given lubricant and base texture, the viscoelastic properties of different pad materials will affect the squeezing process, and this effect can be quantitatively measured.

Other methods of measuring the downward acceleration of the plate include a linear potentiometer (to measure displacement) mounted alongside the spindle, an inductive coil, or perhaps a linear variable differential transformer. Whatever measurement techniques are

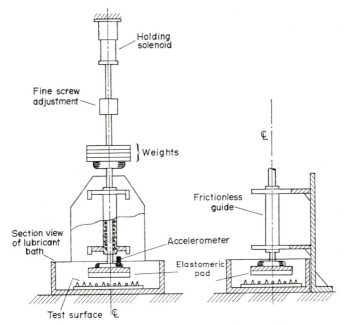

Fig. 11.7. Impact and dynamic draping model.

used, the instrumentation should preserve a high order of accuracy consistent with the given application. Simple models of the type illustrated in Fig. 11.7 provide an exceedingly powerful tool for investigating fundamental tribological phenomena, and we emphasize their contribution in any experimental programme.

11.7 Cross-cylinder Apparatus

We now describe briefly a cross-fibre or cross-cylinder apparatus for measuring polymer friction,[1] as shown in Fig. 11.8. The lower cylinder or fibre is held taut in a carriage which slides to the right at a slow linear speed of about 2 mm/min. The upper fibre is held by a fixed cantilever at right angles to it, and is pressed downwards against the lower fibre by flexing in a vertical plane. When the lower carriage is set in motion, the upper fibre is dragged with it and the horizontal deflection is a measure of the frictional force. We may

observe the deflections using a microscope through a glass window mounted on the surrounding enclosure. The motion is generally intermittent and exhibits "stick–slip" behaviour.

For a given upper fibre, the load range is relatively small (perhaps 50 to 1). By using
fibres of different thicknesses, however, the load range can be considerably extended. Thus
for very thin fibres the lightest load is 10^{-6} g, and for the thickest fibres the maximum
loading is about 10 g. Both drawn and undrawn fibres may be used (usually polytetrafluoroethylene (teflon), nylon, or polythene), and the results show that the coefficient of friction

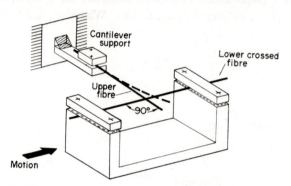

FIG. 11.8. Cantilever apparatus for measuring friction between crossed fibres.

between the fibres decreases non-linearly with increasing normal load. The contact area
between the cylinders is circular, and is thus comparable with that between a sphere and a
flat plane. The whole apparatus in Fig. 11.8 is normally surrounded by an airtight enclosure
so that measurements can be made *in vacuo* or in controlled atmospheres.

11.8 Four-ball Rolling Machine

Most experimental methods of measuring friction use relative tangential sliding as their
mode of operation. In contrast with this, the four-ball machine sketched in Fig. 11.9 causes
continuous rolling between the surfaces. It is known that a common type of failure in rolling

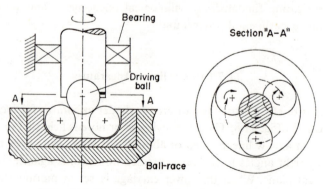

FIG. 11.9. The four-ball rolling friction machine.

elements is pitting fatigue, and this can be investigated satisfactorily with the four-ball device. In its simplest form, the machine consists of three lower balls which are allowed to rotate freely in a ball-race and a fourth ball held in a chuck and loaded against them. Rotation of the chuck in one direction causes the lower balls to rotate in the opposite direction and at the same time to precess or migrate in a circular orbit, as indicated in the right-hand sectional view.

It can be shown that a slight increase in radial loading on the spinning balls will cause a large reduction in fatigue life[46] because of increased stresses in the contact areas. The presence of small cracks or pits (which act as stress raisers) will therefore have the effect of reducing life considerably. The use of the basic apparatus in Fig. 11.9 can be extended by using bath lubrication or by surrounding the device with a simple induction heater for elevated temperature tests.

11.9 Interferometric Equipment

The most precise data on lubricant film thicknesses can be obtained with the optical interferometric apparatus which is shown schematically in Fig. 11.10. An optically smooth rubber surface with spherical curvature is held in a fixed position with its convex surface

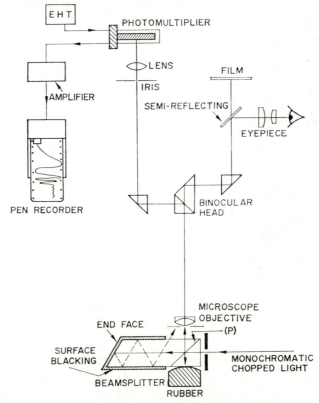

Fig. 11.10. Schematic of optical interferometric equipment.

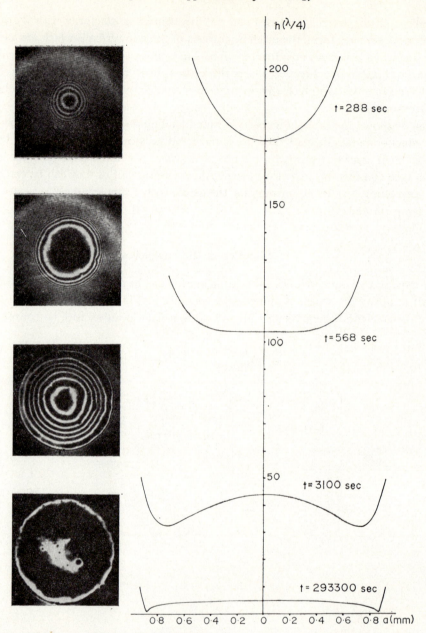

FIG. 11.11. Interferograms and profiles of liquid entrapment: relative normal approach of glass plate and spherical rubber surface. [2]

uppermost, as indicated in the bottom of the figure. After suitable lubrication, a flat glass surface (which is, in fact, the underside of a prism beamsplitter) is gently loaded in a horizontal position on to the rubber surface by means of a lever arm (not shown). The contact area is then examined in monochromatic light through a microscope. In dry contact, strong

Newton's rings are visible between rubber and glass without the need for silvering either surface, but their intensity is greatly diminished in the presence of a liquid. The prism beam-splitter in Fig. 11.10 is used to ensure that the fringes remain visible for liquid contact. It is designed[2] to reject that fraction of incident monochromatic light not directed on to the contact area, so that the background light is kept to a minimum. The glass used in the beam-splitter has a 1.51 refractive index, and is semi-silvered in the plane P to give 50/50 beamsplit-ting. The unwanted half of the incident beam is lost gradually by multiple internal reflec-tions. The absolute separation between glass and rubber surfaces is determined by counting the number of interference fringes created at the centre of contact during the squeezing process. The count is monitored by a photomultiplier and recorded as an electrical signal, while the contact area is simultaneously photographed.

Figure 11.11 shows interferograms of liquid entrapment produced between a spherical rubber surface and glass plate during relative normal approach.[2] The following data are valid for this experiment:

Load = 5 g
Refractive index = 1.40 } dimethyl silicone lubricant
Absolute viscosity, μ = 10^6 cp
Radius of curvature, R = 2 cm
Elastic modulus, E = 6×10^6 dyn/cm^2 } rubber surface
Wavelength, λ = 5461 Å for mercury green light

The illustrations show clearly the rapid formation and slow collapse of the "bell" entrap-ment of lubricant with time. The corresponding profiles of the rubber surface and their distances from the bottom surface are indicated on the right-hand side of the figure. For the low rates of shear attained by the dimethyl silicone in these experiments, it can be assumed to behave as a Newtonian liquid. There are three distinct advantages to be gained by using this fluid:

(a) No swelling or distortion of the rubber surface.
(b) A refractive index sufficiently different from rubber or glass to give reasonably strong interference fringes.
(c) A viscosity sufficiently high to slow the squeeze process such that the early stages of liquid entrapment can be clearly seen and photographed.

Mercury green light has been used to photograph the interferograms in Fig. 11.11. Films squeezed to thicknesses less than 1000 Å generally become too thin to show interference fringes, and a black tint appears as the film approaches zero thickness.

The interferometric method gives a degree of precision which cannot be approached by the other experimental methods listed. However, the method is limited at present to the specialized case of very smooth surfaces. Furthermore, it is distinct from the other methods in so far as it records precisely the elastohydrodynamic interaction effects between surfaces, but does not measure frictional force or wear. Interferometry is a relatively precise tool of interest to the research scientist and applied physicist, and it has contributed substantially to a fundamental understanding of lubricated contact conditions between elastic surfaces. When a finer order of precision is required, we must turn our attention to electron-beam methods.

11.10 Electron Diffraction and Microscopy

The physicist will regard the methods outlined in this chapter as useful in measuring and recording gross surface properties in a frictional interaction, but he will also conclude that such macroscopic approaches are severely limited without direct study of surfaces on an atomic scale. The modern approach therefore recognizes the need for observing the structure of well-defined, atomically clean surfaces of solids and the study of phenomena occurring at interfaces. Such work was not possible until about 10 years ago because of two experimental limitations:

(a) The lack of means for preparing atomically clean surfaces and the lack of a sufficiently high vacuum to prevent the adsorption of environmental impurities. At an ambient pressure of 10^{-6} torr, it takes about one second for a surface to be completely covered with a layer of foreign atoms, assuming that every incident molecule adheres to the surface. At 10^{-10} torr this time is increased to about 3 hours. Great advances in vacuum technology during the last 15 years have made it possible to attain and routinely sustain pressures of 10^{-9} and 10^{-10} torr.

(b) The lack of satisfactory means for observing surface structures on a microscopic scale. Several techniques are now available for such observation. The most widely applied of these are *low-energy electron diffraction, field-ion/field-emission microscopy,* and *molecular beam techniques.*[3]

Low-energy Electron Diffraction

This technique dates back to 1927 when the wavy nature of matter was experimentally verified for the first time by the observation of diffraction patterns obtained by sending electrons through a metal crystal.[3] In that experiment, diffraction patterns were obtained even at electron energies far too low to penetrate the crystal. The cause of the patterns was attributed to surface structures arising from gases adsorbed on the crystal surfaces. The method received new impetus following development work during 1959 at the Bell Telephone Laboratories.

In simplest form, an incident electron beam projected from an electron gun is chopped into pulses before reaching the crystal. Discrete diffracted electron beams coming from the crystal are detected by a movable detector device and fed to an oscilloscope, which then displays either a diffraction spot pattern or a plot of diffraction beam intensity versus electron energy. The most significant result from the low-energy electron diffraction method has been the discovery of a phenomenon described as "reconstructive rearrangement". Thus, for most surfaces, the adsorption of a gas causes surface atoms to rearrange themselves into different structures to accommodate the gas being adsorbed. It is usually necessary to supplement electron diffraction patterns with information gained from other sources (such as kinetic data, infrared spectroscopy, etc.).

Field-emission and Field-ion Microscopy

Field-emission microscopy, invented in 1936 by Müller, made it possible for the first time to scan a surface on a scale approaching atomic dimensions. Observations in a field-emission microscope are dependent on the emission of electrons from a metal into a high electric field. This simple device consists of a vacuum system containing a fluorescent screen and the emitter specimen. The latter has the shape of a sharply pointed needle with a radius of curvature near the tip of about 1000 Å. Images of the molecules and atoms on the tip appear on the screen with a resolution of about 20 Å. The field-ion microscope is similar to the field-emission microscope except that a small quantity of helium gas is admitted to the space between the emitter specimen and the screen and a higher electric voltage is used. The resolving power of this instrument is about 2.5 Å, which is sufficiently fine to directly observe surface defects and interstitial foreign atoms as well as adsorbed gas atoms.

Molecular-beam Methods

A molecular beam is defined as a directed beam of electrically neutral molecules traversing a region in which the pressure is so low that unwanted molecular collisions are negligible. Such beams, when directed at surfaces, may be used for studying such phenomena as reflection, condensation, nucleation, energy exchange, and adsorption.

We conclude by observing that field-emission and field-ion microscopy are best suited for studying the adsorption and diffusion of atoms on metal surfaces, whereas low-energy electron diffraction is to be preferred in the case of crystalline materials. Molecular beam methods are used for studying chemical reactions at surfaces under highly controlled conditions. All of these methods yield valuable and necessary information on the nature of surfaces which may subsequently form part of a frictional interface.

11.11 Miscellaneous

Several experimental methods not listed in this chapter have been described elsewhere in the text and they deserve brief mention. Figure 2.2 illustrates a simple profile measuring device which measures mechanically and records electrically the roughness features of an engineering surface, as described in Section 2.2. The outflow meter in Fig. 2.6 is a hydraulic method of obtaining the same result, namely a measure of the mean texture depth of a rough surface. The principle underlying the outflow meter when used as a hydraulic device may be used in designing a pneumatic instrument for finer surfaces. In this case, air replaces water as the operating fluid (see Fig. 2.9). All of these methods are of interest to the practical engineer since they measure simply the macroscopic features of a given surface.

Figure 3.16 is a schematic diagram of an ultra-high-speed deceleration friction apparatus, and a full description is given in Section 3.14. The method consists in suspending and rotating a steel ball in a magnetic field and then decelerating the spinning ball with the aid of friction pads. An ingenious method of separating the adhesion and hysteresis components of elastomeric friction is made possible by the tape and shaker assembly in Fig. 4.17. The Fitzgerald apparatus described in Section 10.4 and sketched in Fig. 10.5 is one of the most

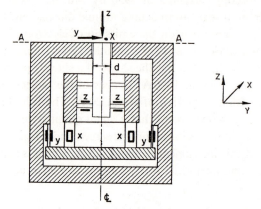

FIG. 11.12. Three-dimensional surface force transducer.

highly developed and frequently used devices for measuring internal friction in elastomeric samples.

Finally, an ingenious device for measuring the components of a three-dimensional surface force is shown in Fig. 11.12. Let us assume for simplicity that a body slides randomly across the interface AA in the figure. As the body crosses the centre line, components X and Y of the frictional force and the normal load Z are recorded instantly from the corresponding strain gauge outputs. The strain gauges are mounted on steel support strips to record bending stresses as shown. If the sliding body is rigid, its dimensions must be such that the contact area is less than $\pi d^2/4$, where d is the diameter of the transducer element appearing at the surface. This will ensure that the X and Y force components recorded correspond to the true frictional force when the sliding body is completely supported by the transducer element. We observe that there are no size limitations in the case of a highly elastic sliding body. The transducer is an extremely useful instrument in friction measurement.

We note in conclusion that a given example involving tribological interactions may, in fact, be a good deal more complex than some of the experimental techniques in this chapter might suggest. In such cases, a given complex event may be broken into a number of simple events, each with perhaps its own weighting factor. Each of the simple events can be simulated by one or other of the rigs described earlier, so that a full comprehension of that event finally emerges from the test programme. We are required to have a broad familiarity with the complex event before pursuing this technique, and both experience and judgement are required to assemble the component test results and interpret them in terms of the phenomenon as a whole. We note that the same number of simple events with a different weighting distribution (or a different number and selection of events with similar weighting) may, in fact, constitute an entirely different complex phenomenon than the one under consideration. The assembly process or synthesis is therefore crucial. The concept of using simplified models or techniques to simulate a given dynamic complex event provides information and an understanding that cannot otherwise be obtained, and it deserves special emphasis as an experimental tool.

PART II

APPLICATIONS

CHAPTER 12

MANUFACTURING PROCESSES

12.1 Metal Machining

The history of metal cutting dates from the latter part of the eighteenth century, which is about the time when James Watt built the first successful steam engine. It is reported that one of his greatest difficulties in developing this machine was the boring of the cylinder casting. The problem was eventually solved when John Wilkinson[61] invented the horizontal boring mill, which consisted of a cutting tool mounted on a boring bar, the latter mounted on bearings outside the cylinder in such a way that it could be fed into and rotated within the cylinder workpiece. Metal cutting as we know it today started with the introduction of this tool, and research into the physics of machining operations commenced about 70 years later. Today, machine tools and metal cutting form the basis of our industry, and metal-cutting features at some stage in the manufacture of industrial products.

All metal-cutting operations are likened to the fundamental process illustrated in Fig. 12.1 in which a wedge-shaped tool with a straight cutting edge (making an angle with the plane of the paper) is constrained to move relative to the workpiece in such a way that a layer of metal is removed in the form of a chip. If the cutting edge is at right angles to the plane of the diagram, the cutting mechanism is said to be orthogonal, otherwise the term oblique cutting is used. The surface along which the chip flows is known as the rake face (or, more simply, as the face) of the tool, and it intersects the tool flank (or flank) to form the cutting

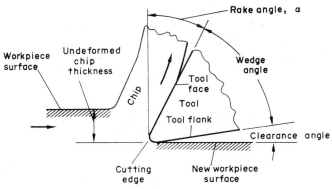

FIG. 12.1. Terminology used in metal cutting.

249

edge as shown. The depth of material removed from the workpiece is known as the undeformed chip thickness and it may vary in size in practical cutting operations. Three angles
are identified in Fig. 12.1 of which the most important is the rake angle α. This is essentially
the angle which the tool face makes with the vertical in the case of orthogonal cutting. The
wedge angle is the included angle between face and flank in the cutting tool, and the clearance
angle that between the tool flank and new workpiece surface. The sum of the rake, wedge,
and clearance angles is 90°.

Metal-cutting operations may be classified in any or all of several ways, as indicated in a
later section. Methods of classification may be based on whether the mode of operation of
the tool is rotary or reciprocating, continuous or intermittent, etc., or whether the tool or
workpiece or both move, or whether single-point or multiple-point tooling is used. In the
case of single-point tools, Fig. 12.2 shows additional terminology in common use today. The
major cutting edge is mainly responsible for chip removal, and it usually takes the greatest

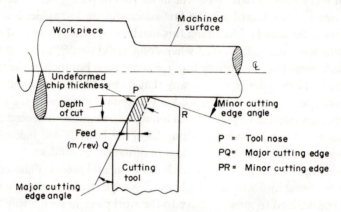

FIG. 12.2. Additional terminology for single-point cutting operations.

share of the cutting load. The minor cutting edge is mainly responsible for producing the
finished workpiece surface. The angles which these edges make with the tool shank are known
as major and minor cutting edge angles as shown. The radius of the tool nose plays an
important part in determining the surface finish of the machined surface. The depth-of-cut
in Fig. 12.2 is self-explanatory, and the feed is the distance advanced axially by the tool
per revolution of the workpiece. Multiple-point tools (such as twist drills or milling teeth)
can be regarded as a series of two or more single-point cutting tools solidly connected to a
common body.

12.2 Friction in Metal Cutting

Early conceptions of the chip formation mechanism during cutting presupposed that a
crack was propagated ahead of the wedge-shaped tool, thus permitting ready separation
of the chip from the workpiece, as indicated in Fig. 12.3(a). Modern theories have since

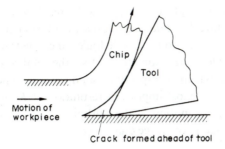

Chip

Tool

Motion of
workpiece

Crack formed ahead of tool

(a) Earlier conception of cutting

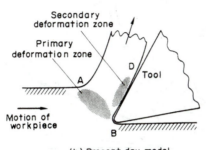

Secondary
deformation zone

Primary
deformation zone

D

A

Tool

Motion of
workpiece

B

(b) Present-day model

FIG. 12.3. Earlier and existing models of the cutting process.

disproved this concept, and it is now generally agreed that there are two distinct regions which can be identified during the cutting operation:

(a) A *primary* deformation zone at the base of the chip, where a continuous shearing action persists. This region can be considered a region of maximum internal friction (see *AB* in Fig. 12.3(b)).

(b) A *secondary* deformation zone occurring between tool and chip characterized by interfacial friction along the plane *BD*.

If we now examine the interfacial friction plane *BD* in more detail, we observe from Fig. 12.4 that it comprises a "stiction" region *BC* and an adjacent length *CD* in which gross sliding occurs. The relatively high normal pressures acting within the stiction zone are responsible for a reduced coefficient of sliding friction for reasons that will appear later. Here the large degree of normal loading promotes a greater actual contact area where local welding between the major asperities of the surfaces will occur in accordance with the welding–shearing–ploughing theory of Chapter 3. Figure 12.4 shows the relative distributions of normal and shear loading along the friction plane *BD*, and Q_F represents the resultant force acting on the tool as a result of these distributions N and F respectively. We must also include a ploughing force P acting on the tool edge or nose within the relatively small work-tool interface. The total resultant force R acting on the tool is the vector sum of Q_F and P, as shown in the inset of Fig. 12.4, and the angle τ which N makes with Q_F is called the friction angle.

We observe that the ploughing force P may also be resolved into components along and perpendicular to the rake face BC, and these could then be added to F and N respectively. This procedure, however, appears to have little logic, since the force P acts along the work-tool interface rather than along BD. Sometimes, too, the total resultant force R is resolved into components F_S and F_N, which lie respectively along and perpendicular to the shear plane AB in Fig. 12.3. However, the most important resolution of R appears to be along and at right angles to the direction of motion of the workpiece, thus defining the cutting force F_C and thrust force F_T illustrated in the force diagram of Fig. 12.4.

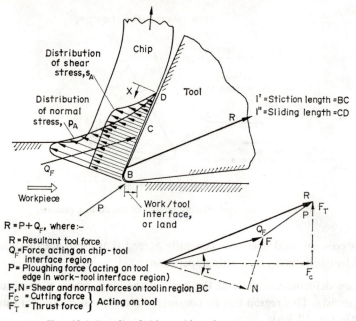

$$R = P + Q_F, \text{ where:} -$$

R = Resultant tool force
Q_F = Force acting on chip-tool interface region
P = Ploughing force (acting on tool edge in work-tool interface region)
F, N = Shear and normal forces on tool in region BC
F_C = Cutting force ⎫ Acting on tool
F_T = Thrust force ⎭

l' = Stiction length = BC
l'' = Sliding length = CD

FIG. 12.4. Details of chip-tool interface and tool forces.

During metal cutting it has generally been observed that the mean coefficient of friction between chip and tool can vary considerably, and is affected by such factors as cutting speed, feed rate, rake angle, etc. The main reason for this variation is the very high value of normal pressure which exists at the chip-tool interface, as explained previously. For example, these normal pressures when machining steel can be as high as 34,000 atm, and can cause the real area of contact to approach the apparent area over a portion of the chip-tool interface ($A_{act}/A_{app} \rightarrow 1$). Consider now the frictional force F from eqn. (3.17) and the coefficient of friction f written in two forms thus:

$$F = A_{act}s, \tag{3.17}$$

$$\begin{cases} f = \dfrac{F}{W} = \dfrac{A_{act}s}{A_{act}p^*} = \dfrac{s}{p^*}, & (3.20) \\[3mm] f = \dfrac{F}{W} = \dfrac{A_{app}s_A}{A_{app}p_A} = \dfrac{s_A}{p_A}, & (12.1) \end{cases}$$

where s_A and p_A are the apparent shear stress and pressure respectively taken over a small area of contact. Equation (3.20) describes the simple theory of adhesion in Chapter 3 in the absence of work-hardening effects, and the ratio (s/p^*) on the right-hand side is constant. The simple theory applies, of course, for $A_{\mathrm{act}} \ll A_{\mathrm{app}}$, and in this case we observe that F is directly proportional to W. The second form of the coefficient of friction therefore implies that the apparent stresses s_A and p_A are proportional to each other for $A_{\mathrm{act}}/A_{\mathrm{app}} \ll 1$, and this is seen to be the case along CD in Fig. 12.4.

We now imagine W to be increased to the point where the actual area of contact approaches the limiting value of the apparent area (i.e. $A_{\mathrm{act}}/A_{\mathrm{app}} \rightarrow 1$). It is clear from eqns. (3.20) and (12.1) that under these conditions $s_A \rightarrow s$, where s (in the absence of work hardening) has an invariant value characteristic of the workpiece. Thus F now assumes a constant value, and we see at once from eqn. (12.1) that the coefficient of friction varies inversely with the apparent pressure p_A. The stress distributions in Fig. 12.4 over the chip-tool interface BC show clearly that s_A is reasonably constant, whereas p_A increases towards the nose B of the tool. The coefficient of adhesional friction therefore varies in accordance with eqn. (12.1) along BC, while having a constant value along CD.

The following analysis of the chip-tool interface conditions during the formation of a chip is a guide to the previous discussion. Let the variation in p_A within the region CD of Fig. 12.4 be given by the expression

$$p_A = K_1 x^{K_2}, \tag{12.2A}$$

where x is the distance along the tool face (starting from the point where the chip loses contact with the tool), and K_1, K_2 are constants. Since $p_A = p_{\max}$ at $x = CD = l$,

$$p_A = p_{\max}(x/l)^{K_2}, \tag{12.2B}$$

and by integrating this expression, the normal load N becomes

$$N = b \int p_A \, \mathrm{d}x = b p_{\max} \int \left(\frac{x}{l}\right)^{K_2} \mathrm{d}x = \frac{p_{\max} b l}{1 + K_2}, \tag{12.3}$$

where b is a length perpendicular to the plane of the paper.

Within the sliding region CD, the coefficient of friction f is constant, so that from eqns. (12.1) and (12.2B)

$$s_A = f p_A = f p_{\max}(x/l)^{K_2}. \tag{12.4}$$

The shear stress $s_A \rightarrow s$ within the adhesion region as we have seen, and we can now write for the total frictional force F

$$F = bsl' + \frac{bf p_{\max} l''^{K_2+1}}{l^{K_2}(1 + K_2)} = bsl' + \frac{bsl''}{1 + K_2} \tag{12.5}$$

with s replacing $f p_{\max} (l''/l)^{K_2}$ from eqn. (12.4). From eqns. (12.3) and (12.5) the angle of friction τ is defined by

$$\tan \tau = \frac{F}{N} = \frac{s}{p_{\max}} \left(1 + K_2 \frac{l''}{l}\right). \tag{12.6}$$

It is convenient to replace the maximum apparent pressure p_{max} in this equation by the mean value $\bar{p}$, where

$$\bar{p} = \frac{N}{bl} = \frac{p_{max}}{1+K_2}$$

from eqn. (12.3). Thus

$$\tan \tau = \frac{s}{\bar{p}} \frac{[1+K_2(l''/l)]}{[1+K_2]}. \tag{12.7}$$

Since the bracketed terms in eqn. (12.7) are largely constant, the mean angle of friction τ for the basic machining operation in Fig. 12.4 becomes

$$\tau = \tan^{-1}(K/\bar{p}) \tag{12.8}$$

where K is a constant and $\bar{p}$ can be shown to be a function of the rake angle α.

12.3 Heat Generation

High temperatures are generated during metal cutting near the nose of the tool, and these temperatures have an important bearing both on the rate of wear of the cutting tool and on the friction which develops at the chip-tool interface. We will therefore consider briefly the mechanism of heat generation during the basic machining process illustrated in the previous figures, and the resulting temperature distribution in chip and workpiece.

Let Q denote the total heat generated during cutting as a result of providing a cutting component F_c of resultant tool force (see Fig. 12.4). Then

$$Q = F_c V_c = Q_s + Q_f, \tag{12.9}$$

where V_c is the speed of cutting, Q_s the heat generated in the primary deformation zone (shearing heat), and Q_f the heat generated in the secondary deformation zone (frictional heat). Since $Q_f = FV_f$, where F is the frictional force and V_f the velocity of chip flow, the shearing heat component can be obtained directly from eqn. (12.9). The total heat generated, of course, must be removed by the chip, tool, and workpiece thus:

$$Q = Q_C + Q_W + Q_T, \tag{12.10}$$

where the suffixes C, W, and T refer to chip, workpiece, and tool respectively. Because the chip material near the tool face flows rapidly, it has a much greater capacity for the removal of heat than the tool. Thus Q_T is very small, and Q_C appears to be the largest component of heat removal. Consider the experimentally determined distribution of temperature within chip and workpiece shown in Fig. 12.5 for the case of orthogonal cutting.[61] For a point X in the material moving towards the cutting tool, heating occurs during passage through the primary deformation zone, and this heat is carried away by the chip. Point Y, however, passes through both deformation zones, and heating is continued until it has left the secondary deformation zone; it is then cooled somewhat as heat is conducted into the body of the

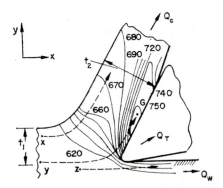

FIG. 12.5. Temperature distribution in workpiece and chip for orthogonal cutting. [61] Material: mild steel. Cutting speed: 22.5 m/min. Width of cut: 6.35 mm. Tool rake angle: 30°. Workpiece temperature: 611°C. $t_1 = 0.6$ mm. $t_2 = 0.9$ mm. All temperatures in °C.

chip, the latter finally achieving a uniform temperature throughout its length. We note that the maximum temperature occurs some distance from the cutting edge along the face of the tool. The point Z (which remains within the workpiece) is heated as it passes below the tool cutting-edge by the conduction of heat from the primary deformation zone.

Since the maximum temperature rise ΔT_{max} occurs at the point G in Fig. 12.5 where the material leaves the secondary deformation zone, we can write

$$\Delta T_{max} = \Delta T_s + \Delta T_m, \tag{12.11}$$

where ΔT_s is the temperature rise which occurs within the primary deformation zone due to shearing, and ΔT_m is the corresponding temperature rise within the secondary deformation zone BD due to chip-tool friction. We can also regard ΔT_m as the maximum temperature rise within the chip itself, where ΔT_f for distinguishing purposes is the mean temperature rise in the chip. Then

$$\Delta T_s = \frac{Q_s(1-\beta)}{\varrho C_p V_c t_1 b}, \tag{12.12}$$

$$\Delta T_f = \frac{Q_f}{\varrho C_p V_f t_2 b}, \tag{12.13}$$

where β is the proportion of Q_s conducted into the workpiece, ϱ the density of the material, C_p its specific heat at constant pressure, t_1 the undeformed chip thickness, t_2 the actual chip thickness, and b the width of cut.

To obtain a value of β in eqn. (12.12) and also to establish a relationship between the mean and maximum temperature increases in the chip, we must solve the differential equation for heat conduction within a two-dimensional solid. [61] In metal-cutting operations, this equation assumes the following simplified form

$$\frac{\partial^2 T}{\partial y^2} - \left(\frac{R_T}{t_1}\right)\frac{\partial T}{\partial x} = 0, \tag{12.14}$$

where the x- and y-directions are shown in Fig. 12.5. Equation (12.14) assumes an idealized model for the metal-cutting process, where both deformation zones are represented by plane heat sources of uniform strength, as shown in Fig. 12.6. A further assumption is that no heat is lost from the surfaces of workpiece and chip. The "thermal number" R_T in eqn. (12.14) is defined by $\varrho C_p V_c t_1/k$, where k is the thermal conductivity of the material.

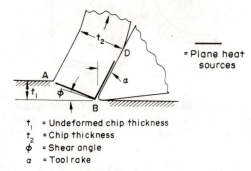

t_1 = Undeformed chip thickness
t_2 = Chip thickness
ϕ = Shear angle
a = Tool rake

FIG. 12.6. Idealized model of cutting process employed in theoretical work on cutting temperatures.

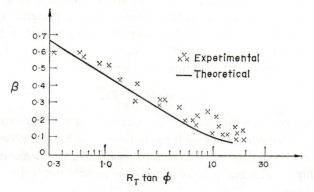

FIG. 12.7. Division of shear zone heating between chip and workpiece as function of thermal number and shear angle.

A solution to eqn. (12.14) has been shown[61] to provide an expression relating β to R_T tan φ, where φ is the shear angle defined in Fig. 12.6. The expression is plotted in Fig. 12.7 and there is close agreement with experiment over a wide speed range. Solutions to eqn. (12.14) also show a relationship between the ratio $(\Delta T_m/\Delta T_f)$ and a new parameter γ, defined as the width of the secondary deformation zone per unit chip thickness, and these results appear in Fig. 12.8. Here α_1 is a numerical factor which when multiplied by the actual chip thickness t_2 gives the length of contact BD between chip and tool (see Fig. 12.4). In practice, α_1 can be estimated from the wear on the tool face, and the width of the secondary deformation zone from photomicrographs of the chip cross-section. If we now combine eqns. (12.11), (12.9), (12.12), and (12.13) with the information in Fig. 12.8, we obtain for the maximum temperature rise ΔT_{max} in the chip

$$\Delta T_{max} = \frac{F_c(1-\beta)+f(\gamma)\,F(V_f/V_c)}{\varrho C_p t_1 b}, \qquad (12.15)$$

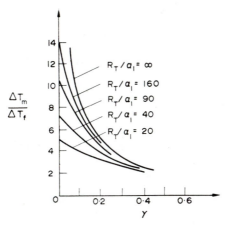

FIG. 12.8. Effect of width of secondary deformation zone on chip temperature.

where $V_f t_2 = V_c t_1$, $f(\gamma)$ denotes the functional relationship between $(\Delta T_m / \Delta T_f)$ and γ from Fig. 12.8, and β is obtained from Fig. 12.7.

Let us now briefly consider a typical example where we desire to estimate as accurately as possible the maximum temperature rise in the chip:

Tool rake angle	$\alpha = 0$
Cutting force	$F_c = 91$ kg
Width of secondary deformation zone chip thickness	$\gamma = 0.2$
Thrust force	$F_T = 80$ kg
Cutting speed	$V_c = 2.2$ m/s
Undeformed chip thickness	$t_1 = 0.25$ mm
Width of cut	$b = 2.5$ mm
Velocity of chip flow	$V_f = 0.7$ m/s
Length of chip–tool contact BD	$\alpha_1 t_2 = 0.75$ mm

The total heat generated is given by eqn. (12.9):

$$Q = F_c V_c = 91 \times 2.2 = 200 \text{ kg m/s}$$

and the heat due to friction along the chip–tool interface (since $F_T = F$) is

$$Q_f = F_T V_f = 80 \times 0.7 = 56 \text{ kg m/s.}$$

Thus $$Q_s = Q - Q_f = 144 \text{ kg m/s.}$$

Letting $\varrho = 6.1$ g/cm³, $k = 4.44$ kg/s °C, and $C_p = 50.4$ m/°C for mild steel,

$$R_T = \frac{6.1 \text{ g/cm}^3 \times 50.4 \text{ m/°C} \times 2.2 \text{ m/s} \times 0.25 \text{ mm}}{4.44 \text{ kg/s°C}} = 38.1.$$

Now, since $\alpha = 0$, $\tan \varphi = V_f/V_c$, and

$$R_T \tan \varphi = 38.1 \times 0.7/2.2 = 12.1.$$

From Fig. 12.7 we therefore obtain a value of about 0.1 for β. Thus,

$$\Delta T_s = \frac{144 \text{ kg m/s} \times 0.9}{4.44 \text{ kg/s} \,^\circ\text{C} \times 38.1 \times 2.5 \text{ mm}} = 306^\circ\text{C}$$

and

$$\Delta T_f = \frac{56 \text{ kg m/s}}{4.44 \text{ kg/s} \,^\circ\text{C} \times 38.1 \times 2.5 \text{ mm}} = 132^\circ\text{C}.$$

We can write

$$\alpha_1 = \frac{0.75}{t_2} = \frac{0.75 \tan \varphi}{t_1} = \frac{0.75 \times 0.7}{2.2 \times 0.25} = 0.955,$$

so that

$$R_T/\alpha_1 = \frac{38.1}{0.955} \doteq 40.$$

From Fig. 12.8, with $R_T/\alpha_1 = 40$ and $\gamma = 0.2$, we find that $\Delta T_m/\Delta T_f = 4.1 = f(\gamma)$. Thus

$$\Delta T_m = 4.1 \times 132 = 541^\circ\text{C},$$

and from eqn. (12.11) the maximum temperature rise in the chip becomes

$$\Delta T_{\max} = 306 + 541 = 847^\circ\text{C}.$$

Equation (12.15) can also be used to find $\Delta T_{\max}$.

12.4 Tool Wear

The fundamental nature of wear varies with operating conditions, as we have seen in Chapter 9. In metal cutting, four main forms have been identified, namely adhesional, abrasive, diffusive, and electrochemical. Adhesional wear is caused by the fracture of welded asperity junctions between two metals according to the welding–shearing–ploughing theory outlined in Chapter 3. In metal cutting, the fracture of junctions may tear out tiny fragments of the tool material which adhere to the chip or workpiece. Abrasive wear occurs when hard particles on the underside of the chip pass over the face of the tool and remove tool material by gouging or mechanical action. These hard particles may be highly strain-hardened fragments of an unstable built-up edge on the tool. Diffusive wear is a result of the movement of atoms from the tool to the workpiece, this being a consequence of intimate contact and high temperatures. The process takes place within a narrow reaction zone between the two materials and causes a weakening of the surface structure of the tool. Electrochemical wear may occur in metal cutting when the passage of ions between tool and work causes scaling of the tool surface. Due to the high temperatures existing during cutting, a thermoelectric e.m.f. is generated at the work–tool junction, causing large electric currents to circulate. This results in the passage of ions and a resulting breakdown of the tool material at the tool–chip interface.

Figure 12.9 depicts how progressive wear occurs in a metal-cutting tool. A crater is formed along the face of the tool as a result of the flowing action of the chip. At the same time, a flat wear land of length l_W appears on the flank due to the rubbing action of the

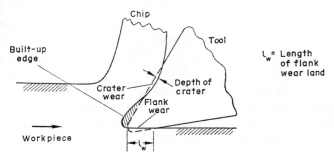

FIG. 12.9. Regions of progressive wear during metal-cutting.

newly formed workpiece surface. A built-up edge may also appear at the nose of the tool due to metal transfer from the softer workpiece. The cratering effect is most likely a consequence of diffusive wear at high speeds since it occurs in the region of maximum temperature rise. At low sliding speeds it is less pronounced, and is due to adhesive and abrasive (or ploughing) wear. The life of a cutting tool under very high-speed conditions is largely determined by crater wear.

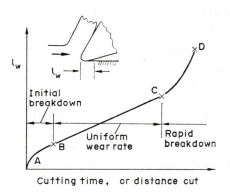

FIG. 12.10. Development of flank wear with time or distance cut.

Under all cutting conditions, wear occurs along the flank of a tool as shown in Fig. 12.9. The usual criterion for the extent of this form of wear is the length of the wear land between tool flank and workpiece. Figure 12.10 shows a typical graph of the progress of flank wear with time or distance cut. The curve can be divided into three regions:

(a) The region *AB*, where the sharp cutting edge is quickly broken down, and a finite wear land is established.
(b) The region *BC*, where uniform wear occurs.
(c) The region *CD*, where wear takes place at an accelerating rate—perhaps due to increased tool temperatures (the latter due to the presence of such large wear lands).

Usually, tools are re-ground before the end of the uniform wear section is attained.

The presence of a built-up edge on the tool face during cutting can affect tool wear in various ways, sometimes decreasing and at other times increasing the life of a cutting tool. We have seen that the built-up edge (see Fig. 12.9) can result in grooving or slow speed cratering because highly strain-hardened fragments from the built-up edge adhere to both chip and workpiece surfaces and abrade the tool faces. When cutting very hard materials, a stable built-up edge can be beneficial by protecting the tool surface entirely from wear and performing the cutting action itself. Built-up edges may also be responsible for sudden tool failures when using carbide-tipped cutting tools.

12.5 The Action of Lubricants

Cutting fluids (usually in the form of a liquid) are applied to the work–tool system to improve conditions of dry cutting. The cutting fluid can act either as a lubricant, or a coolant, or both. Mineral oils diluted with water are applied in the form of an emulsion with the objective of providing a large coolant capacity. When lubricating action is more important than cooling, oils are used; this is normally confined to slow-speed cutting operations such as screw cutting, broaching, and gear cutting.

Cooling provides the following three advantages:

(a) an increase in tool life by virtue of a reduction in temperature;
(b) easier handling of the finished workpiece; and
(c) a reduction in thermal distortion (caused by severe thermal gradients generated within the workpiece during machining).

In the case of grinding operations, the last two factors are important.

The action of lubricants in metal cutting appears to be in the form of a boundary lubrication mechanism. This is because the extremely high pressures existing in the region of the chip–tool interface during machining do not permit complete hydrodynamic lubrication (where chip and tool would be separated by a thin film of fluid). Indeed, the lubricating action of cutting fluids is mainly of a chemical rather than physical nature. Thus carbon tetrachloride (CCl_4), although not normally used in engineering applications, can have startlingly beneficial effects on the cutting process—reducing the specific power consumption by 60% in certain cases.[61] Pure chemical compounds should have the following three properties in order to act as an efficient lubricant in metal cutting:

(a) a small molecular size to allow rapid diffusion and penetration to the chip–tool interface;

(b) a suitable reactive ingredient which promotes a boundary lubrication action between tool and chip; and

(c) the ability to be broken down at the temperatures and pressures existing at the chip–tool interface.

Figure 12.11 shows the effect of cutting speed on the lubricating action of carbon tetra-chloride, and it appears that this compound possesses the three desirable properties listed. At low cutting speeds the chip–tool friction is greatly reduced by the application of the fluid, due largely to the formation of a low shear-strength film of copper chloride which acts as a boundary lubricant. The carbon tetrachloride also eliminates a small region of "sticking" friction on the tool flank during dry cutting. The effectiveness of all cutting lubricants diminishes as the cutting speed increases in a manner similar to that shown in Fig. 12.11 for CCl_4. This is partly explained by a loss of penetration of the fluid to the chip–tool interface at these higher speeds and partly by the accompanying increased temperatures which reduce the effectiveness of the solid boundary lubricant formed.

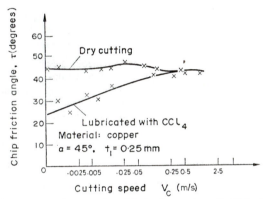

FIG. 12.11. Effect of cutting speed on lubricating action of carbon tetrachloride.

One very significant parameter which affects not only lubricating action in metal cutting but also every aspect of machining operations, is surface roughness of the newly formed workpiece. The built-up edge near the nose of the tool may be continually breaking away in fragments and re-forming, and the fractured particles are carried away on the surfaces of chip and workpiece. Such roughness effects are aggravated by large built-up edges on the cutting tool and also by the occurrence of chatter or vibration of the machine tool, in-accuracies in machine tool movements, irregularities in feed rate, defects in the structure of the workpiece material, surface damage caused by chip flow, and discontinuous chip formation when machining brittle solids.

12.6 Classification of Machining Processes

Having examined in detail the mechanics of the basic metal-cutting operation with special attention to its tribological effects, we now attempt to classify most of the common machining operations used today. Table 12.1 shows an extremely broad tribological classification of

TABLE 12.1. TRIBOLOGICAL CLASSIFICATION OF MANUFACTURING INDUSTRIES

Tribological classification	Processing operation	Associated production equipment	Related manufacturing industry
Metal-on-metal (rigid–rigid) surface pairing	Forging Stamping Grinding Milling Lapping Reaming Guillotining Facing Boring Turning Tapping Broaching Drawing Extrusion Forming, etc.	Bearings Lathes Drill presses Drop hammers Miscellaneous rotary and reciprocating machinery	Wire manufacture Iron and steel industries Light engineering Heavy engineering Metal processing Mechanical components Tool design
Plastic[a]-on-metal (flexible–rigid) surface pairing	Injection and blow moulding Thermo forming Extrusion Drawing Vacuum forming Coating Laminating Friction welding	Miscellaneous die and moulding equipment Also, lathe bearings, drill presses, etc.	Tyre manufacture Plastics industry Building and construction industry Shoe manufacture Mechanical components Flooring industry Miscellaneous Wire manufacture
Fibre-on-fibre (flexible–flexible) pairing	Spinning Weaving Threading Carding, etc.	Miscellaneous spinning and weaving equipment	Textile manufacture Rope industry Plastics Hosiery and knitwear Cable manufacture

[a] Includes elastomeric materials, solid lubricants, and rubbers.

manufacturing industries. While not purporting to be complete, this listing shows the extremely wide relevance of tribology and friction to most manufacturing operations. The processing techniques listed in the second column involve, of course, other considerations besides surface interactions which must receive attention prior to any design procedure. However, the tribological aspect appears to be a common and unifying factor in all cases. The processing operations listed can be grouped together in many different forms, such as, for example, according to whether the motion involved is rotary or reciprocating, or according to whether a single-point or multiple-point tool is used, or perhaps dependent on the relative speed of the event. We will now examine briefly the nature of some common processing methods.

Single-point Tools

Five of the most common machining operations with single-point tools are shaping, planing, turning, facing, and boring. In the case of shaping, a flat surface is generated on the workpiece by reciprocating the tool and feeding the workpiece in a direction at right angles to the reciprocating motion. In planing, the workpiece is reciprocated and the work fed (Fig. 12.12). Two cutting edges are involved in both operations:

(a) the major cutting edge is responsible for chip removal and takes most of the cutting load; and

(b) the minor cutting edge produces the finished workpiece surface.

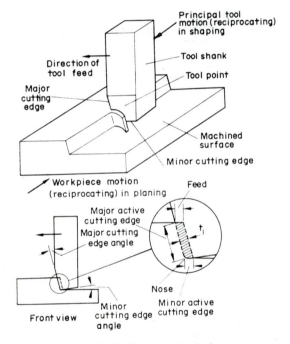

FIG. 12.12. Shaping and planing.

Both cutting edges intersect to form the nose of the tool whose radius has a very pronounced effect in deciding the surface finish of the machined surface. The turning operation shown in Fig. 12.13 is similar to shaping and planing except that the workpiece rotates about an axis of rotational symmetry while the tool is fed in a direction parallel to the axis of rotation. Major and minor active cutting edges are again defined in the figure. Facing and boring operations are sketched schematically in Fig. 12.14, and we observe that rotation of the workpiece in each case is accomplished on a lathe, as in the case of turning. In facing, a flat surface is generated by feeding the tool across the end of a rotating workpiece in a direction normal to the axis of rotation. In the case of boring, an internal cylindrical surface is ground by a single point tool known as a "boring bar".[61]

18*

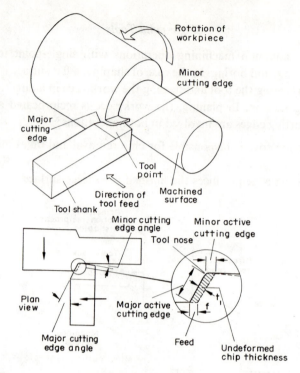

FIG. 12.13. Turning.

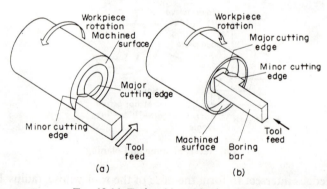

FIG. 12.14. Facing (a) and boring (b).

Multi-point Tools

With multi-point tools the relative motion between tool and workpiece is such that each tool point contributes to the removal of workpiece material. For example, a twist drill can be regarded as two single-point cutting tools fixed to a common shank which is rotated about its own axis, as depicted in Fig. 12.15(a). As the drill is fed in a direction parallel to its axis, each cutting edge should remove an equal share of the workpiece material. We note that the rake angle in a twist drill varies along the cutting edge from positive at the outer radius

(a) Drilling

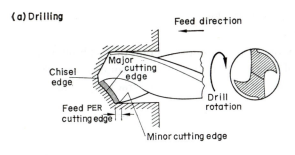

(b) Tapping

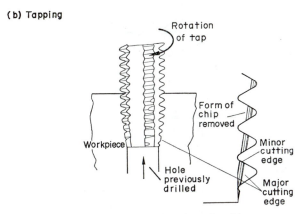

FIG. 12.15. Drilling (a) and tapping (b).

to negative near the centre of rotation. In tapping, each cutting edge of a helical broach removes a small layer of metal, so that the desired thread shape is obtained as shown in Fig. 12.15(b). The fully shaped thread finally obtained serves to burnish the work surface and clear away fragments of chips which tend to collect in the machined threads.

Figure 12.16 illustrates the cutting action in face milling. The geometry of the cutting edge is relatively complex, although major and minor cutting edges can be identified as

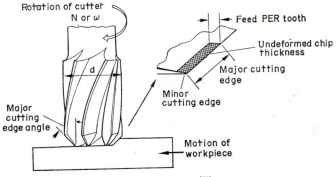

FIG. 12.16. Face milling.

shown. The undeformed chip thickness may vary during a cut, and is also dependent on the major cutting-edge angle.

Figure 12.17 shows the equally complex action of a slab-milling cutter. Here the cutter, which has a number of long helical cutting edges, is rotated while the workpiece is fed past it. Each cutting edge removes a layer of material of gradually increasing thickness.

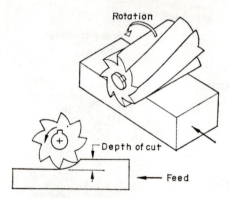

FIG. 12.17. Slab milling.

In broaching the tool is forced past the stationary workpiece at low speed, and each broach tooth removes a small layer of the work material. Cutting conditions are close to orthogonal and the process is similar to that found in shaping except that the work of cutting is shared between several cutting edges (Fig. 12.18).

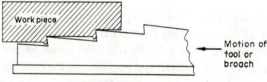

FIG. 12.18. Broaching.

The final process which we consider here and which uses a multi-point tool is grinding. Grinding wheels usually consist of a large number of abrasive particles called grains, which are held together by suitable bonding. The action of grinding is very similar to that of slab milling except that the cutting points are irregularly shaped and randomly distributed over the active face of the tool. Figure 12.19 shows (a) the microscopic structure of a typical grinding wheel, and (b) a schematic of grinding action. Each active grain removes a short chip of gradually increasing thickness. Because of the random nature of grain shape, there is considerable interference or ploughing action between each active grain and the new work surface. This results in the formation of worn areas at the tips of the active grains due to progressive wear. As grinding proceeds, these worn areas increase, thus increasing the interference or friction and also increasing the force on each grain. Eventually, this force is sufficiently large either to tear or pluck out the worn grain from the wheel or to fracture

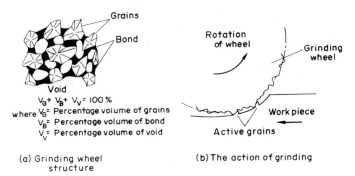

where V_G = Percentage volume of grains
V_B = Percentage volume of bond
V_V = Percentage volume of void

$V_G + V_B + V_V = 100\%$

(a) Grinding wheel structure

(b) The action of grinding

FIG. 12.19. Grinding.

the worn grain. In this manner, new unworn grains and new cutting surfaces are being constantly exposed, so that the process exhibits a distinct self-sharpening characteristic.

The hardest type of grain used in grinding wheels is diamond, and this is widely used for grinding very hard materials such as cemented carbides. Grains of aluminium oxide or silicon carbide are more commonly used in the manufacture of grinding wheels. Wear characteristics in grinding are similar to the general model depicted in Fig. 12.10 for flank wear in metal cutting. Thus there is an initial transient or breakdown period followed by a region of uniform wear rate, and, finally, a region of rapid breakdown. Generally, the grinding wheel should be dressed or replaced before the final region begins.

While this treatment of metal-cutting processes is far from complete, it shows that the more common methods all remove metal in basically the same manner. Thus the action of the basic wedge-shaped cutting tool as treated in earlier sections has wide applicability to virtually all metal-cutting operations. Furthermore, since each cutting process involves the shearing of a softer metal by a harder metal, we can identify a distinct frictional mechanism in each case. Two very significant variables in all metal-cutting operations are the speed of cutting and the temperature attained in tool, chip, and workpiece.

12.7 Friction Welding

The first known use of friction to produce heat occurred when early man used a fire stick to light a fire. It was 1891 before an American patent proposed that friction-induced heat could be used to promote welding between a tube and a vee-shaped die. During World War II, friction welding was used in Germany to butt-weld plastic pipe and in the United States to assemble plastic components. In recent years, Russian investigators have been foremost in developing and applying the process. Some common examples today are the friction welding of plastics during the assembly of lenses, instrument knobs, container caps, aerosol bottles, and tool handles, and the friction welding of metals in such assembly operations as axle shafts, shanks on drills, tool joints, etc.

Figure 12.20 illustrates the simplest and most common scheme for friction welding. Here, two cylindrical bars are aligned axially while one of the bars is rotated and the other held

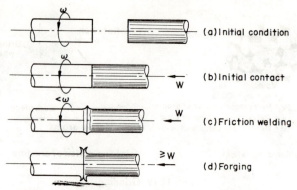

FIG. 12.20. Basic process during friction welding.[62]

stationary. The angular speed ω is sufficiently great that when the non-rotating member is advanced against the rotating member and pressed against it with a force W, enough frictional heat is dissipated at the interface to promote welding during a short time-interval. The final forging phase occurs when the rotation is stopped and the load W maintained or increased for a short period. Trimming of the newly formed joint can be carried out for this example in a lathe, and the total operation (excluding trimming) may last from 2 to 30 s.

The effective coefficient of friction between the contacting bars in stages (b) to (c) of Fig. 12.20 is far from constant. For a given rotational speed ω, the sliding velocity at any point of the interface varies directly with distance r from the centre of rotation or axis of the bars. As a consequence, the coefficient of friction varies across the section being welded. The action of rubbing and normal loading at the interface breaks down oxide film formations if the bars are metallic, and subsequent increase of loading in the forging phase causes the actual area of contact to approach the apparent area, so that a very strong diffusive bond is formed between the specimens. We note carefully that if the same normal loading is initially applied without tangential or spinning motion, the bars would contact and become welded together only at isolated high spots since $A_{act} \ll A_{app}$. The superposition of a rapid relative spinning action increases local temperatures to high values in a very short time period, thus permitting intimate contact to take place not only between the asperities of the surfaces in question but also in the void spacing between asperities, as revealed in Fig. 12.21.

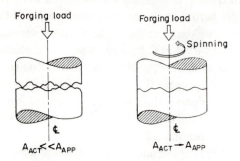

FIG. 12.21. Effect of spinning on mechanism of friction welding.

The heating effect supplied by spinning one of the specimens to be welded is sufficient to cause large-scale deformation of opposing asperities, so that a strong weld is formed over the whole of the apparent contact area; however, at the same time, such heating is not sufficient to melt the surfaces and cause molten liquid films to form at the interface as in the case of conventional welding.

The rotary friction welding process described by Fig. 12.20 has several limitations which must be considered:

(a) An axis of symmetry is required for at least one of the members being welded (as in the case of two rods, or two tubes, or a rod and a sheet).
(b) Very accurate alignment of the rods is required.
(c) The "upset" or bulge created at the interface must be subsequently removed by machining.
(d) The apparatus in general must be rugged and sturdy to withstand high pressure and rapid spinning.
(e) Since the rate of heat generation FV at a particular rotative speed ω is proportional to distance from the axis of symmetry r, it is not possible to have optimum welding conditions at all points of the section simultaneously.

Many of these deficiencies are removed by the vibratory sheet welding equipment shown in Fig. 12.22. An unbalanced flywheel F directly coupled to a motor M is mounted in a cage C. Initially, the cage is held rigidly in support frame S, using four stops A and a spherical bearing B. One workpiece W_2 is clamped to the piston of ram R, and the other workpiece

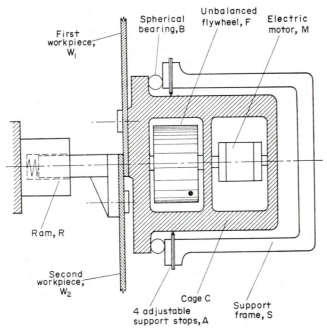

Fig. 12.22. Versatile friction welding apparatus.

W_1 to the cage C. After this clamping operation, F is run up to the desired speed, whereupon the ram R forces the workpieces together. The stops A are then opened, and the cage C (now with the attached workpiece W_1) oscillates about bearing B due to the unbalance in flywheel F. This motion is opposed by a rotary frictional force between the workpiece, which produces the heating effect necessary for welding.

By bringing the flywheel rapidly to rest (this can be accomplished by regenerative braking or mechanical damping), the frictional welding effect between the workpieces is accomplished. As an alternative to arresting flywheel motion, a clutch arrangement can be used to separate a continuously spinning flywheel from the first workpiece at the appropriate instant.

The parameters which must be controlled with this apparatus are:

(a) the amount of flywheel unbalance;
(b) the flywheel speed versus time cycle (in the case of flywheel deceleration) or the de-clutching versus time cycle (for constant flywheel speed); and
(c) the ram loading versus time cycle.

Optimum values of these parameters depend on the geometry and composition of the work-pieces, and can be obtained experimentally. The friction-welding apparatus in Fig. 12.22 is extremely versatile since the workpieces need not have rotational symmetry nor accurate alignment. Furthermore, the relative slip speed at the interface is constant over the section, thus permitting a more uniform weld to be formed.

The coefficient of friction f during friction welding varies in a complex way with velocity. Besides the shearing of junctions, there are other sources of heat such as the deformation of the contact spots and the material behind them (ploughing). As the temperature is increased to 200–300°C, dry friction obtains in the case of metals due to the evaporation of lubricants and absorbed layers, and f increases, producing a further increase in temperature. This is accompanied by seizure or pressure welding on a relatively large scale, with a further increase in temperature and plastic deformation predominating across the section (Fig. 12.23). By the time seizure occurs, most of the adsorbed films have been destroyed by heat, as we have seen, but the final disappearance of oxide films requires large-scale plastic deformation with a steep rise in friction torque. The average temperature during this rise may

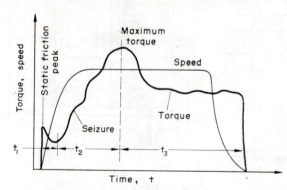

FIG. 12.23. Torque and speed curves during friction welding of low carbon steel.[63] t_1 is the normal kinetic friction; t_2 the gross seizure; and t_3 the shearing of highly plastic material.

be only 100 or 200°C. The final reduction in torque beyond the maximum point in Fig. 12.23 is due to the shearing of a highly plastic (and sometimes molten) material; here the temperature may be in the range 900–1100°C.

Friction welding in general has several advantages over conventional welding, and these are itemized below:

   (a) Welds are highly reproducible and repeatable with uniform quality across the section.

   (b) Very thin heat-affected zone because of a very short heating time.

   (c) Uniform heat generation (in the case of the apparatus in Fig. 12.22) because of a uniform sliding velocity across the interface.

   (d) Not necessary to form a liquid layer at the interface (in the case of metals), thus greatly reducing the probability of forming brittle, intermetallic junctions.

   (e) Very wide range of materials can be welded. Thus strong welds can now be formed between metals which were hitherto impossible or extremely difficult to weld by conventional methods (e.g. refractory metals such as tungsten, brittle metals like beryllium, dissimilar metals including zircalloy–stainless steel combinations, etc.).

   (f) Friction welding can be readily adapted to the welding of plastics as well as metals.

   (g) Need for surface preparation reduced or eliminated. This is because the process automatically breaks up oxide layers on metals and permits welding to take place between pure metals.

   (h) Versatility, lower power requirements, readily adaptable to automation.

12.8 Extrusion Processes

There are, indeed, many types of manufacturing processes today where extrusion of both metallic and plastic materials is the principal mechanism of shaping and sizing. In the case, of metals, rolling contact dies are frequently used to reduce the diameter of steel or aluminium rod and bar specimens, as shown in Fig. 12.24(a). Several pairs of die rolls are used, having a decreasing centre-line spacing d to allow for the reducing diameter of the finished product. The rolls are case-hardened and must withstand severe temperatures at the contact surfaces as the specimen passes through the die with increasing velocity. Furthermore, the centre line of each roll makes an angle of somewhat less than 90° with the centre line of the rod billet to impart a rotary as well as longitudinal motion to the specimen. Plastic flow of the rod occurs at the outside surface in a controlled and uniform manner, and no lubricant is required between the rolls and bar since the mechanism of rolling contact minimizes interfacial adhesion.

Figure 12.24(b) shows the principle of sliding contact extrusion for metals of which the most common example is wire drawing. Because of extremely high adhesion between die and wire, a solid lubricant or soap is applied to the wire before entering the die. Plastic flow occurs readily in the wire specimen by virtue of high temperatures produced by friction at the contact surfaces of the die. The wire is forced through the die by pulling the emerging section of wire, whereas a pushing action is normally used in the rolling contact example of Fig. 12.24(a).

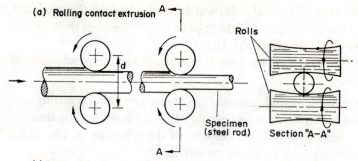

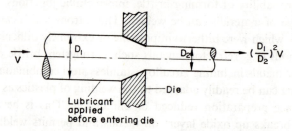

FIG. 12.24. Rolling and sliding contact methods of extrusion.

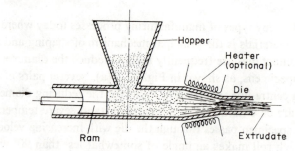

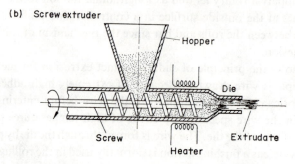

FIG. 12.25. Two common types of plastics extruder.

The extrusion of plastics is normally accomplished with a ram or screw type of extruder, as depicted in Fig. 12.25. In both cases the plastic material is placed in pellet form in a hopper as shown. The ram extruder during its active stroke compresses the plastic and forces it in plasticized form through a forming die. The extrudate during its passage through the die experiences viscous shearing and wall friction forces, of which the magnitude can be regulated with a heater. The screw extruder drags molten plastic along through a narrowing section to a die and has the advantage that continuous action is possible with a relatively simple design. This is not possible with the ram extruder in Fig. 12.25(a) although more complex designs permit ram action on a continuous basis.

The extrusion methods of Figs. 12.24 and 12.25 have the common feature that they are often continuous and of relatively lengthy duration. Impact extrusion, by contrast, occupies a very short time period, during which large forces and deformations are produced in a controlled manner. Figure 12.26 depicts three common impact extrusion methods for aluminium and other alloys. In all cases, pressure is rapidly applied to a metal slug or blank held in a die.

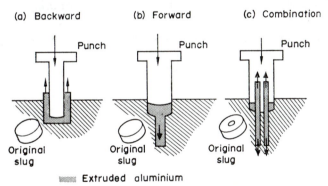

FIG. 12.26. Three impact extrusion methods for metals.

The metal flows plastically upon impact, and the flow can be controlled for an almost unlimited variety of shapes and sizes. Shearing and frictional wall stresses are produced rapidly during the period of flow and determine the extent of the latter. In backward extrusion the metal flows through an orifice formed between the punch and die walls in the reverse direction of punch travel. The metal forms a simple shell about the punch. In forward extrusion the metal flows on impact through an orifice formed by a narrowing section, the direction of flow being the same as the direction of travel of the punch. For the combination extrusion method, metal flows upon impact in both reverse and forward directions.

12.9 Miscellaneous

Because of the complexity of manufacturing processes in today's modern world and the diversity of operations where friction or lubrication in some form plays a vital role, this chapter can at best give a broad survey of manufacturing technology where tribology is used inadvertently or otherwise in the forming and shaping of a product. Several processing operations do not fall conveniently under the headings, and they will therefore be introduced briefly in this section.

Stamping and guillotining are impact processes of relatively short duration in which a rapidly falling mass (having an embossed die in the one case and a cutting edge in the other) contacts a softer material or "blank" usually of metallic composition. Vacuum forming in plastic materials and cold forming in metals are other processes where frictional interfaces either directly or indirectly determine the final product shape. Shot peening and sand blasting are used to relieve internal friction stresses by directing a sudden blast of air (containing lead shot or coarse sand particles) on to one or several metallic specimens. In the field of lubrication, mist cooling has been successfully used as a more effective replacement for conventional flood or liquid cooling. Here an atomized mist generated in a special nozzle design is directed on to the work during machining operations. In some cases (such as the cooling of high-speed dies), mist cooling can increase die life by as much as 500% because of its instantaneous action. A particular advantage in drilling, tapping, grinding, and cutting operations is the blowing away of chips from the workpiece during finishing operations. Other manufacturing processes with tribological implications include electrolytic grinding, electromachining, chemical milling, plasma torch cutting, and explosive forming.

CHAPTER 13

AUTOMOTIVE APPLICATIONS

13.1 Pneumatic Tyre Performance

Perhaps the most significant application of elastomeric friction following the principles outlined in Chapter 4 is the pneumatic tyre, both in its rolling and sliding modes of behaviour. In general terms, the pneumatic tyre fulfils six functions as follows:

(a) Allows a comparatively free and frictionless motion of the vehicle by means of rolling.
(b) Distributes vehicle weight over a substantial area of ground surface, thus avoiding excessive stresses both on the wheel and road.
(c) Cushions the vehicle against road shocks.
(d) Transmits engine torque to the road surface with a low power consumption.
(e) Permits, through tyre adhesion, the generation of substantial braking, driving, and steering loads.
(f) Ensures lateral and directional stability.

No other device exists today which can fulfil these varied functions in as efficient a manner as the pneumatic tyre. In its normal rolling mode the tyre exhibits both internal friction as a result of the continuous flexing of tread and sidewalls, and external friction because of tread "squirming" and micro-movement within the contact patch between wheel and road. The tyre as a whole presents a complex design problem, and the final form which emerges must inevitably be a compromise between conflicting requirements. Thus increased ride comfort means greater shock absorption but also increased power consumption in transmitting engine torque with perhaps overheating and premature fatigue failure in severe manoeuvring conditions. From a tribological viewpoint we require a fundamental understanding of the frictional mechanism which develops in the tyre-to-ground contact region under different driving conditions. Such an understanding is of prime importance from a safety viewpoint, and other considerations (such as ride comfort and noise control, while perhaps of equal significance in the overall design procedure) will be regarded as secondary.

Tread Pattern

The incorporation of a distinct tread pattern into the running band of automobile tyres is another example of how a compromise must be reached in the overall design procedure. If there were no rain or dust deposits on a road surface, the highest coefficients of sliding friction (perhaps as great as 5 in special cases) would be attained with a perfectly smooth tread since the adhesion contribution to friction is maximized by a large available contact area (see eqn. (4.19)). However, the existence of the slightest water film in this case would suppress the adhesion, and produce an extremely hazardous condition (the coefficient of friction may be as low as 0.1 or less). The compromise reached, of course, is to provide an adequate tread pattern on the surface of the tyre. This reduces the coefficient of adhesional friction under dry conditions to values less than unity since the area of contact is reduced because of the grooves. However, the effective coefficient of friction under wet conditions is considerably increased (a mean value of $f \doteq 0.4$ for locked-wheel skidding on wetted road surfaces). In physical terms, the existence of grooving between tread elements permits drainage of excess water from the tyre footprint, so that the adhesional mechanism is largely restored. Thus as regards frictional coefficient an adequate tread pattern offers a compromise between the higher and lower coefficients that would be obtained with a smooth tyre under dry and wet conditions respectively.

From a design viewpoint, the grooving or channeling in the tread pattern should be capable of discharging a reasonable volume of water from the tyre footprint during the very short time available at high speeds of rolling. Although road surface texture also contributes to drainage from the contact area, the bulk water is removed by the tread design. The three basic tread patterns used today may be classified as zigzag, ribbed, and block as shown in Fig. 13.1, and experiments indicate that there is very little to choose between the best examples of each type. In addition to the basic function of bulk water removal, the pat-

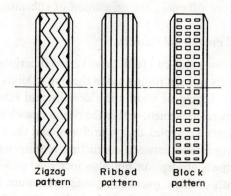

Zigzag   Ribbed   Block
pattern   pattern   pattern

Fig. 13.1. The three basic tread pattern designs.

terned tyre must also permit localized tread movement or "wiping" to assist in the squeezing out of thin water films on the road macro-texture. This necessitates the provision of sipes or cuts leading into the grooving. Modern tread designs therefore exhibit the following:

(a) *Channels* or *grooves* (with the basic arrangement as in Fig. 13.1, or with variations in these designs). The volume of grooving varies little from one tread pattern to the next. The grooves are approximately 3 mm wide and 10 mm deep, and lead continuously outward from the centre of the tyre footprint.

(b) *Transverse slots* or *feeder channels*. While generally of smaller cross-sectional dimensions than the main channels which they serve, the transverse slots are not continuous but end abruptly within the tread rubber. They assist in displacing bulk water from the tyre footprint, and also permit gross or macro-movement of the tread during wiping action.

(c) *Sipes* or *miniature cuts* leading into the channels or feeder channels. These permit local tread "squirming" or micro-movement, which is characteristic of the rolling process. Sipes do not, however, contribute to drainage from the footprint directly.

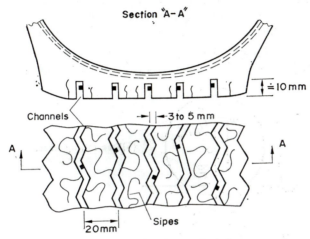

FIG. 13.2. Channels and sipes in a typical zigzag pattern.

Figure 13.2 shows a typical zigzag commercial tread pattern having channels and sipes. We observe that in this design, feeder channels are not used directly because their effectiveness is otherwise obtained by the zigzag pattern.

Rolling and Sliding

The pneumatic tyre can experience either rolling or sliding; no other possibilities exist. Sliding is characterized by its relative simplicity, although the locked-wheel sliding condition rarely occurs except in cases of flooding by heavy rainfall. Since every element of the sliding tyre has no motion relative to the frame of the vehicle, the velocity of slip of each tread element in the contact area relative to the road surface is the same, being identical with the forward velocity of tyre or vehicle. Furthermore, the same tread elements are subjected to the frictional retardation force, so that noticeable wear of the tread occurs unevenly along portions of the tyre circumference.

The rolling tyre by contrast distributes the wear resulting from severe brake application in a uniform manner about the tyre circumference since different tread elements continually enter the contact patch. In addition, the extent of such wear is less, since the mean velocity of slip of the tread relative to the road is considerably lower. The rolling tyre includes free

278   *Principles and Applications of Tribology*

rolling, braking, accelerating (or driving), cornering or any combination of these fundamental modes. Figure 13.3 shows the forces and moments acting on (a) a free rolling, (b) a braked rolling, and (c) a driven rolling tyre. In each case, longitudinal tractive forces are produced locally in the contact area, giving rise to net forces F_R, F_B, or F_D acting on the tread, and the load reaction vector W acts at a small distance a ahead of the centre of contact.

For the freely-rolling tyre there can be no net moment about the wheel centre, and the resultant force $\sqrt{(W^2+F_R^2)}$ therefore passes through O as shown. When braking is applied, the rolling resistance force F_R increases many times to the braking force value F_B, and we now observe that the resultant force $\sqrt{(W^2+F_B^2)}$ has a moment arm b about the wheel centre O. Thus a moment equal to $b\sqrt{(W^2+F_B^2)}$ is created to oppose the clockwise braking torque T_B as shown in Fig. 13.3(b). In the case of driving, a similar reasoning applies except that the net longitudinal force F_D in the contact patch now acts in the direction of travel, and the moment $b\sqrt{(W^2+F_D^2)}$ opposes the counter clockwise driving torque T_D.

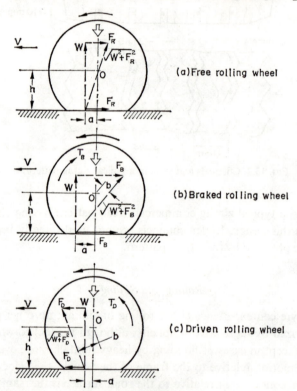

FIG. 13.3. Forces and moments acting on tyre for different rolling modes.

Assuming that steady-state conditions prevail, we can take moments about O for each of the three characteristic rolling modes in Fig. 13.3 thus:

$$\left.\begin{aligned} F_R &= W(a/h), \\ F_B &= T_B/h + W(a/h), \\ F_D &= T_D/h - W(a/h), \end{aligned}\right\} \quad (13.1)$$

and

where h is the axle-to-ground height. We observe from these equations that F_B and F_D are modified by the algebraic addition of a load effect due to the eccentricity of the ground reaction force. If we now consider that the wheel is subjected to load transfer effects in braking or driving because of the tendency of a vehicle to pitch downwards on the front wheels during braking and downwards on the rear wheels during driving, the normal load W is modified by the bracketed term in the following equations:

$$F_B = \frac{T_B}{h} + W\frac{a}{h}\left[1 \pm 4\frac{\ddot{x}}{g}\left(\frac{h_{cg}}{L}\right)\right],$$

and

$$F_D = \frac{T_D}{h} - W\frac{a}{h}\left[1 \mp 4\frac{\ddot{x}}{g}\left(\frac{h_{cg}}{L}\right)\right],$$

(13.2)

where h_{cg} is the height of the centre of gravity of the vehicle above ground level, L is the vehicle wheelbase,[†] and $\ddot{x}$ the acceleration or deceleration of the vehicle. Within the bracketed terms in these equations, the first sign refers to the front wheels and the second sign to the rear wheels.

Consider now the contact area for a rolling tyre under conditions of braking, driving, and cornering. In each case virtually no slip takes place within the forward part of the contact patch, whereas appreciable slip occurs towards the rear of contact (see Fig. 13.4 for the

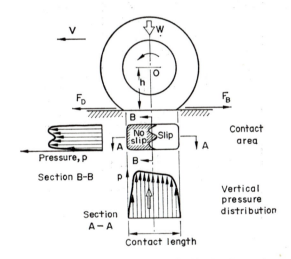

FIG. 13.4. Contact area and pressure distribution for rolling tyre.

cases of braking or driving). Details of the distribution of slip velocity for a braked, driven, or cornering tyre in the rolling mode are given in Fig. 13.5. It is convenient in visualizing the results in Fig. 13.5 to consider the wheel centre fixed and the road appearing to move to the right with a velocity

$$V = \omega R_e$$

(13.3)

[†] It is assumed in eqns. (13.2) that each wheel of the vehicle carries an equal load W when at rest, and that the centre of gravity of the vehicle lies midway between front and rear wheels. See Section 13.2.

19*

equal and opposite in sign to the original forward velocity of the wheel and tyre. The angular velocity of the wheel is denoted by ω in eqn. (13.3) and R_e is the effective rolling radius defined by the same equation. In the case of braked rolling, the band velocity of the tyre increases to the "road velocity" value ωR_e upon entering the contact patch, and maintains this value until approximately one-half of the contact length has been traversed, as shown clearly in Fig. 13.5(a). At this point the tyre band velocity decreases non-linearly towards the rear of contact, thereby producing a variable longitudinal slip velocity in the forward direction. The magnitude of this slip velocity increases rapidly with speed, and has particular importance in promoting skidding on wet surfaces, as we shall see in the next section. A similar slip velocity distribution, but this time in a backward direction, is obtained within the rear of the contact length for a driven tyre, as shown in Fig. 13.5(b). For the case of cornering, the tread elements deform linearly along CE in a lateral direction until the elastic or strain energy stored within them exceeds the local tyre–ground adhesion value, whereupon the same elements slip sideways in the rear of the contact area, as shown in Fig. 13.5(c).

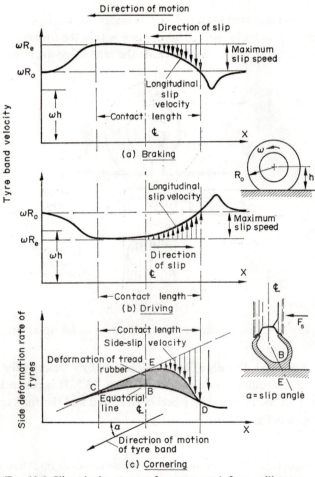

FIG. 13.5. Slip velocity at rear of contact patch for a rolling tyre.

The distribution of slip velocity within the last half of the contact zone is very similar for cornering, braking, or driving, as Fig. 13.5 indicates, and in all of these cases there is virtually no relative motion[†] within the forward contact patch. In the case of braking or driving, the longitudinal slip velocity has a tendency to overshoot the rearmost point in the contact patch, and this is interpreted physically as a discontinuity as the tread rubber suddenly adapts from longitudinal to circumferential motion during the process of rolling.

Brake and Drive Slip

We have seen how a pneumatic tyre in a given rolling situation exhibits a non-linear distribution of slip at the rear of the contact patch, but as yet we have no overall method of determining how severe our braking, cornering, or driving may have been in a particular instance. As a guide in this direction, the brake slip ratio s_B and drive slip ratio s_D give estimates of the severity of braking and driving thus:

$$s_B = \left.\frac{\omega_{R_0} - \omega_{br}}{\omega_{R_0}}\right|_{V = const} \tag{13.4}$$

$$s_D = \left.\frac{\omega_{dr} - \omega_{R_0}}{\omega_{dr}}\right|_{V = const} \tag{13.5}$$

where ω_{R_0} is the angular velocity of a wheel in free rolling and ω_{br} and ω_{dr} are the angular velocities during braking and driving respectively for the same forward travel speed V. We

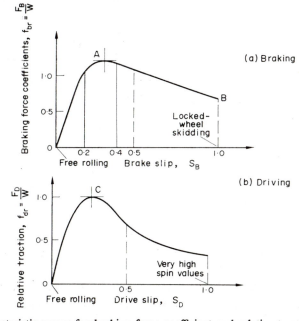

FIG. 13.6. Characteristic curves for braking force coefficient and relative traction during rolling.

[†] Excepting tread squirming motion as described later.

observe at once that for free rolling, $\omega_{br} = \omega_{dr} = \omega_{R_0}$ and both s_B and s_D are zero. On the other hand, the most severe form of braking causes a locked-wheel condition, whereupon the wheel ceases to rotate and ω_{br} becomes zero, giving $s_B = 1$. Similarly, the most severe acceleration causing "wheel spin-up" renders ω_{R_0} insignificant compared with ω_{dr}, and $s_D \rightarrow 1$. *The severity of braking and driving is therefore gauged in most practical cases by a range of values of s_B and s_D between 0 and 1.* Sometimes we refer to s_B or s_D as "nominal slip", since in any example a distribution of slip velocities characterizes the area of contact in rolling as we have seen in Fig. 13.5, although a single number is used following the definitions in eqns. (13.4) and (13.5).

Figure 13.6(a) shows how the braking force coefficient f_{br} ($= F_B/W$) varies with brake slip ratio s_B, and Fig. 13.6(b) gives a similar plot of relative traction f_{dr} ($= F_D/W$) vs. drive slip ratio s_D. Both coefficients of friction indicate a maximum at values of s_B or s_D between 0.2 and 0.4. Within this range, the mean velocity of longitudinal slip (see Fig. 13.5) produces the adhesion peak in elastomeric friction as described in Chapter 4. It is desirable to operate in the vicinity of the peaks A and C in Fig. 13.6, since the available friction for braking or accelerating is maximized; in practice, however, it is difficult to identify where the peaks occur. Furthermore, even if this knowledge were available, the portion AB in Fig. 13.6(a) is unstable, and once the point A is reached or exceeded, the value of s_B rapidly attains the locked-wheel value.

The distribution of longitudinal force within the contact length of a rolling tyre for the cases of free rolling, braking, and driving is shown in Fig. 13.7. In the first free-rolling case a virtually symmetrical and low-amplitude sinusoidal variation of shear force is shown,

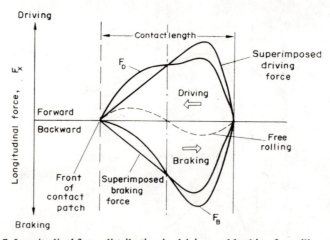

Fig. 13.7. Longitudinal force distribution in driving and braking for rolling tyre.

acting in the forward (or travel) direction in the front of the contact length and backwards within the rear half of contact. Superimposed on this basic distribution is a triangular and "backward" distribution in the case of braking, due to the linearly increasing stretch of tread elements as they traverse the contact length. For the driving mode, the superimposed distribution is also triangular but acts in a forward direction with respect to vehicle motion as shown. By comparing Figs. 13.6 and 13.7 we observe that the peak values of longitudinal force for either braking or driving correspond to a certain slip velocity within the last part of the contact area. This is in accordance with the usual observation that a finite velocity of sliding is required (usually of the order of 1 cm/s or less) to attain maximum adhesion. A similar distribution of transverse shear force is obtained in the case of cornering.

A final remark must be made with regard to the interpretation of regions of zero slip in the contact area of a rolling tyre, as indicated in Figs. 13.4 and 13.5. Within these regions, the phenomenon of tread squirming occurs according as tread elements experience the transition from roundedness to flatness. The orientation of the squirming movement is random and its magnitude very small, so that the term micro-movement is appropriate. By contrast, the rear part of the contact area as described above experiences a directional and non-linear variation in slip velocity of tread relative to road (this has been broadly described as macro-movement).

Wet Tyre Friction

Consider the rolling behaviour of a pneumatic tyre on a wet, rough roadway, and let it be supposed that the contact length be divided into three distinct regions as shown in Fig. 13.8. Let us visualize for simplicity that the centre of the rolling tyre is fixed and that the roadway moves with a velocity V as indicated. We observe that to a first approximation there is no relative motion between tyre and road within the forward contact region (see also Fig. 13.5) as the former traverses the contact length. In the case of heavy flooding, a finite wedge angle is defined between tyre and water in the region just ahead of the contact length as shown in the figure, and a hydrodynamic upward thrust P_H is created by the change in momentum of the water within the wedge. The magnitude of the upward thrust increases with the square of the forward speed V of the tyre relative to the roadway. Within the "contact" length itself, the tread elements must first squeeze their way through the remaining film of water

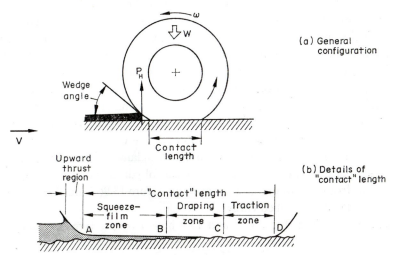

Fig. 13.8. The three zones of contact in wet rolling.

before contacting the major asperities of the road surface at the point B in Fig. 13.8(b). The normal load acting on the tread elements throughout the contact length AD is due to tyre inflation pressure p_i. From B to C, a vertical draping of the tread occurs about the larger asperities of the surface, and rubber properties (such as hardness, hysteresis, and resilience) play a significant role in determining the extent and rate of penetration of the tread by road asperities. At the point C, the tread has completed its draping action, and vertical equilibrium is attained.

We observe that true contact between tyre and road is established only within the region CD in Fig. 13.8(b) under wet conditions. If the squeezing process is accomplished in a minimum time by a suitable choice of road texture (see Figs. 6.15 and 6.16) the length AB is minimized for a given rolling speed V, and if BC remains unaffected it is clear that a maximum length CD remains for developing traction. An increase of speed will eventually increase the squeeze-film region AB to the point where it occupies almost the whole length AD, and very low traction forces are developed. This critical speed is defined as the viscous hydroplaning limit, and it is determined chiefly by the squeeze-film characteristics of the front part of the contact path.[†] At such a speed, the hydrodynamic upward thrust P_H is large but still less than the normal wheel load W. A second higher speed exists such that $P_H = W$, and this is called the dynamic hydroplaning limit. In practice, the latter is significant only for the landing of aircraft, where the dynamic limit is approached from higher velocities. The viscous hydroplaning limit represents a critical rolling speed for all road vehicles, and the squeeze-film length AB extends substantially throughout the normal contact region AD. Postponement of this critical limit to rolling speeds which exceed the normal maximum on given roads can only be satisfactorily achieved by an optimum selection of texture in the road surface.

Let us now consider briefly the nature of events within the traction zone in Fig. 13.8(b) under wet conditions. For braking, driving, or cornering, we have seen from Fig. 13.5 that the rear of the "contact" area is characterized by an increasing velocity of relative slip between tyre and road. The direction of the slip velocity is, of course, longitudinal in the case of braking or accelerating, and lateral for the cornering condition. Under wet conditions, this velocity generates hydrodynamic pressures on the leading slopes of road surface asperities, which tend to separate tyre and surface (see Fig. 8.15). The separating effect increases with slip velocity, so that according as forward speed is increased, contact is lost at the rearmost part of the tyre footprint. According as speed is further increased, the separating effect rapidly spreads forward from the rear of the contact patch. At the same time, the front part of the "contact" length is rapidly eroded by a backward-penetrating squeeze-film separation. Figure 13.9(b) shows the simultaneous erosion of both front and rear parts of the contact length just prior to the viscous hydroplaning limit. An instant later, when $V = V_{HV}$, the separating effect at the rear due to elastohydrodynamic action and the squeeze-film effect in the front of the contact zone overlap. Figure 13.9(a) illustrates the condition of dynamic hydroplaning described earlier.

We have seen clearly in Figs. 13.8 and 13.9 how a rolling tyre may attain the hydro-

---

[†] Elastohydrodynamic effects at the rear of contact also play a lesser role, as seen later.

planing limits at critical speeds, and we now take a look at what happens to a sliding or locked-wheel tyre under otherwise identical conditions. From lubrication theory (see Chapter 6), it can be shown that twice the speed is necessary under sliding conditions compared with rolling to achieve the dynamic hydroplaning condition $P_H = W$. This can be physically explained on the basis that both surfaces defining the wedge in the rolling con-

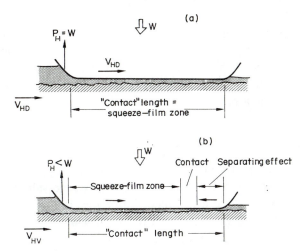

FIG.13.9. The onset of skidding for rolling tyre: (a) dynamic hydroplaning; (b) viscous hydroplaning. Note: $V_{HD} > V_{HV}$.

dition attempt to entrain lubricant into a narrowing passage, whereas only one of the surfaces (namely, the lower road surface) is active in this manner for sliding (see Fig. 13.10 for a comparison of rolling and sliding behaviour of tyres at the dynamic hydroplaning limit). Let us examine briefly what happens as speed increases in the case of a rolling tyre on a wet road surface. At a certain speed V_{HV} the viscous hydroplaning limit is attained (see Fig. 13.9), and the tyre ceases to rotate because the force F_B (or F_D or F_R; see Fig. 13.3) in the tyre–ground plane decreases instantaneously to a low value which is just sufficient to

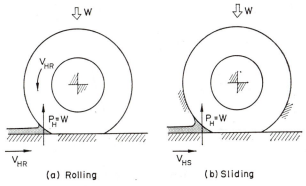

FIG. 13.10. Dynamic hydroplaning in (a) rolling and (b) sliding. Note: $V_{HS} = 2V_{HR}$.

balance the load reaction eccentricity torque. We now effectively have a locked-wheel sliding condition at a speed V_{HV} which is much lower than the dynamic hydroplaning limit V_{HS} in Fig. 13.10(b). Indeed, if the viscous hydroplaning limit V_{HV} did not exist and rolling speed were further increased, the condition in Fig. 13.10(a) would eventually occur, namely the onset of dynamic hydroplaning in rolling.

In practical terms we must concern ourselves primarily with the viscous hydroplaning limit in automobile applications. We note that it is not at all necessary to have a flooded road surface for this phenomenon to occur, and the slightest film of water may be sufficient to induce viscous hydroplaning. The relative importance of the rearward elastohydrodynamic separating effect increases as film thickness reduces. We shall see in the next chapter that the viscous hydroplaning effect can be avoided only by a suitable choice of road surface asperities and micro-roughness at asperity peaks.

Abrasion and Wear

The three types of wear pertinent to elastomeric materials appear to be present in the rolling or sliding pneumatic tyre (see Chapter 9). An average value of the thickness of rubber abraded on a road surface with rounded asperities is about 0.25 mm for a travel distance of 1000 km. The main property of tread rubber which determines its resistance to abrasion is fatigue life, i.e. the number of deformation cycles which the surface layers in the tread can withstand without failure under given operating conditions. These conditions, of course, include a state of complex stress, high temperature, and various other factors which apply in the contact area. When the road surface exhibits asperities with sharp peaks and rough edges, the tread surface shows relatively large tears and lengthwise scratches and cuts. Under these conditions, abrasive wear predominates although the fatigue mechanism is also active. Sudden braking, cornering, and swerving increase the intensity of abrasive wear

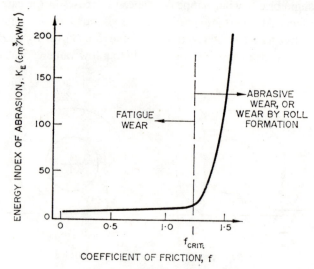

FIG. 13.11. Tread wear and frictional coefficient for different rubbers.

considerably, and there is a high probability that wear by roll formation takes place to a limited extent in these cases. Thus the pneumatic tyre in service generally exhibits all three types of wear.

Let it be assumed that there exists a certain critical shear stress τ_{crit} for each tread rubber. Tears and cracks will occur in the surface layers when the actual shearing stress τ exceeds this value. For $\tau < \tau_{crit}$ the fatigue-type wear predominates, whereas for $\tau > \tau_{crit}$ either wear through roll formation (on fairly smooth surfaces) or abrasive wear (on rough surfaces with sharp projections) occurs. The shearing stress can be related to the product of the coefficient of friction and normal pressure as follows:

$$\tau = fp \quad \text{and} \quad \tau_{crit} = f_{crit}p. \tag{13.6}$$

Thus for $f < f_{crit}$, the wear is due to surface fatigue, and for $f > f_{crit}$ the other forms of wear predominate. Figure 13.11 shows a plot of energy index of abrasion for various tread rubbers against the corresponding measured coefficient of friction, and it is seen at once that f_{crit} has a value of about 1.25. Above this value, the wear is very severe and the rubber surface exhibits ridges similar to that depicted in Chapter 9. On the other hand, for values of f less than unity, the energy index of abrasion is low and the surface has a smooth appearance.

Another property of tread rubber which affects the type and intensity of wear is its hardness, as indicated by Fig. 13.12 for SBR[†] tread rubber. On abrasive paper, an increase in

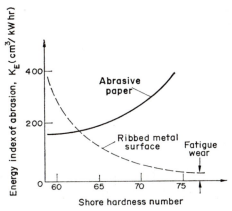

FIG. 13.12. Effect of hardness on wear of SBR tread rubber.

hardness increases the stress concentration on the sharp projections, and conditions become more favourable for abrasive wear and micro-cutting to exist. Thus the intensity of wear is seen to increase. On the other hand, the abrasion of a soft rubber on the smooth projections of a ribbed metal surface occurs as a consequence of roll formation. According as the hardness increases, the probability of roll formation decreases, and at a certain value of hardness (about 75 shore hardness) the prevailing type of wear is due to fatigue.

† Styrene-butadiene.

Abrasion occurs within the regions of gross slip at the rear of the contact zone (see Fig. 13.5) particularly under dry conditions, and also to a lesser extent in the front part of the contact area where micro-slip of the tread rubber takes place. When the direction of abrasion coincides with the direction of slip, the abrasion can be regarded as useful, since it contributes directly to establishing a coefficient of friction which opposes the slip motion. On the other hand, when the direction of abrasion differs from the direction of tyre slip, the abrasion can be regarded as parasitic, and it contributes nothing to the effective friction at the tyre–ground interface. Consider as an example[2] the transverse shear stresses which occur in a tyre during free rolling, as shown in Fig. 13.13. These shear stresses are directed outwards from the longitudinal centre line in the contact area, and they contribute to side forces of considerable magnitude which are exactly equal and opposite. The generation of these side

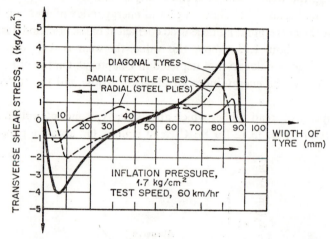

FIG. 13.13. Transverse shear stress distribution across contact width for radial and diagonal-ply tyres during free rolling.

forces produces lateral wear and abrasion of the parasitic type since they contribute nothing to the coefficient of rolling friction while they are a source of considerable abrasion. Figure 13.13 shows clearly that the magnitude of the transverse shear stress (and hence the side forces) is considerably greater for a diagonal-ply tyre[†] than for radial-ply tyres[†] and among the latter the use of steel plies produces minimum side forces for a given rolling speed and inflation pressure.

Now it is known that the life and overall performance of radial-ply tyres are greatly superior to the performance of conventional or diagonal-ply tyres, and Fig. 13.13 may well explain at least in part why this is the case. With a minimum of parasitic abrasion and a reduction in effective heat generation, the radial-ply tyre can therefore produce the same rolling coefficient of friction with substantially lower overall abrasion. From this view-

[†] The construction of diagonal-ply tyres is fundamentally different from the radial-ply type. The diagonal-ply tyre has several layers of rayon or nylon cord laid diagonally within a rubber matrix. The radial-ply tyre, by contrast, has an inextensible or rigid cord structure or "belt" and relatively weak sidewalls. [2]

point, it would appear that an optimum design for a pneumatic tyre would minimize the parasitic abrasion losses by maximizing the percentage of slip in the direction of overall motion.

We finally consider an approximate relationship between volumetric abrasion losses Δv and nominal slip (s_B in braking, or s_D in driving, or s in general terms). Nominal slip is simply defined as the mean velocity of slip for the tyre elements relative to the road surface v, divided by the forward velocity V of the tyre as a whole. If we remove the time-dependence from numerator and denominator,

$$s = v/V = L/L_c, \tag{13.7}$$

where L is the mean slip distance of an element in the contact patch during rolling and L_c is the total contact length. If we also assume in accordance with Fig. 13.6 for small slip values that the coefficient of friction f is proportional to s, and that the abrasion losses Δv are proportional to the frictional energy dissipation, then it follows that

$$\Delta v \sim \underset{\substack{\text{frictional energy} \\ \text{dissipation}}}{FL} \sim fWL$$

$$\sim (sW)s \tag{13.8}$$

$$\sim s^2 \quad \text{for constant wheel load } W,$$

which indicates that the abrasion losses are proportional to the square of the nominal slip. The severity of driving, braking, or cornering in terms of abrasion losses can be gauged from this relationship. Indeed, practical experience suggests that in terms of overall wear, the cornering mode contributes substantially more than the most severe forms of braking and accelerating during the life of a tyre.

13.2 Braking Mechanisms

Two distinct frictional interfaces are involved in the simple act of attempting to brake an automobile, truck or aircraft, thus: (a) between tyre and road (or runway), and (b) between brake linings and drum (or disc).

We have considered exhaustively the first interface in the previous section, and here we confine our attention to the equally important second interface. Both interfaces should play a part in a well-designed braking operation; thus application of the brakes by the driver causes relative slip between brake linings and drum or disc and relative slip between tyre and road. We observe that throughout this operation the rolling mode is preserved, and the vehicle glides smoothly to a halt. Let us now assume that an icy condition reduces the tyre–road available friction to a very low value ($f < 0.1$). In this case, application of the same brake pedal force as that required to preserve rolling in the first example now causes a locked wheel condition to develop. Thus no energy dissipation or relative slip occurs at the brake-lining–drum (disc) interface, and a maximum relative slip appears at the tyre–road boundary. Similarly, if we assume that the brake lining interface is rendered largely ineffective by the presence of engine oil or water, brake application causes a maximum relative slip at this interface with very little effect appearing between tyre and road. Obviously, these last two examples offer very ineffective braking action, since relative slip is confined to one or other of the two frictional interfaces.

Modes of Braking

We now consider briefly several basic modes of brake application and their effect on vehicle motion. The most common modes are:

(a) Front-wheel braking.
(b) Rear-wheel braking.
(c) Four-wheel braking.
(d) Diagonal braking.
(e) Anti-skid pulsed braking.

Figure 13.14 depicts the general nature of front, rear, diagonal, and all-wheel braking, the activated wheels in each case being denoted by shading. We observe that rear-wheel braking represents an inherently unstable mode if a locked-wheel condition develops. Under these conditions, the vehicle undergoes an uncontrollable spin motion of 180 degrees or less. By

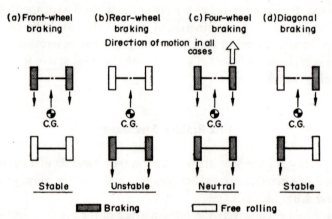

FIG. 13.14. Different modes of vehicle braking.

contrast, front-wheel and diagonal braking are stable under locked-wheel conditions, since the braked vehicle does not have a pronounced tendency to spin. Four-wheel braking may be described as neutral from this viewpoint, being characterized by a limited spin of the vehicle.

It is, of course, neither necessary nor desirable that the degree of braking should be sufficiently severe to create a locked-wheel or skidding condition for some or all of the wheels of the vehicle. Directional control or steering of the vehicle virtually disappears when all four wheels are in a locked-wheel mode, and, furthermore, the degree of vehicle retardation is by no means maximized (see Fig. 13.6). Control of the vehicle has been maintained in

Fig. 13.14(a), (b), and (d) at the expense of retardation effectiveness by preserving two of the wheels in the free-rolling mode. Indeed, front-wheel braking causes severe load transfer under locked-wheel conditions, and we have already seen that rear-wheel braking is unstable, so that the diagonal braking concept has distinct advantages.

We may ask ourselves the logical question: How effective are the various braking modes in Fig. 13.14 for moderate or non-locking wheel braking? In such a case, the rolling mode is preserved at all four wheels, and only the conventional four-wheel braking mode is fully effective. Braking is then sacrificed in the other modes at the freely rolling wheels for no apparent reason. However, there is a very high probability that some of the braked wheels will progress rapidly from a rolling to a locked-wheel condition on slippery road surfaces even for light brake application, and the advantages of the diagonal system again appear.

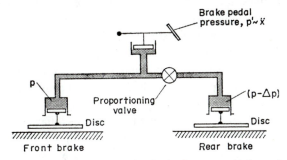

FIG. 13.15. Pressure attenuation by brake proportioning valve.

Load transfer in braking presents a difficult problem from the designer's viewpoint. Because the vehicle centre of gravity is placed at a distance h_{cg} above the road surface, the application of brake torque to all four wheels causes the inertia force of the vehicle to react as an additional load increment ΔW on each of the front wheels, while a diminution of load by the same amount[†] takes place at each of the rear wheels. It can be shown simply that

$$\Delta W = 4W(\ddot{x}/g)(h_{cg}/L),\qquad(13.9)$$

where W is the load per wheel, L the vehicle wheelbase, and $\ddot{x}$ the vehicle deceleration as a result of braking action. Since the friction force F_B at the tyre-to-ground interface is equal to the coefficient of friction f_{TR} times the normal load, we can write

$$F_B = f_{TR}(W \pm \Delta W) = f_{TR}W[1 \pm 4(\ddot{x}/g)(h_{cg}/L)],\qquad(13.10)$$

where the suffix *TR* refers, of course, to the tyre–road interface. The positive sign in the bracketed terms in eqn. (13.10) refers to the front wheels and the negative sign to the rear wheels. The coefficient of friction f_{TR} is a function of braking torque between disc and brake lining as well as tyre–road conditions, and we see at once the dilemma posed by eqn. (13.10). The effective tyre–ground braking force F_B increases at the front wheels and decreases at the rear wheels in a manner which depends directly on vehicle deceleration $\ddot{x}$, and the difference

[†] Assuming equal loading on each wheel and a symmetrically located centre of gravity.

in the values of F_B between front and rear wheels increases with the severity of brake application. A designer is limited to the selection of only one vehicle deceleration $\ddot{x}$ at which to
equate F_B at front and rear, and for all other decelerations the braking forces F_B at the front
and rear wheels are different. In particular, very severe brake application with this system
causes the tyre-to-ground braking force F_B at the front wheels to substantially exceed that
at the rear. Under these circumstances, the same hydraulic pressure p in the brake fluid
may cause rear-wheel lock-up, whereas the front wheels continue to roll. Thus four-wheel
braking is likely to develop effectively into the highly unstable locked rear-wheel mode
because of severe load transfer effects.

One method of simply alleviating this problem is by the use of a simple proportioning
valve in the hydraulic brake circuit between front and rear brakes, as indicated schematically
in Fig. 13.15. In its simplest form the proportioning valve is a constriction in the flow. Let
us assume that the driver applies a pressure p' to the brake pedal, thus initiating a vehicle
deceleration $\ddot{x}$. As a consequence, a pressure p appears in the hydraulic brake fluid which
actuates the front brakes and a reduced pressure $(p-\Delta p)$ in the corresponding hydraulic line
to the rear brakes. We observe that the pressure drop Δp across the proportioning valve
varies directly with brake-pedal pressure p', the latter being proportional to $\ddot{x}$. We shall see
later on page 294 that the applied torque T_B at any wheel is proportional to the actuating
hydraulic pressure p, although the proportionality constants differ for front and rear wheels.
Thus, we may write $T_B = K_1 p$ for the front and $T_B = K_2 p$ for the rear wheels of the vehicle, with $K_2 > K_1$. From a design viewpoint, the relative magnitudes of K_1 and K_2 are
determined by selecting the area over which the pressure p acts and the general dimensions and composition of drum or disc. Applying this information to the first of eqns.
(13.2), we obtain:

$$F_B = \frac{K_1}{h}p + W\frac{a}{h}\left[1 + 4\frac{\ddot{x}}{g}\left(\frac{h_{cg}}{L}\right)\right] = K_0 + K_3 p + K_4 \ddot{x} \qquad (13.11)$$

for the front wheels, where K_0, K_3 and K_4 are constants. Similarly, for the rear wheels
with $\Delta p \sim \ddot{x}$:

$$F_B = \frac{K_2}{h}(p - K\ddot{x}) + W\frac{a}{h}\left[1 - 4\frac{\ddot{x}}{g}\left(\frac{h_{cg}}{L}\right)\right] = K_0 + K_5 p - (K_4 + K_6)\ddot{x} \qquad (13.12)$$

with K, K_5 and K_6 as additional constants. We observe that in the absence of any brake
proportioning device whatever (i.e., $K = K_6 = 0$ in eqn. (13.12)), there exists a critical
value p_c of brake pressure and a corresponding value $\ddot{x}_c$ at which the brake forces F_B are
numerically equal at the front and rear wheels.

Perhaps the simplest means of visualizing the action of a brake proportioning valve is
to consider that its chief purpose is to match braking effectiveness at the tyre/road and
brake/drum frictional interfaces. Thus, severe braking as we have seen creates a sizeable
load increment ΔW which adds to the front wheel load and subtracts from the rear. For
given conditions at the tyre/road interface, the coefficient f_{TR} is constant at all four wheels,
and hence the braking force F_B is proportional to normal load. The front wheels can and
should therefore transmit a much greater braking force to the decelerating vehicle than the
rear wheels in the case of severe brake application. The proportioning valve monitors

these conditions by effectively increasing the actuating pressure at the front wheel brake/ drum interfaces and reducing it at the rear. In this manner, the braking capability of *both* interfaces is matched and rear wheel lock-up avoided.

Another possibility frequently made use of with proportioning valve systems is the cut-off of hydraulic pressure from the rear wheel brakes when the brake-pedal pressure exceeds a certain maximum value. The proportioning valve then takes the form of a spring-loaded piston in the hydraulic line between the brake master cylinder and the rear brakes, as shown in Fig. 13.16. For very severe brake applications, the hydraulic brake line pressure from

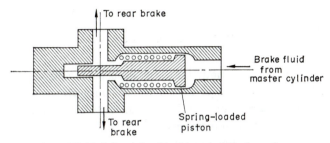

FIG. 13.16. Schematic of brake proportioning valve.

the master cylinder which would cause rear-wheel lock-up is automatically cut off by the system in Fig. 13.16, and the rear wheels therefore continue to roll freely during the braking period.

We can conclude generally that the proportioning valve, with all its simplicity, compensates for the effects of load transfer in a braking vehicle, so that the probability of individual wheel lock-up is greatly reduced.

Uniform Disc Wear

Consider briefly the action of a friction disc and actuator during braking. We are interested specifically in obtaining the axial force F_A necessary to produce a certain friction torque T_B and pressure p. Two methods are in general use, dependent on the properties and support of the disc. Let us suppose in the first case that the discs are rigid, so that the distribution of pressure across the face of the disc is reasonably uniform, as shown at left in Fig. 13.17.

Now the rate at which work is done by friction over any elementary area dA on the disc surface is given by

$$d\dot{W} = \underbrace{fp\,dA}_{\substack{\text{Elementary}\\\text{friction force}}} \times \underbrace{\omega r}_{\text{Velocity}} = K(pr), \tag{13.13}$$

which is greatest in the outer areas (i.e. as $r \rightarrow D_2$). Thus the greatest amount of wear takes place at the largest radii, and here the pressure rapidly decreases according to the pressure distribution sketched on the right-hand side of Fig. 13.17. It can easily be shown that the rate of work done by the friction force is now constant at all radii between $\frac{1}{2}D_1$ and $\frac{1}{2}D_2$. This can be seen from eqn. (13.13), where the local pressure p is no longer uniform (as it

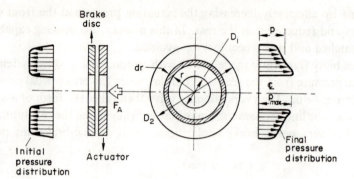

FIG. 13.17. The action of a disc brake.

was during the initial running-in period), but varies in such a manner as to preserve the product pr constant. Indeed, any other condition would cause $d\dot{W}$ to vary in some manner across the face of the disc, and a differential wear mechanism would continue to operate until pr became constant. Thus a uniform wear condition results.

Let the maximum pressure for the uniform wear condition occur at the inside edge of the disc (i.e., $p = p_{max}$ at $r = \frac{1}{2}D_1$; see Fig. 13.17). The uniform wear criterion can then be described by

$$pr = \text{const} = p_{max}(\tfrac{1}{2}D_1). \tag{13.14}$$

The axial force F_A is the pressure integral over the face of the disc thus:

$$F_A = \int 2\pi pr\, dr = \pi p_{max} D_1 \int_{\frac{1}{2}D_1}^{\frac{1}{2}D_2} dr$$
$$= \tfrac{1}{2}\pi p_{max} D_1(D_2 - D_1) = (\pi/4)\,\bar{p}(D_2^2 - D_1^2), \tag{13.15}$$

where the average pressure $\bar{p}$ on the face of the disc can be shown to be related to p_{max} as follows:

$$\bar{p} = \left(\frac{2D_1}{D_1 + D_2}\right) p_{max}. \tag{13.16}$$

The brake torque T_B is obtained by integrating the product of the frictional force and radius thus:

$$T_B = \int f_{BD}(2\pi r\, dr)\, pr = \pi f_{BD}\, p_{max} D_1 \int_{\frac{1}{2}D_1}^{\frac{1}{2}D_2} r\, dr,$$

$$T_B = (\pi/8) f_{BD}\, p_{max} D_1(D_2^2 - D_1^2) = (\pi/16) f_{BD}\, \bar{p}(D_1 + D_2)^2(D_2 - D_1) = K'\bar{p}, \tag{13.17}$$

which shows that T_B is proportional to mean pressure $\bar{p}$ (or p) as we assumed earlier provided the coefficient of friction f_{BD} between brake disc and actuator remains constant. We can eliminate pressure entirely from the right-hand side of eqn. (13.17) by substitution from eqn. (13.15) thus:

$$T_B = \frac{F_A f_{BD}}{4}(D_1 + D_2). \tag{13.18}$$

Equation (13.18) represents an extremely useful design criterion, since we obtain the available torque capacity in terms of the dimensions of the disc, the proposed axial force required for operation, and material properties as reflected in the coefficient of friction. The dimension $\frac{1}{4}(D_1+D_2)$ is called the friction radius, having a value of about 1.5 for automotive brakes and clutches.

Anti-skid Pulsed Braking

We have seen that application of even moderate braking effort on slippery road surfaces may cause wheel lock-up (or locked-wheel skidding), and the brake-slip ratio s_B rapidly attains the value unity (see Fig. 13.6; also the insert in Fig. 13.18). This condition is highly undesirable because of the loss of directional control of the vehicle when all four wheels lock simultaneously. Furthermore, the effective retarding force due to the locked-wheel coefficient f_{LW} is not as large as might obtain by permitting a partial slip condition. We would like to operate continuously at the peak A of the friction–slip curve, since we would then have the dual advantage of maximum retardation force and directional control. However, the peak in the insert of Fig. 13.18 is highly unstable, and once attained it inevitably degenerates into the locked-wheel mode at the point B. Apart from this consideration, we

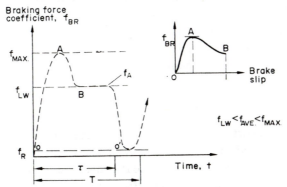

FIG. 13.18. Operating characteristics of anti-skid pulsed braking system. τ = time of brake application per pulse; T = pulsing period.

must bear in mind that friction versus slip curves of the type illustrated in Figs. 13.6 and 13.18 are instantaneous characteristics—a moment later both the magnitude and position of the peak and the slope of the unstable portion AB will in general be different. It is therefore not only extremely difficult to measure the friction–slip curve OAB in practice, but because of its continual variation in shape, position, and size, the instantaneous measured value has little practical significance.

Anti-skid braking systems of the on–off type provide a simple and effective means of continuously monitoring the friction versus slip characteristic for a braked rolling tyre. In principle, the brakes are applied and released in a rapid pulsing action, so that the braking force coefficient f_{br} traverses the curve OAB rapidly in both directions. Thus the mean

20*

value of the coefficient lies somewhere between the maximum and locked-wheel values, as indicated in Fig. (13.18), i.e.

$$f_{LW} < f_{ave} < f_{max}.$$

For a given brake application, anti-skid pulsing devices therefore provide an additional increment $\Delta f = f_{ave} - f_{LW}$ in braking coefficient, but the time of application during braking is reduced in the ratio (τ/T). It has been found in practice that the application of a higher coefficient f_{ave} for a shorter time period $\Sigma\tau$ is approximately equivalent to applying the steady, locked-wheel value, f_{LW}. *There is therefore no distinct gain in net retarding force and little or no reduction in stopping distance from the use of a pulsed, anti-skid braking system. The great advantage of the latter is, of course, the retention of steering or directional control during the off-brake or rolling periods* $\Sigma(T-\tau)$. In the final analysis, and considering safety as a primary objective, the ability to steer around and avoid collision objects in a panic braking manoeuvre is far more important than a reduction in stopping distance without directional control of the vehicle.

The development of anti-locking braking devices proceeded rapidly during the fifties because of the need to minimize landing hazards for aircraft on flooded runway surfaces. Many variations of the on–off type of brake control evolved, and details are beyond the scope of this book. In the simplest form, the rate of wheel "spin-up" at touchdown is used as an indication of the state of the runway. If this is less than a critical value, a hydroplaning condition can be assumed to exist between tyre and runway (see Section 13.1), and the on–off braking system automatically overrides the pilot's brake control of the aircraft to induce the free-rolling mode of operation. In recent years, anti-lock braking devices have been incorporated in automobiles of the more expensive class.

Brake Friction Materials

Perhaps the most important property of brake friction materials is their durability or wear resistance, because of the need to rapidly dissipate part of the kinetic energy of the vehicle. This energy is rapidly transformed into frictional heat energy which raises the temperature of the brake. The maximum temperature reached depends on the time and severity of brake application and on the heat-dissipation capacity of the brake system, and lies approximately in the range 200–260°C for automotive asbestos brake linings.

The nature of the contact interface between brake linings and disc or drum surfaces is such that it cannot be classified either as a metal–metal or metal–elastomer combination. Wear phenomena associated with brake linings in use therefore reflect mechanisms which are characteristic of both types of interface. Five basic wear mechanisms[64] may be distinguished:

  (a) Thermal wear.
  (b) Abrasive wear.
  (c) Adhesion–tearing wear.
  (d) Fatigue wear.
  (e) Macro-shear wear.

We observe that the abrasive and fatigue types of wear normally accompany elastomer–metal combinations, whereas the remaining types of wear are typical of metal-on-metal interactions (see Chapter 9).

Thermal wear deserves special attention because it encompasses a group of physical and chemical reactions in the course of which inter-atomic bonds are continuously broken. These reactions include pyrolysis (or thermal decomposition), oxidation, particulation, explosion, melting, evaporation, and sublimation. The rates at which they occur increase exponentially with temperature. Pyrolysis occurs predominantly at the centres of linings and pads, and to a lesser extent at corners and edges. Oxidation, on the other hand, predominates at corners and edges, being less severe near the centre. Explosive reactions can occur under highly abusive braking conditions, where the rate of heat input is so high that solids are converted into gases well beneath the surface; because these gases are greater in volume than the solids which they displace, they create instantaneous pressures which rupture the lining in an explosive manner. Flash temperatures created by the sudden rupture of welded asperity tips may be as high as 760°C, and this causes rapid decomposition of the organic compounds used in friction materials. Thus fibrous asbestos in the brake lining is converted to powdery olivine, and the cast-iron mating surface is transformed from pearlite to martensite.

Of the remaining types of wear in brake-lining materials, abrasive wear includes ploughing and grinding by wear debris and foreign particles. Fatigue wear occurs in two forms, namely thermal and mechanical. Thermal fatigue is caused by repeated heating and cooling, which induce cyclic stresses in the surface material and steep thermal gradients. A special case of thermal fatigue is called thermal-shock cracking, occurring as a result of a single, abusive loading. Mechanical fatigue is caused by repeated mechanical stressing of the lining material. Macro-shear is a relatively sudden failure of a friction material which has been previously weakened by heat, and it is most likely to occur at elevated temperatures and under severe braking conditions.

All these forms of wear can be controlled or minimized by the rapid removal of frictional heat, so that the sliding interface is kept as cool as possible. Thermal wear in particular can also be controlled by the use of polymers, fibres, and friction modifiers that have high thermal stabilities. Brake design should eliminate if possible high local pressures, especially at the corners of linings and pads; it should also block the passage of road dirt into the interface and yet permit the removal of wear debris.

13.3 Engine Friction

The purpose of lubrication between piston rings and the inside surfaces of engine cylinders is fourfold; thus

(a) The reduction of friction and wear.
(b) Sealing the annular gap between piston ring and cylinder by means of a lubricant film.

(c) Cooling of frictional surface between piston ring and cylinder.

(d) Cleaning of rubbing surfaces by washing away carbon and metal wear particles.

The first objective is by far the most important one, and we observe that this can be fully achieved only by completing and satisfying the remaining objectives. For high-speed engines, some form of splash lubrication is adequate, and the speed of the moving parts is sufficient to carry lubricant from the crankcase to the pistons and rings in the form of oil mists. Pistons of low- and medium-speed engines, however, require lubrication by some positive feed devices, such as mechanical oilers. In all cases care must be exercised to supply the minimum quantity of lubricating oil consistent with maintaining a thin film of oil on the internal cylindrical surfaces. Such problems as sticking piston rings, excess carbon deposits in the circulating oil, and even excessive wear of the liner and piston rings, can be traced to excess lubrication of the cylinders.

Other friction surfaces requiring lubrication in an internal combustion engine are:

(a) Crankshaft and main bearings.

(b) Crankpins and their bearings.

(c) Crossheads and guides.

(d) Wristpins and their bearings.

(e) Valve gear.

Main bearings in small stationary engines are lubricated by ring oilers, and in larger size engines by positive-feed lubricators. Pressure lubrication using gear pumps is also used to circulate large quantities of lubricating oil in engines having enclosed crankcases. Oil seals are widely used to prevent oil leakage along a shaft, and the elastohydrodynamic effects which characterize their performance are dealt with in Chapter 16.

Friction Mean Effective Pressure

The power absorbed by mechanical friction in an internal combustion engine is equal to the indicated power developed in the cylinders less the shaft (or output, or brake) power, thus

$$P_F = P_I - P_B. \tag{13.19}$$

As we will see later, most of the "friction power" P_F is absorbed by friction between piston rings and cylinders. Let P_F' denote the fraction of P_F appearing as piston ring friction so that we may write

$$P_F' = f_{Pc} W V, \tag{13.20}$$

where f_{Pc} is the effective coefficient of friction between piston and cylinder, W is the load[†] between piston rings and cylinder lining, and V the mean piston speed. The friction mean effective pressure (f_{mep}) is then defined as follows:

---

† The load W is such that the mean pressure between piston rings and cylinder lies in the range 0.2–0.4 bar.

$$f_{\text{mep}} = \frac{CP_F'}{A_p \bar{V}} = \left(\frac{CW}{A_p}\right) f_{Pc}, \tag{13.21}$$

which is independent of mean piston speed $\bar{V}$, and for a given design is proportional to the frictional coefficient f_{pc}. The constant C in eqn. (13.21) has the value 2 for two-stroke engines and 4 for four-stroke engines.

Figure 13.19 shows a plot of the percentage f_{mep} contributed by pistons and rings and the percentage due to bearings and valve gear versus piston speed in metres per minute. We observe that friction mean effective pressure is independent of mean piston speed, and for a

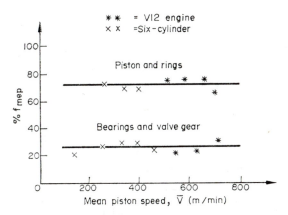

FIG. 13.19. Friction mean effective pressure as a function of mean piston speed.[65]

multi-cylinder engine the piston ring friction contributes approximately three-quarters of the total mechanical friction. The most common method of measuring friction in engines is by motoring, that is determining the power required to drive a non-firing engine by an outside power source. There is an obvious objection in that since the engine is not firing, neither the pressure nor the temperature acting at the bearing surfaces are representative of conditions under normal operation.

The actual friction as measured in two-stroke engines is approximately twice as great as that measured in four-stroke engines, so that from eqn. (13.21) the friction mean effective pressure is about the same. Two-stroke engines have longer and heavier pistons than four-stroke engines, and they also have more piston rings. Furthermore, the motoring method of obtaining frictional data usually permits cylinder heads to be left in place for two-stroke engines, so that there is gas loading on the piston during the compression stroke. Both of these factors contribute to the substantially increased friction forces measured in two-stroke engines.

Figure 13.20 shows piston-ring friction as a function of piston displacement during the induction, compression, power, and exhaust strokes which typify the four-stroke cycle. We observe in particular from these curves that:

(a) The average force of friction during the compression and exhaust strokes is about the same.

(b) The average force of friction during the power stroke is about twice that during the induction stroke.

(c) Friction forces are high just after TDC and BDC probably because of inadequate lubrication between piston rings and cylinders.

In general, piston friction increases with increases in oil viscosity, piston speed, gas pressure, and (to a lesser extent) compression ratio. An aircraft engine appears to have an especially

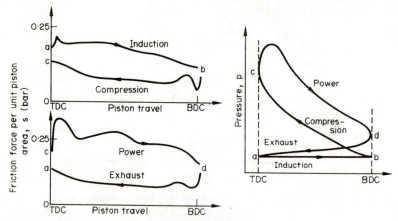

FIG. 13.20. Piston-ring friction measurements in special internal combustion engine.[65] TDC = top dead centre; BDC = bottom dead centre; s = friction force on piston per unit piston area.

low value of mechanical friction due to light reciprocating parts which minimize loads and relatively large piston-to-cylinder clearances. In addition, the pistons of earlier aircraft engines were relatively short in axial length, with the non-thrust faces cut away for weight-saving.

13.4 Miscellaneous

Windscreen Wiper Design

Typical operating speeds for automobile windscreen wipers are about 70 cm/s, and the average frictional coefficient is about 0.2. Experiments have shown[66] that the observed friction of lubricated wiper blades far exceeds the contribution from viscous shear forces alone, and there is strong evidence that small asperities exist along the tips of the blades. These asperities affect the flow of lubricant between the wiper blades and the glass surface of the windscreen. When liquid passes under a blade, it is broken into fine threads by the asperities with the result that streaks of liquid are left behind the path of the wiper. These in turn tend to break into droplets and to obscure driver vision.

The size and stiffness of the asperities in the blade surface are dependent on the particle size of carbon filler incorporated into the rubber matrix of the wiper. Carbon filler is necessary in the first place to prevent undue blade wear, and this can be minimized by selecting the smallest carbon particles. However, after continued use, these blades have the largest and sharpest asperities. Experiments have shown that windscreen wipers operate below the hydrodynamic régime and within what has been previously described as the boundary lubrication or mixed-friction range (see earlier) where contact between blade and glass is part liquid and part solid. In fact, because of filler particles in the blade or dirt deposits on the glass, elastohydrodynamic effects predominate in the mixed friction range following the theory outlined in Chapter 8.

Ideally, the windscreen wiper blade should exhibit both low friction and good wiping properties. The latter implies that not only should dirt particles be effectively removed by scraping action, but that existing water on the windscreen be spread into a thin, even film. Provided an even film is formed, it is not necessary to remove every water droplet for clear vision. A poor wiper will not only leave streaks and patches of water, but it may also fail to scrape off dirt particles because of elastohydrodynamic action. A good wiper will not only be satisfactory in scraping and wiping, but it must give long life. It has been suggested[66] that the optimum wiper design will have two separate functions, e.g. scraping and wiping. *At present the evidence suggests that elastohydrodynamic action is highly beneficial to the wiping action*, and it is responsible for a low coefficient of sliding friction.

Clutch Design

We have dealt extensively with braking mechanisms in Section 13.2, and much of the theory pertaining to contact conditions at the frictional interface between brake linings and drum or disc can be applied directly to the case of automotive friction clutches. It is apparent that the principle of operation of both a friction clutch during engagement and of a friction brake during braking is to bring two members having relative motion to a state where no relative motion exists. The operation of a clutch is therefore essentially the same as that of a brake, although there are structural differences in the two units because of control requirements and the necessity of providing for heat absorption or dissipation in braking systems. Plate clutches are similar in construction and operation to disc brakes according to Fig. 13.17, and several pairs of discs or plates may be used to transmit large torques. Other common clutch types which depend for their operation on the existence of relative motion at a frictional interface include cone, band, and block clutches.

We conclude this chapter by observing that automotive bearings and seals have not been dealt with, even though their tribological implications are widespread. The subject of bearing friction will, in fact, be dealt with exclusively in Chapter 15 and the action of mechanical and lip seals will receive special treatment in Chapter 16. The omission will therefore be rectified to some extent by the subject matter of successive chapters.

CHAPTER 14

TRANSPORTATION AND LOCOMOTION

14.1 Surface Texture in Roads and Runways

Perhaps the major application of elastomeric friction is the rolling or sliding performance of pneumatic tyres on pavement surfaces. In the last chapter we have dealt with the behaviour of pneumatic tyres in some detail, and here we confine our attention to the nature of surface texture in roads and runways. We must bear in mind, of course, that the frictional interface between tyre and road is determined by complex interaction events during free rolling, braking, driving, cornering, skidding, or any combination of these modes, and its properties reflect the result of this interaction. Nevertheless, it is extremely useful to separate the individual contributions of tyre and road in order to understand the fundamental events which subsequently determine the frictional coupling in the contact area.

It is now a well-established fact that the coefficient of sliding friction on a wet road surface decreases with increase of speed,[†] and that the rate of decay is a function primarily of drainage ability. This drainage is normally provided both by the tread grooving in the tyre and by the void spacing between asperities of the pavement surface. Let us imagine a perfectly smooth tyre (i.e. having no tread pattern whatever), so that only the road texture is capable of providing drainage under wet conditions. Figure 14.1 then shows four characteristic friction vs. velocity curves for road surfaces having different sharpness at asperity peaks and different void spacing. Two general laws emerge from the data in this figure:

(a) the magnitude of the coefficient of sliding friction at any given velocity depends only on the sharpness of the texture; and

(b) the slope of the friction versus velocity characteristic at any given velocity depends only on the mean void spacing of the texture.[67]

We observe from Fig. 14.1 that the curves AA' and CC' have about the same void spacing, so that their slopes are approximately equal. This also applies to the curves BB' and DD'. Furthermore, in comparing either AA' and BB' (or CC' and DD'), the surface having the larger void spacing in each case has a lower rate of decay of frictional coefficient with speed. This is because drainage takes place more readily at the higher speeds on surfaces having a larger void spacing, so that intimate contact between tyre-tread elements and road surface is more likely to occur. We also observe from Fig. 14.1 that the curves AA' and BB' have

---

[†] This decrease is effective over a very wide speed range.

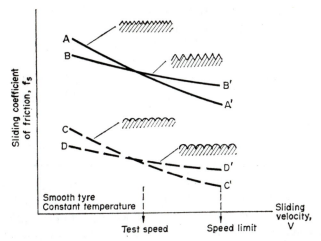

Fɪɢ. 14.1. Characteristic friction vs. velocity curves on road surfaces with different sharpness and void spacing.

approximately the same frictional coefficient since the road textures are pointed and therefore similar. In like manner, the curves CC' and DD' have a lower but almost identical coefficient of friction because the textures are rounded and therefore similar.

We can see at a glance from Fig. 14.1 the fallacy of attempting to characterize the skid-resistance (or frictional coefficient) of a particular pavement by performing one test measurement at a preselected speed. Let us suppose that in Fig. 14.1 the speed of testing corresponds to the cross-over point between the curves AA' and BB' or between CC' and DD' as shown. We will further suppose that the average maximum speed (or speed limit, if any) on a particular road network is somewhat higher than the test speed. If follows that the testing device will rate pavements AA' and BB' as identical in frictional performance, whereas at the higher speeds, surface BB' is obviously superior: similarly for surfaces CC' and DD'. The situation is not serious if the test speed and maximum average speed are the same or virtually the same, since the road or runway is then rated at its worst friction value. In the general case where these speeds are considerably different, however, it is clear that one test speed is not sufficient. Either one or more additional speeds (clearly different from the original test speed) must be selected at which to measure skid resistance, or otherwise the gradient or slope of the friction versus velocity curve for that pavement must be established from drainage considerations. The latter information combined with the measurement of skid resistance at one sliding speed is sufficient to determine frictional performance over a range of speeds. The outflow meter described in Section 2.6 is the most suitable instrument in use today for determining the slope of the friction versus velocity curves from texture measurements.

Optimum Surface Texture

Consider the relationship $\lambda = V/\omega$, where λ is the mean wavelength of asperities in the road or runway texture, V is the sliding speed or speed of travel for a tyre, and ω the "frequency" of indentation of tread elements by road asperities, giving rise to the hysteresis component of friction (see Chapter 4). As we have seen earlier, there exists a certain value of ω for which the hysteresis component of friction is a maximum, and in designing an optimum texture we would like to select λ in accordance with the above equation so that this maximum is approached or realized. For typical road surfaces and rubber materials, however, the speed V so obtained normally exceeds average maximum speeds of driving, even in Europe. It is also necessary that the mean wavelength provides asperities which are sufficiently large to ensure adequate drainage of water into neighbouring voids at some typical maximum speed (see Fig. 14.1). In practice, the mean wavelength selected may be a compromise between drainage requirements (which suggest a lower size limit) and hysteresis[68] requirements (which indicate an upper limit) at the speed limit or average maximum speed for the pavement under consideration. However, since the hysteresis requirement pertains to the sliding mode alone and drainage requirements apply to both rolling and sliding, *it is certain that the drainage criterion predominates in the final selection of wavelength.*

The adhesion contribution to friction can be maximized by providing in addition a sufficiently sharp texture. Figure 14.2(a) shows one form of an idealized random road surface, where the individual asperities are pointed to provide sufficiently high localized

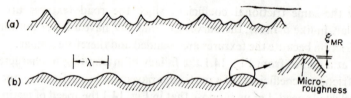

FIG. 14.2. Idealized (a) and typical (b) road surfaces.

pressures between tread and road surface. This permits a ready break through water films resting on the surface and facilitates contact between tread and road. Actual road surfaces have profiles more in accordance with Fig. 14.2(b). Here, again, the hysteresis and drainage conditions may be satisfied by choosing the wavelength λ at a maximum design skid speed, but we observe that since the asperity tips are predominantly rounded rather than pointed, there is a tendency to entrain water by elastohydrodynamic action over these peaks (see Chapter 8) during normal rolling, and thereby to promote viscous hydroplaning. To avoid such a possibility it is necessary to carefully select an adequate micro-roughness ε_{MR} at asperity tips, to ensure the disruption of the fluid entrainment mechanism. This micro-roughness is provided by the surface finish on road aggregate materials, and its magnitude lies somewhere in the range 10–100 μ.[69]

We must emphasize the need for providing micro-roughness when the degree of roundedness has a certain critical value. Figure 14.3 shows the pressure distributions which result when a cone and sphere are pressed in turn against an elastic rubber plane. The cone and sphere can, in fact, be considered as extreme cases of sharpness and roundedness, and they therefore define limits for sharp and rounded aggregate materials in road construction. Let p^* be a hydrodynamic pressure defined as the minimum value required to break through the squeeze-film thickness on a particular surface. In the case of the cone, the maximum elastic pressure occurring at its tip easily exceeds p^*, even though the average value of the pressure distribution $\bar{p}$ is considerably lower than p^*. For the sphere in Fig. 14.3(b), both p_{max} and $\bar{p}$ are less than p^*, so that no penetration of surface water films can be expected. The cone, of course, would be subject to severe wear at its peak, and heavy tearing and penetration of tread rubber would take place if such asperities were placed in an actual test surface. It

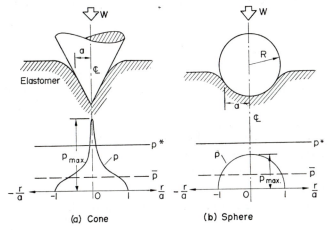

(a) Cone   (b) Sphere

FIG. 14.3. Pressure distributions for (a) cone, and (b) sphere, pressed into an elastomeric plane.

cannot therefore be seriously considered as a desirable shape or as an ideal prototype asperity in road surfaces. The spherical or hemispherical asperity does not suffer from this wear problem, but its inability to penetrate through squeeze films is such a serious limitation that we must reject it as a practical asperity shape. There is obviously some compromise between the two examples in Fig. 14.3, and this may take the form of a conical projection

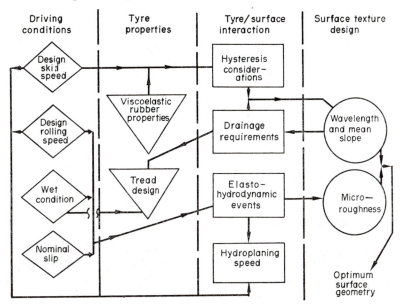

FIG. 14.4. Sequence of events in selecting optimum surface geometry.

with a rounded tip. Presumably this shape of asperity would just penetrate through surface films, while there would be a limit on wear and tear in the mating elastomer surface. The criterion for the degree of roundedness at the tip of this asperity would be given by the equality $p_{max} = p^*$. Even if such an asperity shape were readily available, there would still exist asperities of the rounded or flattened type, and the need for micro-roughness would again arise.

The selection of optimum surface texture follows a rational procedure which is depicted in Fig. 14.4. The mean wavelength and slope are dictated by hysteresis–drainage factors, whereas the micro-roughness is selected to ensure the existence of an adequate adhesional mechanism at asperity peaks. The various factors involved are grouped for convenience in the following categories:

(a) *Driving conditions* (forward speed, rolling–sliding, nominal slip and wet–dry).
(b) *Tyre properties* (viscoelasticity, tread design, etc.).
(c) *Interaction events* (hysteresis, drainage, elastohydrodynamic factors, hydroplaning, etc.).
(d) *Pavement geometry* (wavelength, mean slope, micro-roughness).

Elastohydrodynamic Considerations

We have described both in Chapters 8 and 13 the mechanism of elastohydrodynamic separation for pneumatic tyres rolling (and sometimes sliding) on a wet textured pavement surface. Figure 14.5 outlines schematically how the various factors determine a compatibility

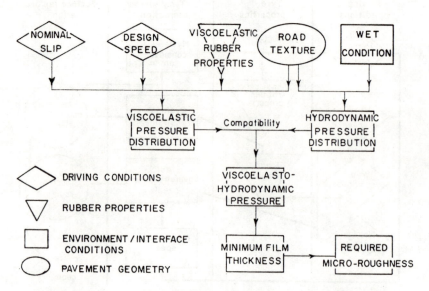

FIG. 14.5. Selection of optimum micro-roughness for given pavement.

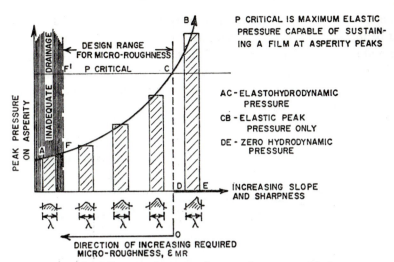

FIG. 14.6. Micro-roughness requirements for pavement design.

between the viscoelastic and hydrodynamic pressure distributions about each typical asperity in the road surface. The resulting elastohydrodynamic or viscoelastohydrodynamic pressure can then be used to determine the minimum film thickness from lubrication theory, and hence the amplitude of the minimum required micro-roughness at asperity peaks is known.

Let us now examine closely the micro-roughness requirements for different asperity shapes, as illustrated in Fig. 14.6. We assume that the mean wavelength λ has already been selected from drainage–hysteresis requirements, but the mean slope of the asperities is still variable within prescribed limits. We also assume that increasing slope and sharpness are related in some manner. Figure 14.6 shows clearly that the peak pressure on any asperity increases rapidly and non-linearly with mean slope to a critical value at the point C, but the elastohydrodynamic pressure distribution can be sustained only in the region AC. As the sharpness increases beyond the value at C, the corresponding elastic pressure is too great to permit the existence of a water film at asperity tips, and thus the hydrodynamic pressure component has zero value along DE. The elastic or viscoelastic pressure continues to increase indefinitely along CB.

It is apparent that no micro-roughness is needed along CB since no water film exists in this range at asperity tips. Also, for very small slopes, there is inadequate drainage in the void spacing between asperities (e.g. along AF). *The design limits for the selection of micro-roughness are therefore clearly specified in the range F'C in Fig.* 14.6. The extent of the range $F'C$ remains to be determined from further research, but it is certain that λ has an order of magnitude of 5–15 mm and ε_{MR} ranges from 10–100 μ.

Flooded and Damp Conditions

We have hitherto considered the case of wet road and runway surfaces, since texture is required to establish contact between tyre-tread and pavement only under wet conditions, and we now distinguish between two forms of wetness, i.e. flooded and damp conditions.

When flooding occurs, the tread grooving of the tyre and the mean void width of the pavement must be together capable of discharging sufficient water from the contact patch between tyre and road at the average maximum speed for that particular road section. In practice, of course, the average maximum speed may vary from perhaps 45 km/hr in urban areas to 120 km/hr on motorways, and the texture requirements differ widely depending on the type of road under consideration.

In Fig. 14.7 we consider the case of three roads A, B, and C corresponding respectively to average maximum speeds of 45, 75, and 115 km/hr respectively. The maximum coefficient of friction[†] when plotted versus vehicle speed for each of these roads shows a rate of

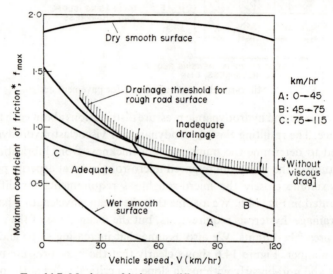

FIG. 14.7. Maximum friction on different flooded road surfaces.

decrease which is distinctly greater at speeds in excess of the maximum rating for each of the surfaces. Thus each curve has a point of discontinuity corresponding to its maximum speed rating as shown in the figure. Furthermore, we observe that as the design limit is increased (i.e. by proceeding from roadway A to roadways B and C), there is a loss of braking capacity at speeds below the design limit. Thus it is clear that pavement B is designed for a higher speed range than A, but the performance of B is inferior to that of A below the design speed limit of the former. A similar argument applies to surface C when compared with either A or B. In practical terms we would like to have as constant a frictional coefficient as possible over the design speed range for any pavement, and we therefore ensure that the point of discontinuity lies outside this range. Each point of discontinuity of slope in Fig. 14.7 represents the limiting speed for which the design macro-roughness provides adequate drainage. Thus by joining the discontinuities occurring in each curve, it is possible to establish a drainage threshold as shown.

---

† Obtained during braked rolling.

With damp pavement surfaces, the phenomenon of viscous hydroplaning as described in Chapter 13 may occur. Here we must ensure that an adequate micro-roughness ε_{MR} exists at road surface asperity peaks to counteract the mechanism of elastohydrodynamic film entrainment, and thus guarantee effective adhesion even in the presence of thin films. The "feel" of the surface texture is the best qualitative measure of micro-roughness; thus surfaces which are sufficiently harsh to the touch may be deemed to have an effective micro-roughness amplitude. We note that whereas the flooded condition is comparatively rare on roads and runways, the damp or thin-film situation occurs whenever there is any precipitation what-soever, or even a high humidity. Furthermore, the viscous hydroplaning phenomenon occurs at the rear of the contact patch in wet rolling as we have seen, when the front part experi-ences dynamic hydroplaning under flooded conditions. The adhesion-generating mechanism is therefore the principal contributor to braking effectiveness in wet rolling, irrespective of the degree of precipitation or rainfall.

Pavement Grooving

One of the more effective means of eliminating flooded conditions on roads and runways is by providing longitudinal or lateral grooving within the pavement. This has found partic-ular favour in runway design for two reasons. In the first place, aircraft tyres for the most part have little or no pattern in the tread,[†] and they therefore depend almost completely on the runway to provide the necessary drainage from the tyre footprint. Furthermore, the cost of grinding grooves in existing concrete runways is generally regarded as excessive, but the cost of incorporating such grooves during the construction of new runway strips is almost negligible.

Transverse grooving is undoubtedly more effective in terms of drainage ability than longitudinal grooving of the same design and spacing, since the escape path for the water under a pneumatic tyre has the same direction as the maximum pressure gradient in the contact area. It is therefore not surprising to find transverse grooving used widely in runway design where drainage ability is more critical as a consequence of high landing and touch-down speeds.

One disadvantage of transverse grooving is excessive tread wear during landing, which necessitates more frequent changes in tyres during service. It can be shown that the in-creased wear is due to a combination of normal draping action of the tyre tread within the grooves and the existence of elastohydrodynamic contact conditions along the flat surface of the runway between grooves. Figure 14.8 shows vertical longitudinal cross-sections taken within the contact area of an aircraft tyre which is (a) rolling, and (b) sliding on a surface with transverse grooving. The local bending of the tread surface at B is reinforced by the familiar "concave down" elastohydrodynamic effect at A. This promotes an aggravated contact condition along the edges of the grooves with a resulting increase in tyre wear. However, the drainage properties of grooving are far superior to conventional texture, and

---

† High contact pressures in aircraft tyres make a tread impractical in terms of durability, and there is also a weight penalty.

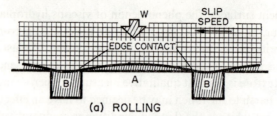

(a) ROLLING

A = ELASTOHYDRODYNAMIC CONTACT
B = DRAPING EFFECT

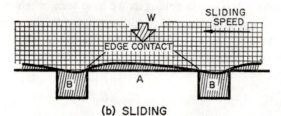

(b) SLIDING

FIG. 14.8. Elastohydrodynamic contact conditions for transverse grooving.

this is undoubtedly the overriding consideration. It is probable that excessive wear can be reduced by either chamfering the edges of the grooves or replacing the rectangular with V-shaped grooving.

14.2 Rail–Wheel Adhesion

The power-to-weight ratio of railway locomotives has increased consistently over the past two decades because of enlarged power plants, the extensive use of lightweight materials, and improved body construction methods. A limited factor to a continued increase in this ratio appears to be the adhesion developed between the driving wheels and rails. The evidence for this is the pronounced tendency for wheel slip to occur especially in starting from standstill and in damp weather.

In attempting to obtain a clearer understanding[†] of the fundamental mechanism of adhesion between wheel and rail, a memory effect known as "secondary conditioning" was discovered to have a profound influence on the adhesion of certain rail sections.[70] This phenomenon arises from the fact that when certain substances, notably oil, are spread upon a rail, secondary effects occur in which a minute quantity of the substance associates so closely with the surface as to virtually form part of the main material. Thus it is possible for two apparently identical steel specimens to reveal widely different coefficients of surface friction in spite of any reasonable amount of cleaning. This memory capability which is generally detrimental for surfaces previously contaminated with oil, can have beneficial effects when the spark-discharge method of improving adhesion is employed, as we shall see later.

[†] Convention on adhesion by Railway Engineering Group of Institution of Mechanical Engineers, London, 1963.

There are generally three methods for improving rail–wheel adhesion, thus:

(a) *Chemical*, by using additives on the rail surface.
(b) *Mechanical*, such as scouring, abrading, and sanding of the railhead.
(c) *Electrical*, including plasma arc or spark discharge between an electrode and the rail.

It appears that both the chemical and mechanical methods used have either consistently failed to provide satisfactory improvements in adhesion or they tend to introduce other undesirable effects. For example:

(a) The use of "Syton" liquids (a colloidal dispersion of silica in water) produces a considerable improvement in the adhesion of dry or wet rails having medium or low values of secondary conditioning, but the results show little improvement on oily rails.
(b) Sodium hydroxide solutions can reduce the effects of oil contaminants by attacking traces of oil which give low secondary conditioning, thereby increasing adhesion. However, excess or long-term use of these additives produces a sludge which may have a deleterious lubricating effect.
(c) Sanding of the rails provides a practical way of obtaining an instantaneous increase in rail–wheel adhesion. However, the sand must be absolutely dry, and there are difficulties with storage and particle-size control. In addition, sanding is not permitted near switching points because of the danger of clogging, and, generally, surface damage occurs due to pitting.
(d) Rails previously treated with silicone fluids tend to give high adhesion values even when subsequently covered with oil, whereas initially clean rails when covered with oil give low readings. However, when the silicone-treated rails are sprayed with water the formation of innumerable droplets brings about a large reduction in adhesion.

The method of spark discharge (by ionizing the air gap between an electrode placed ahead of each driving wheel and the rail) effectively volatilizes the contaminants which normally appear on the rail, and thereby produces systematic improvements in adhesion under the most adverse conditions. Unfortunately, although experiments using spark discharge have been conducted both in the Soviet Union and the United States, the method until recently had found limited application, chiefly because of the disadvantages of high power consumption, additional complexity, electrical noise, and rail deterioration. The problems of high power consumption and electrical noise still persist today.

Before examining further the effects of electrical-spark discharge in improving rail–wheel traction, it is well to consider typical adhesion values under a variety of experimental conditions.

Typical Rail–Wheel Adhesion Values

Table 14.1 shows a listing of rail–wheel adhesion values under dry, wet, or greasy conditions. It is apparent that a large variation in the values of the adhesion coefficient exist, from a minimum of about 0.07 on damp rails to a maximum of 0.35 under dry, clean conditions. Lower adhesion values generally reflect the presence of an oily conta-

21*

mination on the railhead, and the more slippery locations are systematically found on curves, near points, near stations, and at road crossings. Oil contamination from axles and lubricating pads flows on to wheel rims, and both at stations and on curves it finds its way into the contact patch. For points, the source of contamination is obvious, while for road crossings the oil escaping from automobiles is the culprit.

TABLE 14.1. TYPICAL RAIL–WHEEL ADHESION VALUES

Condition of rail	Adhesion coefficient
{ Dry rail (clean)	$0.25 \leqslant f_A \leqslant 0.30$
{ Dry rail (with sand)	$0.25 \leqslant f_A \leqslant 0.33$
{ Wet rail (clean)	$0.18 \leqslant f_A \leqslant 0.20$
{ Wet rail (with sand)	$0.22 \leqslant f_A \leqslant 0.25$
Greasy moist rail	$0.15 \leqslant f_A \leqslant 0.18$
Dew on rail, or foggy damp weather	$0.09 \leqslant f_A \leqslant 0.15$
{ Sleet on rail	$f_A = 0.15$
{ Sleet on rail (with sand)	$f_A = 0.20$
{ Light snow on rail	$f_A = 0.10$
{ Light snow on rail (with sand)	$f_A = 0.15$
Wet leaves on rail	$f_A = 0.07$

As we have seen earlier in Chapter 3, the coefficient of friction or adhesion between two metallic, perfectly clean surfaces at 600–700°C under vacuum conditions can be as high as 10–100. In allowing air contamination, layers of oxides and adsorbed gases will reduce the coefficient to perhaps 1 or 2. With a good lubricating oil it can be further reduced to the range 0.05–0.1. *Every reduction of the coefficient of friction is due to a layer (often monomolecular) of some contaminant adsorbed on at least one of the contacting surfaces.* Such layers must be very solidly attached to the surfaces to be able to resist the enormous pressures which occur between loaded surfaces in relative motion. Paraffinic oils are not good boundary lubricants from this viewpoint because their chemical inertness does not permit them to be readily adsorbed, and direct contact between metallic parts is not prevented. On the other hand, fatty acids are good lubricants because they combine chemically with metals and form as a result very solid and adherent layers on the surface.

The adhesion values listed in Table 14.1 are, of course, approximate and correspond presumably to low speed. When speed is considered as a variable, we obtain typical curves as in Fig. 14.9(a) and (b) showing the systematic decrease in adhesion with increasing locomotive speed. At least thirteen empirical relationships have been proposed[71] between the years 1858 and 1947 to approximate the adhesion versus speed characteristics typified by Fig. 14.9. Usually, these relationships take either of two forms:

$$\left. \begin{aligned} f_A &= k_1 + \frac{k_1}{V + k_3}, \\ f_A &= k_4 - k_5 V^n, \end{aligned} \right\} \tag{14.1}$$

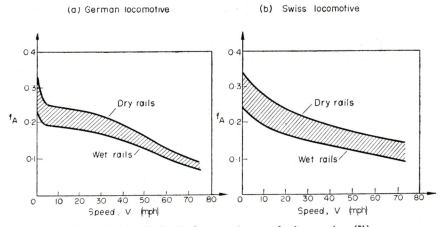

FIG. 14.9. Typical adhesion–speed curves for locomotives.[71]

where k_i are positive constants ($i = 1$–5) and n has the value 1 or 2. From the results of many investigations, the following approximate and simplified equations are proposed for dry rails:

$$f_A = 0.25 \text{ (constant)} \quad 0 < V < 40 \text{ mph}$$

$$f_A = \frac{30}{V+75} \quad 40 < V < 150 \text{ mph}$$

For wet rails we may use $f_A' = 0.6\, f_A$.

The general dependence of f_A on brake-block pressure p (as well as speed V) is shown in Fig. 14.10 taken from actual tests.[71] The decrease of the adhesion coefficient with increasing brake pressure can be expressed by the relationship

$$f_A \sim p^{-0.38}. \tag{14.2}$$

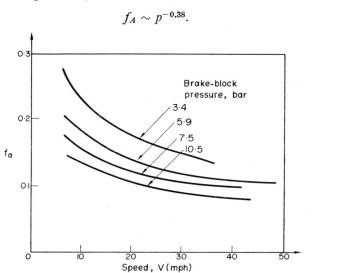

FIG. 14.10. Effect of brake-block pressure on adhesion.

We note that the inverse dependence of f_A on p is in accordance with the simple theory of adhesion for metals as expressed by eqn. (3.20) and, indeed, as shown graphically in Fig. 3.15. Equation (14.2) is valid for pressures up to about 20 atm: *at greater pressure it can be shown that the adhesion coefficient begins to increase again with increasing p, and at the same time becomes independent of speed.* The reason is that probably the high combination of speed and pressure causes overheating of the brake block, softening, and ultimately more intimate contact with the rail. To ensure reasonable wear, the Brinell hardness of brake blocks should be 220–240, and of wheels 240–300.

Other variables which affect rail–wheel adhesion are wheel load, wheel size, and whether braking or driving conditions apply. Wheel load appears to have very little effect on f_A. This is due largely to the fact that for very high loading (as in the case of the wheel–rail situation), the real area of contact approaches the apparent area in size. In this case, the mean pressure in the contact zone is close to the yield limit in compression for steel (about 70 kg/mm²), and the simple theory of adhesion,[†] written as

$$f_A = s/p^*, \qquad (3.20)$$

shows that f_A remains invariant. Increased wheel load, of course, will increase the apparent area of contact, but this affects both numerator and denominator in eqn. (3.20) so that there is no overall change is f_A. For clean surfaces, the value of s/p^* is about 0.80 for steel-on-steel. In most cases, however, the existence of oxides, rain, oil, or other contaminants reduces the effective shear strength s, so that the adhesion values in Table 14.1 are obtained. There is a small reduction in mean pressure and a corresponding increase in apparent contact area according as we increase wheel diameter. Finally, the adhesion coefficient appears to be little different whether braking or driving conditions prevail at the wheel–rail interface.

Mechanics of Braking

Consider the effect of braking a railway wheel from an initial travel speed of 100 km/hr, as shown in Fig. 14.11. The time taken for wheel-lock to develop is approximately 1 s or less from the instant of brake application. During this period, both tractive effort and wheel angular velocity decrease progressively, whereas the velocity of slip between wheel and rail rises non-linearly to attain the final locked-wheel skidding value of 100 km/hr. During this braking action, the contact area between wheel and rail consists of a region of adhesion and a slip zone, as shown in the insert of Fig. 14.12. [70] Theory and experiment indicate a longitudinal shear stress distribution within the area of contact which takes the form of the upper curve in Fig. 14.12. We note that the shear stress is confined to relatively low values within the adhesion zone, and reaches its maximum value within the slip region. This observation satisfies the well-known requirement that a certain relative slip velocity between surfaces is essential for maximum friction. The distribution of longitudinal slip velocity in the contact zone shows a non-linear increase towards the rear of contact according to the lower curve in Fig. 14.12. We note a remarkable similarity between this distribution and that for

[†] This is valid in the case where $A_{real} \doteq A_{app}$; see eqn. (3.20).

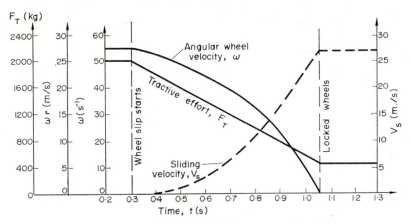

FIG. 14.11. Time history of events for wheel–rail braking.

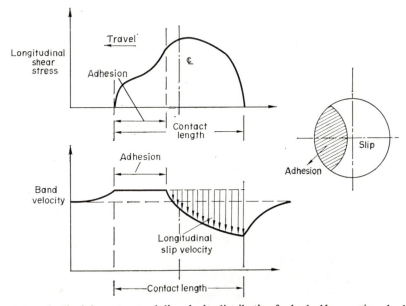

FIG. 14.12. Longitudinal shear stress and slip velocity distribution for braked locomotive wheel on rail.

a braked pneumatic tyre, as shown in Fig. 13.5(a). In both cases, the slip region is followed by an "overshoot" as the band velocity attempts to restore to its undeformed value just outside the contact area. Another remarkable fact is that although the apparent area contact of the tyre situation may be as large as 100 times that for the wheel–rail combination, the actual or real areas of contact are similar in size. In both cases, the longitudinal shear stress distribution has the same overall shape, and the adhesion zone in the front of the contact patch is followed by a slip region towards the rear.

Improvement in Adhesion by Spark Discharge

Extensive laboratory tests using the spark-discharge method of volatilizing rail–wheel contaminants[70] showed a considerable improvement in adhesion values after sparking. Accordingly, an experimental wagon was devised to measure adhesion before and after the spark discharge and to provide the spark discharge itself. Figure 14.13 shows a schematic of a typical experimental wagon. A single electrode of mild steel is mounted ahead of the test wheel and at a height 15–17 mm above the railhead, and a similar electrode is placed to the right of the test-wheel surface. The test wheel is rigidly attached to an experimental axle, on the other end of which a similar wheel is free to roll (not shown in the figure). A hydraulically operated disc brake applies a braking torque to the experimental axle until

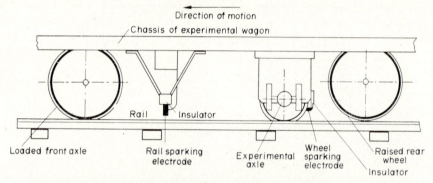

FIG. 14.13. Schematic of experimental wagon for spark-discharge measurements.

FIG. 14.14. Typical pattern on steel surface after sparking.

the test wheel is on the point of slipping. This torque is proportional to the coefficient of adhesion between wheel and rail, and it may be measured with a dynamometric arm which has been suitably calibrated. A high-voltage circuit triggers and sustains a spark discharge simultaneously across the electrode–rail gap and across the electrode–wheel gap at preselected intervals of time.

Figure 14.14 shows a typical pattern left on a steel surface after cleaning by the spark-discharge method. It is believed that no permanent damage (such as pitting, visible surface deterioration, or reduced fatigue life) would occur on rail sections exhibiting an after-sparking pattern similar to that shown in the figure.

The combined results of all tests conducted with the experimental wagon on dry, oily, naturally wetted, and flooded rail sections are presented in Fig. 14.15 as plots of coefficients of adhesion versus a logarithmic-Gaussian scale of cumulative frequency. Here the cumulative frequency corresponding to any coefficient of adhesion under particular test conditions is defined as the percentage of all tests which have been conducted under those conditions, where the measured adhesion was either less than or equal to that value. This method of plotting the experimental data has several advantages. For example, since the recorded data normally appears as a straight line for a particular set of experimental conditions:

(a) its slope is a measure of the spread of the data, and hence some function of standard deviation for a particular choice of scales;
(b) changes in slope indicate special departures from a Gaussian distribution of recorded data; and
(c) improvements in adhesion appear as shifts of the straight line plots to the right.

Figure 14.15 shows clearly that whereas before sparking the adhesion coefficient may be as low as 0.10 (for naturally wetted rail) and ranging up to 0.6, this range is changed to a

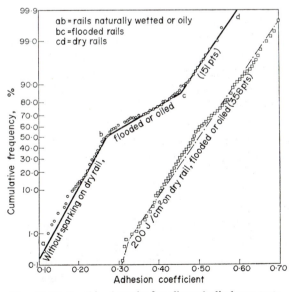

Fig. 14.15. Combined results for all spark-discharge tests.

minimum of 0.30 and a maximum of 0.70 after spark discharge. The results for the un-
sparked rails fall into three separate categories as seen in Fig. 14.15, depending on whether the
rails are naturally wetted, oily, flooded, or dry. We note in particular that the spread of
adhesion values after sparking is much smaller than before, indicating that the spark-discharge method primarily augments the lower adhesion values. If we tabulate the minimum
adhesion values obtained corresponding to 1% cumulative frequency, the percentage increase
in adhesion due to spark discharge can be obtained for each rail condition, as shown in
Table 14.2. For all practical purposes, all measured values exceed the minimum values listed in the table, and thus an indication of the relative slipperiness of the various forms of
surface contamination is given. Two energy levels are used to produce sparking (200 and
400 J/cm²), and the percentage improvement in adhesion is obviously greater at the higher
level. The naturally-wetted and oily rails show the largest percentage improvement after
sparking.

TABLE 14.2. MINIMUM ADHESION VALUES (1% CUMULATIVE FREQUENCY)

Initial rail condition	Coefficient of adhesion			Percentage increase in adhesion after spark discharge	
	Without sparking	After sparking			
		200 J/cm²	400 J/cm²	200 J/cm²	400 J/cm²
Dry	0.296	0.358	0.440	36	54
Naturally wetted	0.108	0.313	0.410	190	280
Flooded	0.214	0.313	0.410	49	92
Oily	0.130	0.286	0.369	120	184

Wear Effects

Our limited discussion of rail–wheel adhesion would not be complete without a brief
mention of wear effects both in the wheel and track. For new wheel and rail profiles, the
contact area is little more than 300 mm² (about the size of a new fivepence piece), and contact pressures approach the yield limit in compression of the railhead material. Plastic
flow of metal occurs both at the surface between wheel and rail, and more readily at a small
depth within the rail profile itself. This accounts for the smooth and shiny appearance of
rails in service compared with the rusted appearance of those which have been disused for
some time. According as wear proceeds in both steel tyre and railhead, the contact area
widens and becomes elliptical in a transverse sense. Figure 14.16 shows the relative profiles
and contact conditions for three combinations of rail and wheel in both cornering and
straight running. We observe that wear of the rail top causes a progressively increasing slope
of its contact surface with respect to the horizontal. The corresponding wear of the steel
tyre is shown clearly in Fig. 14.17. Wear of both tyre and rail increases the slope of the contact interface from about 1 in 20 for the new condition to 1 in 6, while the centre of contact
remains in about the same position (this is indicated by the arrows in Fig. 14.16). If the rail
but not the wheel is now re-ground to the original profile, this slope remains at about the
value 1 in 6, but the centre of contact moves inward with respect to the vehicle or in the direc-

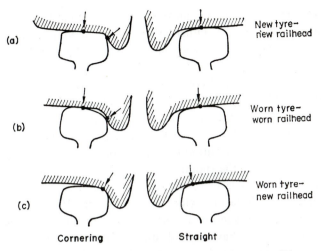

FIG. 14.16. Effects of wheel and rail wear on contact conditions.

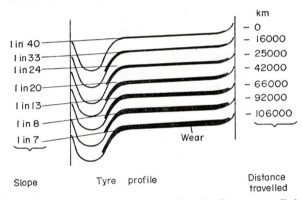

FIG. 14.17. Change in steel-tyre profile with distance travelled.

tion of the wheel flange as shown. At the same time, the cornering condition produces only one contact spot for this combination compared with the normal two contact positions when rail and tyre wear progressively in combination. We conclude that wear has a distinct effect in changing the geometry of the contact area at the wheel–rail interface, and re-grinding of both profiles may be necessary to restore initial conditions. The slope of the contact area and flange wear have also a pronounced effect on the dynamic and hunting performance of rail vehicles and ultimately on cornering stability. Further elaboration here is beyond the scope of this book.

14.3 Land Locomotion

We have considered the tribological aspects of wheel locomotion in this and the preceding chapters, and we have seen that the wheel may consist either of a flexible or rigid structure (compare the pneumatic tyre and steel locomotive wheel). We now regard the wheel in a broader and more fundamental context as perhaps the most significant means of land locomotion in our modern world. Other means of locomotion on land exist[72] such as walking, crawling, or skiing, and the frictional aspects of these methods will be dealt with briefly here.

The exact origin of the wheel, of course, remains uncertain to the present day. It dates at least as far back in history as 3200 BC, when it was used on chariots and carriages to transport warlords and chieftains. The obvious disadvantages of the wheel on soft terrain led to the development of paved roads, which reached a high level of advancement in ancient Roman times. The industrial revolution in Europe found a ready need for the wheel, and it appeared in diverse sizes in the rudimentary machinery of the age.

The locomotion problem of a wheel on deformable terrain (such as soft soil or sand) finds particular application in farm tractors and land rovers. Here the properties of the soil are of primary importance in determining the tractive capability or, indeed, flotation of the vehicle wheels. Soils in general have both plastic and frictional properties, and the shearing stress τ in a soil depends on the coefficient of cohesion c and angle of friction φ[72] according to the equation

$$\tau = c + \bar{p} \tan \varphi, \tag{14.3}$$

where $\bar{p}$ is the mean loading pressure acting at the wheel–soil interface. Plastic masses such as saturated clays or certain types of melting snow can be considered to have zero angle of friction φ, and thus $\tau = c$. On the other hand, the more common granular type of soil exhibits no cohesion or internal bonding ($c = 0$), so that $\tau = \bar{p} \tan \varphi$. One of the complications in attempting to apply eqn. (14.3) to an actual soil is that the relative proportion of c to $\bar{p} \tan \varphi$ is very sensitive to water content. Furthermore, soils generally lack the homogeneity which is characteristic of other deformable materials, so that Coulomb's eqn. (14.3) must be applied with caution.

Consider the traction of a ribbed tyre in soft terrain, as shown in Fig. 14.18. Both theory and experience indicate that the spuds or ribs of elastic tyre-treads clog with soil once they are in contact with the ground, and that they play an insignificant and secondary role in developing tractive effort. We note that this contrasts with the effect of tread pattern on automobile tyres (see Chapter 13). The main effect of the ribs is to increase the effective

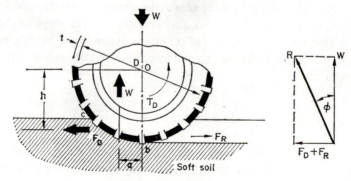

FIG. 14.18. Locomotion of a ribbed tyre on soft soil.

wheel diameter from D to $D+2t$, where D is the undeflected smooth tyre diameter and t the rib depth as shown. Let T_D be the driving torque applied to the wheel and F_D the average value of tractive force developed at distance h below the wheel centre. Then, taking moments about the centre O,

$$T_D = F_D h + Wa, \tag{14.4}$$

where a is the distance forward of the wheel centre at which the load reaction vector W is effective at the tyre–soil interface. Dividing both sides of this equation by h,

$$T_D/h = F_D + F_R, \tag{14.5}$$

where $F_R = W(a/h)$ is the rolling resistance of the wheel. We note that the ratio a/h for locomotion in soft terrain is considerably larger in magnitude than that corresponding to the more usual tyre-on-road situation. This is because the load is effectively supported on soft soil only from the point of initial contact to the maximum penetration position (i.e along the arc bc in Fig. 14.18). If we now divide both sides of eqn. (14.5) by W, we obtain coefficients of friction rather than frictional forces on the right-hand side. Although the F_D and F_R terms in eqn. (14.5) appear to have the same sign, each is taken in a positive sense according to the directions indicated in Fig. 14.18; thus they are, in fact, numerically subtractive. With this in mind, we may divide each term in eqn. (14.5) by the contact area A thus:

$$\frac{T_D h}{A} = \tau - \bar{p} f_R = \bar{p}(\tan \varphi - f_R) \tag{14.6}$$

from eqn. (14.3) for granular type soils ($c = 0$). The shear strength τ of the soil at the shearing boundary is equal to F_D/A, $\bar{p} = W/A$, and the coefficient of rolling resistance f_R $=F_R/W$. Equation (14.6) shows clearly that the application of driving torque T_D to the wheel is a function of soil frictional properties and normal loading. It is important to assess the bearing capacity or "flotation" characteristics of soft soils and terrain,[72] and this is expressed quantitatively by $\bar{p}$ in the last equation.

Skis on Ice

The friction of miniature skis on ice has been dealt with very briefly in Chapter 5, and it has been shown in Chapter 7 that hydrophobic skis give lower sliding coefficients than hydrophilic skis. We now examine the friction of skis on ice and snow in more detail. The lowest coefficients are recorded near the melting point of ice or snow (0°C at atmospheric pressure), and Table 14.3 below indicates the relative values of f_A for different ice and snow conditions in the case of a sliding pneumatic tyre. We observe a remarkable tenfold variation in coefficient from packed snow to wet ice, even though pressure, temperature, and sliding speed remain constant. The magnitude of the coefficient f_A may differ from the values listed in Table 14.3 when skis rather than tyres are involved, but the relative ratings will remain unchanged.

TABLE 14.3. TYPICAL COEFFICIENTS OF FRICTION FOR ICE OR SNOW AT 0°C

Ice–Snow condition	f_A	Speed range	Remarks
Packed snow	0.20	8–65 km/hr	Slight increase of f_A with speed, 0.0007 per km/hr
Rough Ice	0.12	8–32 km/hr	
Glare Ice	0.057	0–32 km/hr	Constant with speed
Wet Ice	0.02	—	

Consider the case of a waxed ski sliding on packed snow, as shown in Fig. 14.19. The effect of waxing is to create a hydrophobic surface condition on the ski, so that there is a tendency to repel water droplets formed by pressure melting. The contact angles between the moving waxed surface and an individual water droplet are approximately 84° and 66° as

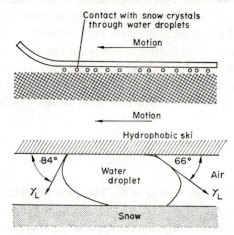

FIG. 14.19. The contact of a hydrophobic ski on snow.[72]

shown, these values being independent of speed above a minimum speed value of 0.4 mm/s.[72] It is apparent that since the values of the advancing and receding contact angles are different, there will be a net retarding force acting on the ski due to surface tension effects. The magnitude of this force F_{ST} may be estimated as follows.

The surface tension force per unit length of drop periphery γ_L is related to the pressure within the droplet p_W by the following well-known equation

$$p_W = 4\gamma_L/d, \tag{14.7}$$

where d is the droplet diameter. If we now consider the horizontal components of the surface tension forces in Fig. 14.19 we obtain the capillary drag F' per unit length of drop periphery, thus:

$$F' = \gamma_L(\cos 66° - \cos 84°) = 0.302\gamma_L$$

and the total horizontal capillary drag F is given by:

$$F = F'd = 0.302\gamma_L d. \tag{14.8}$$

Let there be N_0 droplets (each of diameter d) per unit area of ski surface, and let the contact area of one drop be $(\pi d^2/4)$. Then

$$k = (\pi d^2/4)N_0, \tag{14.9}$$

where k is a measure of the degree of wetness of the ski surface. The total capillary drag per unit area of ski can then be obtained from eqns. (14.8) and (14.9) thus:

$$F_{ST} = N_0 F = 0.386 k \gamma_L/d. \tag{14.10}$$

By substituting for γ_L from eqn. (14.7) into eqn. (14.10),

$$F_{ST} = 0.096 p_W k. \tag{14.11}$$

Finally, by observing that the ski loading per unit area $\bar{p} = k p_W$, we can find from eqn. (14.11) the shear stress τ_{ST} due to capillary drag:

$$\tau_{ST} = F_{ST}/\bar{p} = 0.096 = f_{ST}. \tag{14.12}$$

From eqn. (14.7), the diameter of a water droplet can be estimated to be about 30 μ.

In practice, the size and number of droplets and also the capillary drag do not depend on $\bar{p}$ alone. The water content at the ski–snow interface depends primarily on temperature and on the frictional conditions of the ski and snow surfaces. The latter, of course, through heat generation produce water droplets of a certain size at different locations within the area of contact. Since F_{ST} in eqn. (14.12) is the capillary drag per unit area of ski, the shear stress τ_{ST} can also be regarded as a frictional coefficient. We suspect that the value of 0.096 is perhaps too great to be realistic, partly because of the limiting assumptions made in the analysis (in particular the neglect of temperature effects). This is reflected in the fact that the very small droplet size obtained can scarcely be detected with the naked eye. It is certain that according as the droplet diameter d increases, the capillary drag F_{ST} falls to negligibly small values according to eqn. (14.10).

The contribution of viscous drag to ski retardation is generally considerably greater than that of surface tension effects. From Newton's law of viscosity[†] we can write for the shear stress τ_V due to viscous drag

$$\tau_V = \mu(V/d), \tag{14.13}$$

where V is the forward velocity of the ski and μ the absolute viscosity of water. We can substitute for d in this equation from eqn. (14.7) previously. Thus the coefficient of viscous friction f_V becomes

$$f_V = \frac{\tau_V}{p_W} = \frac{\mu V}{p_W d} = \frac{\mu V}{4\gamma_L}. \tag{14.14}$$

Putting $V = 25$ m/s, $\gamma_L[‡] = 70$ g/s^2, and $\mu[‡] = 1.83$ centipoise, we find that $f_V \doteq 0.163$.

Again, the complexity of the problem may be seen in the fact that we have assumed complete wettability over the surface area of the ski in the above calculation. In practice the

[†] See Chapter 6.   [‡] For water at 0°C.

problem may be compounded by boundary and mixed-contact conditions[†] particularly because of the presence of water vapour. This type of lubrication is produced by extremely thin layers which reduce the friction between ski and snow considerably. Apart from these considerations, conditions will generally be different in the front part of the contact area under the ski compared with the rear end. This is due to the fact that since the melting of interfacial snow requires a certain frictional heat input, the snow particles located at the rear of the ski may find themselves surrounded by more water droplets than those located in the front, due to a longer exposure to the frictional process. Thus the contribution of viscous drag will be more significant towards the rear of contact. At the front of the ski a practically solid friction is to be expected particularly at higher sliding speeds. *Indeed, it is quite probable that the distribution of frictional forces along the contact area of a ski will show a predominance of solid friction towards the front, surface tension in the middle, and viscous drag at the rear.*

We can accomplish more by changing the physical and chemical properties of the runner than by increasing the average pressure $\bar{p}$ under the ski (the objective in the latter case is to generate more water at the frictional interface, and thus produce a better lubricating action). This is illustrated clearly by Fig. 14.20. In most cases, the decrease in f obtained by an increase in $\bar{p}$ is quite small and sometimes zero as shown, whereas a fourfold decrease in sliding friction can be obtained by changing the ski material from dural to bakelite. Other variables which affect ski friction are the surface finish and hardness of the ski material. Hard steel runners produce a greater melting effect than soft runners, particularly at sub-

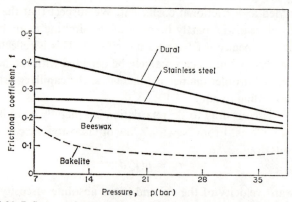

FIG. 14.20. Influence of pressure and ski material on frictional performance.

freezing temperatures. Smooth surfaces may reduce friction close to 0°C compared with rough surfaces, but the opposite effect will be evident at lower temperatures. The reasoning here is that at very low temperatures the actual pressure supporting the load on the ski is greater (see eqn. (4.26)) for a rough ski, and this promotes more efficient pressure-melting of the snow.

Other means of locomotion on snow and ice include sleighs and toboggans, and the same basic mechanism of friction as in the case of the ski is applicable.

[†] See Chapter 7.

The Mechanics of Walking

Walking and crawling are the basic means of locomotion exclusive to the animal kingdom, and their complete study would require not only the application of engineering principles (such as kinematics, dynamics, strength of materials, and energy utilization) but also medical technology, biology, and even anthropology. From the tribologist's viewpoint, the mechanism of walking offers two distinct types of frictional interface which require a fundamental understanding of the variables which make possible the rudimentary act of self-propulsion. These interfaces comprise (a) animal and human joints (shoulder, hip, knee, etc.), and (b) foot-to-ground contact.

We observe that the elementary act of walking would be impossible if either of these interfaces did not exist or failed to behave as required. Thus such crippling diseases as arthritis or paralysis can be regarded as a breakdown of lubrication at the first interface, whereas climatic conditions (such as glare ice) can produce a breakdown of adhesion at the second interface. The friction and lubrication of animal joints will be dealt with in detail in Chapter 16, and we will confine our attention here very briefly to the interface between sole (or heel) and pavement.

Under dry conditions, adhesion between sole and pavement surface is readily developed, and there is little probability of uncontrolled slip. However, when a lubricant such as water or slush is present on pavements, a squeeze film is created between sole and pavement surface which prevents immediate contact between the two. The time of approach of sole (or heel) to pavement is directly proportional to lubricant viscosity[†] and the latter is greatly increased by the presence of dust–water emulsions or melting ice. This approach time must be less than the contact time of shoe on pavement for normal walking on dry surfaces, otherwise slip will occur. By incorporating a surface texture in sole and heel (this usually takes the form of transverse serrations or grooves or sometimes a block-type pattern) we can usually ensure that firm contact is established between shoe and pavement even under the worst climatic conditions. The materials used in sole and heel have viscoelastic properties, so that walking is also concerned with the elastohydrodynamic[‡] effect between sole and pavement under wet conditions.

One interesting design aspect is whether the shoe manufacturers should incorporate leather in the sole and rubber in the heel or vice versa—assuming that cost is not a determining factor. The coefficient of friction of lubricated rubber on a rough surface is almost twice that of lubricated leather on the same surface, so that we may wonder why rubber-soled shoes are not used more widely. However, it is also true that the rate of wear of rubber soles or heels is much greater than that of leather, so that the thickness and weight of a rubber sole are much greater than a leather sole with the same durability. The logic of choosing optimum shoe materials is therefore as follows. At the critical instant when the heel just makes contact with the pavement surface it is essential that the maximum friction requirement be dominant to avoid any possibility of slip; thus the heel should be made of rubber. Here the added thickness required for a reasonable durability is usually regarded as a desir-

[†] See eqn. (6.12).   [‡] See Chapter 8.

able feature. By contrast, aesthetic and weight limitations dictate that the sole be made of leather, and the frictional requirements are obviously less demanding.

Surprisingly, little information is available on the mechanism of crawling in nature. This type of animal locomotion is even more primitive than walking, and it consists essentially of consecutive foldings and unfoldings of extremities. The motion of caterpillars is believed to be of this general nature,[72] and there may be considerable friction between the body and ground.

14.4 Miscellaneous Transportation Topics

The Action of Tyre Studs

We have dealt extensively with the friction of pneumatic tyres in Chapter 13, and here we consider how tyre-to-ground adhesion can be restored under icy conditions. The function of tyre studs is essentially to penetrate or bite into hard ice, so that the tyre as a whole establishes a grip on the surface during rolling. Consider the action of a cylindrical stud in ice as shown in Fig. 14.21(a). The friction mechanism can be likened to the ploughing of a hard metal through a softer matrix, as described in Chapter 3. We can therefore write for the total frictional resistance F experienced by one stud,

$$F = F_{adh} + F_{plough} \qquad (3.15)$$

where the adhesion component F_{adh} is caused by friction between stud and ice along the base of the stud in Fig. 14.21(a). Let us assume that in moving forward with velocity V the ice ahead of the stud is continuously sheared along a plane AB. Thus a segment ABC of ice will move with the stud and will plough its way through the "upstream" ice layer. The ploughing component of friction F_{plough} in eqn. (3.15) can then be described by

$$F_{plough} = \tau_f A \cos \alpha \qquad (14.15)$$

where τ_f is the ultimate shear strength of ice and the area A is given by

$$A = \underbrace{d(h \operatorname{cosec} \alpha)}_{\substack{\text{Area of inclined} \\ \text{shear surface,} \\ AB}} + \underbrace{2(\tfrac{1}{2}h)h \cot \alpha}_{\substack{\text{Area of sides of} \\ \text{inclined shear plane} \\ \text{(twice } \triangle ABC)}} \qquad (14.16)$$

$$= h(d + h \cos \alpha), \operatorname{cosec} \alpha.$$

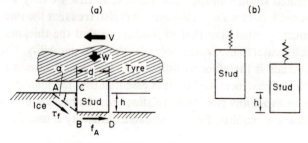

Fig. 14.21. The friction of a tyre stud on ice.

Here d is the stud diameter and h its height of protrusion from the tread matrix. The load W causing penetration of the stud into the ice is provided by the elasticity of the tyre, and we can write

$$W = C_0 - C_1 h, \qquad (14.17)$$

where C_0 and C_1 are constants. We note that the effective spring constant causing penetration of the stud decreases with increasing h: this is indicated schematically in Fig. 14.21(b), and the corresponding reduction in W is noted from eqn. (14.17). Putting $F_{adh} = f_A W$ and making use of eqns. (14.15)–(14.17) in eqn. (3.15), we finally obtain for the total friction force F acting on the stud

$$F = f_A(C_0 - C_1 h) + \tau_f h(d + h \cos \alpha) \cot \alpha$$
$$= f_A C_0 + (\tau_f d \cot \alpha - f_A C_1)\, h$$
$$+ (\tau_f \cos \alpha \cot \alpha)\, h^2$$

which has the form

$$F = C_2 + C_3 h + C_4 h^2. \qquad (14.18A)$$

Typical values of h_{max} and d are 1.5 mm and 2.5 mm respectively. Putting $\tau_f = 7$ bar, $W_{max} = C_0 = 16$ kg, $C_1 = 10.67$ kg/mm, $f_A = 0.10$, and $\alpha = 45°$,

$$F = 1.6 - 0.89h + 0.05h^2, \qquad (14.18B)$$

where F is measured in kilograms and h in millimetres. We note that F is greater at smaller stud protrusion values h.

This analysis neglects temperature effects at the boundary between the stud and the ice, yet important conclusions may be derived from the model in Fig. 14.21. In the first place, let us consider changes in h as being due to varying loading conditions in the contact zone. The spring mounting of the studs within the tyre matrix permits higher values of F_{adh} at maximum compression of the stud (i.e. when h is a minimum). At the same time, the value of F_{plough} is obviously reduced. The analysis shows clearly that the overall effect is to increase the total friction force F on each stud according as h is reduced. Thus the stud performs better in terms of providing an adhesion force in the contact area at points of greater contact pressure.

Let us now briefly examine the effects of stud wear in the model. This can be readily simulated by considering that changes in h which occur do not affect the stud load W. If we therefore neglect eqn. (14.17), the analysis leads to the following relationship between F and h:

$$F = 1.6 + 0.18h + 0.05h^2. \qquad (14.18C)$$

Here it is clear that according as h reduces because of stud wear, the value of frictional force F is correspondingly diminished. Thus if the value of h is reduced by 50% from 1.5 to 0.75 mm, the value of F is reduced by 11%.

In practice, tyre studs have been found to be most effective in restoring tyre–ground adhesion under icy conditions. In the case of soft snow, the use of snow treads and chains is recommended. Tyre studs, however, have one very great disadvantage in that the biting action continues on road surfaces when the ice has disappeared. This causes an accelerated deterioration of road surfaces, and it has become such a serious problem in countries with severe winter conditions that tyre studs have been severely restricted and sometimes banned altogether.

Ball-race Friction

The question of friction in ball-bearing races has particular relevance to transportation, and in most cases the frictional energy dissipated is negligibly small. However, when the speed of rolling of the balls in the races is sufficiently high, the frictional heat generated can create an appreciable temperature rise. This is especially true in jet engines.

Consider the main shaft of a jet engine rotating at high speed in ball-bearing races, as shown schematically in Fig. 14.22. The races themselves experience radial vibration forces in addition to rolling-ball motion, and it is therefore necessary to provide an outer squeeze-

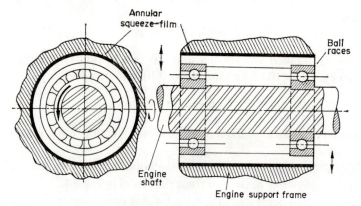

FIG. 14.22. Ball-bearing races and annular squeeze film in jet engine.

film annulus as shown. This effectively absorbs most of the unwanted vibrations from the engine, so that they are not transmitted to the support structure. From a design viewpoint, we desire to know the energy dissipated per unit time because of ball-race friction, and to compare this with the energy dissipated in the squeeze-film annulus in absorbing engine vibrations.

Let the radial thrust[†] T on the balls in the jet engine of a Boeing 707 aircraft be 2750 kg and the rotational speed N of the shaft is 12,000 r.p.m. The diameter D of the bearing is 20 cm and the coefficient of rolling friction f between the balls and the outer race may be taken as 0.0025. The path of the balls per shaft revolution is

$$L = \tfrac{1}{2}\pi D = 10\pi \text{ cm},$$

and the force of friction is given by

$$F = fT = 0.0025 \times 2750 = 6.875 \text{ kg}.$$

[†] A radial thrust is necessary to ensure that the rolling mode of operation is preserved in ball-bearing races. Otherwise there is a possibility of "ball-skid" which causes flat spots and drastically reduces the life of the bearing.

Thus the work done by friction at one rolling interface per shaft revolution is

$$W_F = FL = 2.16 \text{ kJ/m width.}$$

Consider both interfaces (at the inner and outer bearing races); the rate of energy dissipation due to rolling resistance is

$$E_{d_1} = 2W_F N = 864 \text{ kW/m.} \tag{14.19}$$

Now in the case of the squeeze-film damper, let the film thickness h be 3 mm and the absolute viscosity μ of the lubricant 1.75 poise (at the operating temperature of 150°C). From Table A in the Appendix, the damping constant c for a complete journal-type bearing without rotation is given by

$$c = \frac{3\pi\mu}{2(h/D)^3},$$

where the width b is unity, $R = \frac{1}{2} D$, and $H = 0$ for small displacements. Thus

$$c = \frac{3\pi \times 1.75}{2(3/200)^3} = 2.44 \times 10^6 \text{ poise/m width.}$$

The energy dissipated per cycle (assuming a sinusoidal disturbance) is given by $\pi\omega c x_0^2$, where ω is the frequency of rotation in radians per second and x_0 is the amplitude of vibration. For a typical value of $x_0 = 100 \mu$, the total rate of energy dissipated in the squeeze film is given by

$$E_{d_2} = \pi N\omega c x_0^2 = \pi \times 200 \frac{\text{rev.}}{\text{sec.}} \times 400\pi \frac{\text{rad}}{\text{sec.}} \times 2.44 \times 10^6 \frac{\text{poise}}{\text{m-rev.}} \times 10^4 \mu^2 = 1.923 \text{ kW/m.}$$

Thus the rate of energy dissipation within the squeeze film due to the absorption of radial vibrations is generally small compared with ball-race friction. The replacement of the squeeze-film annulus by a high-temperature elastomer can be shown[2] to give little difference in energy absorption rate, and the elastomeric annulus has the advantages of not requiring a pressurizing system to compensate for hydraulic leakage losses. The relatively high rate of energy dissipation by ball friction is partly the result of extremely high rolling speed and is partly due to very large radial loading in the jet-engine example.

Air-cushion Vehicles

We finally consider briefly the general category of air-cushion vehicles, irrespective of whether these constitute land or water transport. The common examples of water transport in mind are hovercraft, captured-air-bubble† boats, or simply conventional ships with an air-film hull. Land transport applications include tracked hovercraft, high-speed "levapad"

---

† Captured-air-bubble boats are conventional hulls with long sideboards or skegs to contain an air bubble which greatly reduces hull-to-water drag.

trains, etc. Most of these vehicle types correspond to a known lift–drag ratio, and they operate in a given speed range so that they appear on a graph such as in Fig. 14.23. The graph includes hydrofoils, helicopters, conventional ships, and high-speed planing boats so that a broad comparison is feasible. The diagonal limit in the right-hand upper corner is the well-known Gabrielli–von Karman envelope line representing the existing state-of-the-art in terms of maximum efficiency. All vehicles fall to the left of this limit, but the closer a given transport system is to the envelope the greater its efficiency.

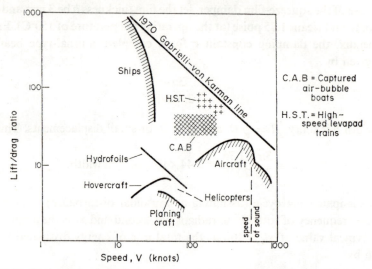

FIG. 14.23. Comparison of lift–drag ratio vs. speed for various land, sea, and air vehicles.

Two well-known modes of transport (ships and aircraft) are close to the Gabrielli line, but they represent extremes of speed in the general transport spectrum. In the intermediate range of velocities, high-speed trains of the levapad air-cushion type and captured-air-bubble vessels have high lift–drag ratios as seen. In the case of high-speed levapad trains and tracked hovercraft, the thickness of the air film is very small compared with general vehicle dimensions, and the tangential force required to shear the film is generally an order of magnitude less than that corresponding to the rolling friction of the finest precision ball-bearings. The drag is therefore considerably reduced, and the lift–drag ratio is high. Part of the savings in frictional power obtained by the use of an air film must be made available as an air cushion to support the vehicle, but in the levapad concept this represents a very small fraction. As we proceed from very small air-film thicknesses to the relatively large air cushions which are characteristic of cross-channel hovercraft, the net power savings at moderate speeds may be largely eroded because of the large power requirement to sustain the cushion. Here the advantage lies in achieving higher forward speeds with little additional expenditure of energy, since the propulsive power is only a small fraction of the support power.

CHAPTER 15

BEARING DESIGN

We HAVE defined viscosity in Chapter 10 as an internal friction mechanism which plays a fundamental role in the various lubrication régimes that we may encounter in engineering practice. These régimes include fully hydrodynamic, boundary, mixed-contact, and elasto-hydrodynamic lubrication as described in Chapters 6, 7, and 8. Bearing design is perhaps the single, outstanding example of the application of hydrodynamic lubrication theory to the design of modern machinery, and in this chapter we will describe in detail the operating principle of the most common bearing types. In general, we can classify bearings in three broad categories according to Table 15.1. This list is by no means complete,[†] but the system of classification appears to be the most logical available.

TABLE 15.1. CLASSIFICATION OF BEARING TYPES

Classification	Description
Fluid-film bearings	Planar motion (plane, curved, step, and composite sliders, pivoted-pad sliders, sector-pad sliders)
	Rotational motion (full and partial journal bearings, floating ring bearings, foil bearings)
Rolling-contact bearings	Roller bearings
	Ball-bearings (radial and thrust loads)
Miscellaneous	Gas bearings, porous bearings, magnetic and electrostatic bearings, hydrostatic bearings.

15.1 Slider Bearings

The velocity distribution under a plane thrust slider has been shown in Chapter 6 to consist of hydraulic and shear components, and these have been superimposed graphically in Fig. 6.3. Let the angle of inclination α of the inclined plane slider be defined by the relationship

$$\alpha \doteq \tan \alpha = \frac{h_2 - h_1}{L} \tag{15.1}$$

† Squeeze-film bearings have not been included.

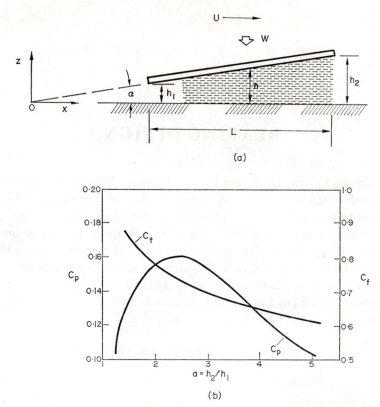

Fɪɢ. 15.1. Plane thrust slider showing (a) coordinate system, and (b) load capacity and friction coefficients.

as indicated in Fig. 15.1, where L is the length of the slider measured horizontally and h_1, h_2 are the minimum and maximum values of lubricant film thickness at the slider extremities. We also define $a = h_2/h_1$ for convenience. From the coordinate system in Fig. 15.1(a), the film thickness h at any location x beneath the slider is given by

$$h = \alpha x = \frac{h_1(a-1)}{L}x. \tag{15.2}$$

Now, Reynolds's equation (6.4) for one-dimensional flow reduces to the following simplified form:

$$d\left(h^3\frac{dp}{dx}\right) = 6\mu U\, dh$$

assuming incompressibility and neglecting the stretch and squeeze terms on the right-hand side of the original equation. By integrating both sides,

$$\frac{dp}{dx} = 6\mu U\left(\frac{h-h_0}{h^3}\right). \tag{15.3}$$

We now substitute for h from eqn. (15.2) into eqn. (15.3) and integrate between the limits $x_2 = h_2/\alpha$ and $x_1 = h_1/\alpha$ thus:

$$p = \frac{6\mu U}{\alpha^3} \frac{(\alpha x - h_1)(h_2 - \alpha x)}{(h_1 + h_2) x^2} \quad \text{for} \quad \frac{h_1}{\alpha} \leqslant x \leqslant \frac{h_2}{\alpha}. \tag{15.4}$$

Since the load W carried by the slider is the integral of pressure over the slider area, we obtain from eqn. (15.4)

$$W = B \int_{h_1/\alpha}^{h_2/\alpha} p \, \mathrm{d}x = C_p \left(\frac{\mu U L^2 B}{h_1^2} \right), \tag{15.5}$$

where the load coefficient C_p is given by

$$C_p = 6 \left(\frac{1}{a-1} \right)^2 \left[\ln a - \frac{2(a-1)}{a+1} \right] = \varphi_1(a).$$

To find a corresponding expression for the friction force F experienced by the slider in shearing the lubricant, we first re-write the expression for the x-component of velocity within the lubricant film from eqn. (6.6) with $U_1 = U$ and $U_2 = 0$ thus:

$$u = \frac{1}{2\mu} \left(\frac{\mathrm{d}p}{\mathrm{d}x} \right) z(z-h) + \frac{z}{h} U.$$

By differentiating this expression with respect to z and then putting $z = h$, we obtain for the shear stress τ at the slider surface

$$\tau = \mu \left(\frac{\mathrm{d}u}{\mathrm{d}z} \right)_{z=h} = \frac{h}{2} \left(\frac{\mathrm{d}p}{\mathrm{d}x} \right) + \frac{\mu U}{h}. \tag{15.6A}$$

We now substitute for $(\mathrm{d}p/\mathrm{d}x)$ from eqn. (15.3) in eqn. (15.6A),[†] and then substitute the resulting expression for τ in the following equation:

$$F = B \int_{h_1/\alpha}^{h_2/\alpha} \tau \, \mathrm{d}x = C_f \left(\frac{\mu U L B}{h_1} \right), \tag{15.7}$$

where the friction factor C_f is given by

$$C_f = \frac{2(a^2 - 1) \ln a - 3(a-1)^2}{3(a+1)} = \varphi_2(a).$$

Both the load capacity coefficient C_p and the friction factor C_f depend only on the value of a, and this is shown diagrammatically in Fig. 15.1(b). We observe that there exists a critical value of a at which load support is a maximum, and this is obtained by setting $dW/da = 0$ in eqn. (15.5). For the plane slider configuration in Fig. 15.1(a), this critical

---

[†] Taking care to substitute for h from eqn. (15.2) in the expression for $(\mathrm{d}p/\mathrm{d}x)$.

value is 2.2. We can deduce intuitively that since load support within the bearing must necessarily be zero at the extreme positions where the slider is parallel ($\alpha = 0$) and perpendicular ($\alpha = \frac{1}{2}\pi$) to the base surface, then there must exist a maximum value of C_p at some intermediate angle of inclination. The friction factor C_f continues to decrease progressively as the value of a rises or as the angle α increases. This implies simply a degeneration of pure frictional drag in the parallel slider position to the general case of combined frictional and pressure drag for finite angles of inclination α. Furthermore, the percentage of frictional drag reduces with increasing inclination. Putting $a = 2.2$ in the expressions for C_p and C_f,

$$\left.\begin{aligned} W_{max} &= 0.1602 \frac{\mu U B L^2}{h_1^2}, \\ F_{max} &= 0.753 \frac{\mu U B L}{h_1}, \end{aligned}\right\} \tag{15.8}$$

from which the coefficient of friction f is given by

$$f = \frac{F_{max}}{W_{max}} = 4.7 \left(\frac{h_1}{L}\right). \tag{15.9}$$

Furthermore, the value of the integration constant[†] h_0 in eqn. (15.3) is given by

$$h_0 = 1.37 h_1,$$

where h_1, as we have seen, is the minimum film thickness.

Let us now suppose that we replace the plane with a curved slider, where the nature of the curvature is given either by the power law relationship

$$h = mx^n \tag{15.10}$$

or by the exponential function

$$h = h_1 e^{\beta x}, \tag{15.11}$$

where m, n, and β are constants. These last two equations are now used in turn to replace eqn. (15.2) for the plane slider, and the same procedure is followed. Table 15.2 shows a

TABLE 15.2. LOAD CAPACITY OF PLANE AND CURVED SLIDERS

Type of slider		Equation for film thickness	Critical value of a	$C_{p_{max}}$
Plane slider		$h = \frac{h_1(a-1)x}{L}$	2.2	0.1602
Curved	Power-law slider	$h = mx^2$	2.3	0.163
	Exponential slider	$h = h_1 e^{\beta x}$	2.3	0.165

[†] This is the value of film thickness at which $dp/dx = 0$.

comparison of plane and curved sliders in terms of maximum load capacity. We observe that the numerical results are almost identical. *Thus, having fixed the magnitudes of the minimum and maximum film thicknesses h_1 and h_2 and the dimensions of the slider, the exact shape of the lubricant film is relatively unimportant.* This important conclusion greatly simplifies the solution of elastohydrodynamic film problems where the precise film-thickness distribution is unknown.

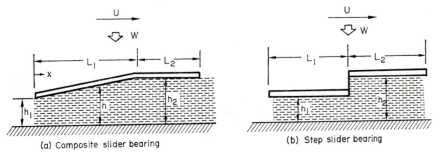

(a) Composite slider bearing   (b) Step slider bearing

FIG. 15.2. Composite slider and step slider bearings.

Two other types of slider bearing are the composite and step bearings shown in Fig. 15.2. The composite slider bearing is a combination of inclined and parallel plane sliders for which the following film-thickness distribution is valid:

$$h = \alpha x \text{ (region } L_1) \quad \text{and} \quad h = h_2 \text{ (region } L_2).$$

Two parallel plane sliders with a step change in film thickness near the centre comprise the step slider in Fig. 15.2(b). Here we may write for the film thickness:

$$h = h_1 \text{ (region } L_1) \quad \text{and} \quad h = h_2 \text{ (region } L_2).$$

By carefully matching boundary conditions at the transition point in the film thickness for each of these bearing types, we can evaluate the load capacity in each case as before.

Pivoted-shoe Bearings

Let us assume that the plane slider in Fig. 15.1(a) instead of maintaining a fixed inclination α is supported by a pivot as shown in Fig. 15.3, and is thus free to assume any angle. The resultant force on the slider must act through the pivotal point, but if the latter coincides

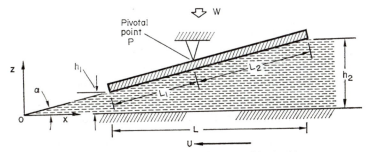

FIG. 15.3. Pivoted-shoe or tilting-pad slider bearing.

with the centre of pressure for the fixed-pad system in Fig. 15.1(a), then there is no essential difference between the analysis for both types of bearing. For convenience, we assume the pivot support is fixed and that the lower surface moves to the left with velocity U as shown. This is dynamically equivalent to the slider moving with velocity U to the right relative to a fixed base as treated earlier.

The centre of pressure for the tilted pad in either Fig. 15.1(a) or Fig. 15.3 is given by the relationship

$$\bar{x}W = B \int_{h_1}^{h_2} px \, dx, \tag{15.12}$$

where B is the width of the shoe perpendicular to the paper and $\bar{x} = OP$ is the distance from the origin of coordinates to the pivotal point P. By using integration by parts on the right-hand side of this equation,

$$\bar{x} = \frac{B}{2W} \left(px^2 \Big|_{h_1}^{h_2} - \int_{h_1}^{h_2} x^2 \frac{dp}{dx} \, dx \right),$$

where the first term disappears because of zero pressure at the extremities of the slider. We now substitute for dp/dx from eqn. (15.3) thus:

$$\bar{x} = \frac{3\mu UB}{W} \left\{ \int_{h_1}^{h_2} \frac{x^2 \, dx}{h^2} - h_0 \int_{h_1}^{h_2} \frac{x^2 \, dx}{h^3} \right\}. \tag{15.13}$$

Before the integration in eqn. (15.13) can be carried out, it is necessary to substitute for h from eqn. (15.2). Finally, the expression for W from eqn. (15.5) is used and from eqn. (15.13)

$$\bar{x} = \frac{L(a^2 - 1 - 2a \ln a)}{2(a^2 - 1) \ln a - 4(a - 1)^2} = \varphi_3(a).$$

This expression shows that for any values of the height ratio a, the pivotal point P is located such that $L_2 > L_1$ in Fig. 15.3. Generally speaking, $L_1 \doteq 5L/9$ and $L_2 \doteq 4L/9$,[†] so that the pivotal point is placed approximately five-ninths of the length of the shoe from the leading edge. Should we desire at any stage to reverse the direction of sliding, the pivot position must be at the midpoint of the shoe. It is an experimental fact that the bearing will operate satisfactorily and assume an angle of inclination even with the pivot located at the centre, although this is denied by the theoretical analysis. The most obvious explanation of this behaviour is that the pad surface is never quite flat both because of geometrical imperfections and the loading stresses which attempt to deform the pad surface into a circular or parabolic shape. It is also possible that the leading edge may be slightly rounded so as to force a centrally pivoted shoe to attain its initial tilt attitude as sliding begins.

Another important phenomenon which must be mentioned briefly is the thermal wedge effect. We have seen that the slightest convexity in the pad surface will create an angle of tilt for a centrally pivoted slider. Let us now suppose that no such convexity or leading-edge roundedness exists so that there is no reason for the slider to assume an angle of inclination.

[†] Since α is a relatively small angle, $L = L_1 + L_2$ in Fig. 15.3.

Let us further ensure that parallelism between slider and base exists by locking the slider in the parallel position. Although the Reynolds equation (see Chapter 6) predicts no load support for sliding parallel surfaces which are both rigid and smooth, there is evidence that some load-carrying effect in fact develops. The theory is that the lubricant that is passing between the parallel surfaces experiences a rise in temperature due to viscous shearing effects. This rise in temperature produces a slight expansion of the lubricant, which attempts to increase its volume as it flows through the clearance space. The result is a pressure build-up and load-carrying capacity, so that a thermal wedge effect is developed where no geometrical wedge exists. The magnitude of the effect is generally small.

Side Leakage Effects

The pressure gradient dp/dx in eqn. (15.3) has been derived from the Reynolds equation by neglecting sideflow in the y-direction. This analysis is valid when the width B of the slider is much larger than its length L. The neglect of sideflow or side leakage means that there is no pressure gradient in the y-direction (i.e. $dp/dy = 0$). As a practical guide, the one-dimensional analysis presented earlier in this section can be applied with sufficient accuracy to real slider bearings if $B \geqslant 4L$. If the width B is less than $4L$, we must allow for the loss in load-bearing capacity resulting from side leakage. Figure 15.4 depicts the nature of the

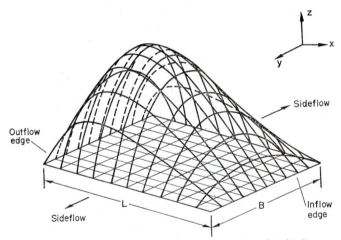

FIG. 15.4. Pressure distribution in slider bearing with sideflow.

pressure distribution beneath a slider where the dimensions B and L are comparable. We must now replace eqn. (15.3) by the more complex relationship

$$\frac{\partial}{\partial x}\left(h^3 \frac{\partial p}{\partial x}\right) + \frac{\partial}{\partial y}\left(h^3 \frac{\partial p}{\partial y}\right) = 6\mu U \frac{dh}{dx}. \qquad (15.14)$$

We cannot proceed to solve this equation directly, and we usually, therefore, obtain approximate solutions mathematically or by the use of analogies,[34] or more commonly we make use of numerical methods. One very approximate method involves the assumption of the following pressure distribution:

$$p = \frac{36W}{L^3B^3} xy(L-x)(B-y),$$ (15.15)

where the origin of coordinates is taken at one of the plate corners and x and y are measured along adjacent edges. Equation (15.15) has the form of a three-dimensional parabola, and it satisfies the boundary conditions that $p = 0$ along the four edges of the slider (i.e. at $x = 0$, L and at $y = 0$, B). Furthermore, it obeys the condition that the pressure integral taken over the entire area of the pad is identical with the load W. In using eqn. (15.15), however, we neglect the fact that maximum pressure is not attained at the geometric plate centre, and the edges and corners of the pad introduce a distortion effect in the true pressure distribution which is not accounted for in the equation.

When exact numerical methods are used to solve eqn. (15.14), the loss of load-carrying capacity and film thickness compared with the simple case of no side leakage is given by a numerical factor η less than unity. For zero sideflow, $\eta = 1$, and eqn. (15.5) gives the load capacity for the case $B/L \to \infty$. Table 15.3 lists values of η for different breadth-to-length ratios.

TABLE 15.3. EFFECTS OF SIDE LEAKAGE ON LOAD

Ratio B/L	η	Ratio B/L	η
0.25	0.060	1.33	0.550
0.33	0.090	2.00	0.680
0.50	0.185	4.00	0.835
0.67	0.278	5.75	0.920
1.00	0.440	∞	1.000

Rotational Fixed-pad Bearings

Several of the fixed-pad bearings shown in Fig. 15.1(a) may be grouped together to form a multiple fixed-pad bearing as shown in Fig. 15.5(a). More commonly, the multiple pads are arranged to form a circle as sketched in Fig. 15.5(b). Here the thrust is supplied by immersing the system in a lubricant and by the rotation of an upper disc or runner about the axis of symmetry of the pad sectors.

Let the mean length of any individual pad be L/n, where L is the total length of all the pads measured along a circle, and n is the number of pads. We also suppose that the width B of a pad is given by $(R_2 - R_1)$ in the figure, and that the angle of inclination or taper of each pad is α. If ΔW is the load support contributed by each pad sector, W_{tot} the total load supported by the system in Fig. 15.5(b), and W the loading of a single pad having the same width B, angle α, and total length L as the system, we can write

$$W_{tot} = n \, \Delta W$$

and, from eqn. (15.5),

$$W = n^2 \, \Delta W.$$

From these last two equations, we conclude that

$$W_{tot} = (1/n) \, W, \qquad\qquad (15.16)$$

so that the bearing capacity of the segmented fixed-pad system in Fig. 15.5 is $1/n$ times smaller than that of an equivalent unbroken surface.

Both the fixed-pad rotational thrust bearing and hydrostatic bearings[†] having the same general dimensions find application in the support of large industrial rotors such as waterwheel generators. We note also that they are extremely competitive in such applications.

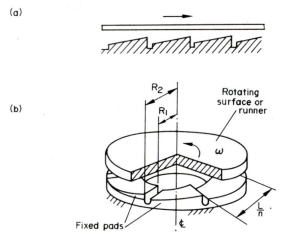

FIG. 15.5. Diagrammatic sketches of (a) multiple fixed-pad linear bearing, and (b) fixed-pad rotational thrust bearing.

The fixed-pad rotational thrust bearing generates its own load support internally, but difficulties arise during stopping and starting, and it is here that substantial wear of the bearing surfaces may occur as a result of solid contact. The hydrostatic bearing is free from these difficulties, but an externally pressurized lubricant is required to support the load during normal operation, and this necessitates additional power. There is also the difficulty of a fluctuating film thickness due to vibrational instability of the bearing surfaces, and control problems arise when two or more hydrostatic bearings are operated from the same pressurized system. On the other hand, the fixed-pad rotational system generates considerably higher temperatures within the lubricant due to the combined action of wedge formation and shearing, and this may be a limiting factor. The advantages and disadvantages of each system must be carefully considered before making a selection in a particular industrial application.

---

† See Section 6.6.

15.2 Journal Bearings

The journal bearing is perhaps the most familiar and useful of all bearing types. It consists fundamentally of a sleeve of bearing material wrapped partially or completely around a rotating shaft or journal, and designed to support a radial load in the presence of a lubricant. In its earliest historical form, the radial load was either zero or very small in magnitude. Thus Newton's concentric cylinder experiment (see Chapter 10) for the measurement of viscosity, and Petroff's bearing (see Chapter 6) constituted perhaps the first journal bearings that we are consciously aware of in history.

Figure 15.6 shows the essential features of a journal bearing. A fixed cylinder or bearing of radius R contains a rapidly rotating journal or shaft having an outer radius r. The space between journal and bearing is filled with an incompressible lubricant of absolute viscosity

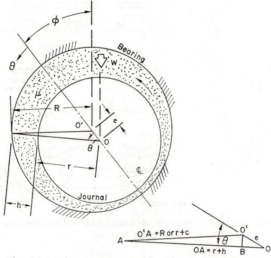

Fig. 15.6. Diagrammatic sketch of journal bearing.

μ, and a load W is supported by the journal. During normal rotation, the centre line OO' of journal and bearing makes an angle φ with the vertical as shown, and its magnitude varies with W. This angle is called the attitude of the journal, and the centre distance OO' is the eccentricity e. For zero loading, $e = 0$ and φ has no meaning: this corresponds to the Newtonian or Petroff bearing previously discussed. On the other hand, $e = 2(R-r)$ and $\varphi = 0$ when W is finite and the journal is at rest. In normal operation, a converging wedge of lubricant generates sufficient pressure to support the load W hydrodynamically.

Let θ define in angular coordinates any position within the lubricant film, and let $c = R-r$ denote the average clearance between journal and bearing. From the triangle OAO' we can then write

$$OA = r+h = OB+BA = e \cos \theta + \sqrt{[(r+c)^2 - e^2 \sin^2 \theta]}$$

$$= e \cos \theta + (r+c) \sqrt{\left[1 - \left(\frac{e}{r+c}\right)^2 \sin^2 \theta\right]}.$$

Now $R = 10^3 e$ and $e \doteq c$, so that the ratio $e^2/(r+c)^2 = 10^{-6}$ and is therefore negligible. Thus

$$h/c = 1+(e/c) \cos \theta. \qquad (15.17\text{A})$$

The ratio (e/c) or ε is called the percentage eccentricity of the bearing. It follows from eqn. (15.17A) that the dimensionless film thickness (h/c) attains maximum and minimum values when $\theta = 0$ and $180°$ respectively.

We assume that the width B of the bearing in Fig. 15.6 is large compared with its radius R, so that end leakage is negligible. Putting $x = r\theta$ in eqn. (15.3), therefore

$$\frac{dp}{d\theta} = 6\mu Ur \left(\frac{h-h_0}{h^3}\right), \qquad (15.18)$$

where h_0 is the value of film thickness at which the hydrodynamic pressure p in the bearing is a maximum, and U is the peripheral speed of the journal. Substituting for h from eqn. (15.17A) into eqn. (15.18) and integrating, we obtain the following expression for p:

$$p = \frac{6\mu Ur}{c^2} \left(\int \frac{d\theta}{(1+\varepsilon \cos \theta)^2} - \frac{h_0}{c} \int \frac{d\theta}{(1+\varepsilon \cos \theta)^3} \right) + K. \qquad (15.19)$$

If we assume that the pressure $p = p_a$ at $\theta = 0$,[†] then the constant K in eqn. (15.19) is numerically equal to p_a. We can then either regard p in eqn. (15.19) as the actual pressure in the bearing as a function of position θ or we can transpose K to the left-hand side of the equation, so that we have a pressure rise term $(p-p_a)$ or Δp. Confusion inevitably arises

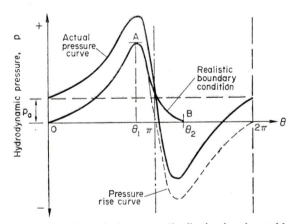

Fig. 15.7. Actual and theoretical pressure distribution in a journal bearing.

when the pressure rise term is frequently given the symbol p rather than perhaps Δp, and the two curves differ by the constant amount p_a as shown in Fig. 15.7. We note that the pressure curves in the region $\pi \leqslant \theta \leqslant 2\pi$ have been depicted by dashed rather than solid lines to indicate that although these now obey the theoretical distribution predicted by

[†] p_a = atmospheric pressure.

eqn. (15.19), no allowance has been made as yet in the theory for cavitation effects.[†] Since a lubricant can in fact sustain very limited negative pressure effects without emulsification and the formation of countless air bubbles, it is usual to superimpose a realistic boundary condition on the theory in the region $\pi \leqslant \theta \leqslant 2\pi$. This takes the form $dp/d\theta = 0$ at $\theta = \theta_2$, where $\theta_2 > \pi$. Indeed, if cavitation effects did not exist, the bearing as such would have no net load support since the converging wedge effect in one part of the film would be exactly matched by a diverging wedge effect elsewhere in the film. The amended curve OAB in Fig. 15.7 is an entirely positive pressure distribution which can be integrated to give the load support W of the bearing. We note the boundary conditions for the pressure curve OAB thus:

$$p = 0 \quad \text{at} \quad \theta = 0 \quad \text{and} \quad \theta_2,$$

$$dp/d\theta = 0 \quad \text{at} \quad \theta = \theta_1 \quad \text{and} \quad \theta_2.$$

By using a change of variable technique,[(73)] the integrals in eqn. (15.19) can be evaluated, so that the pressure rise p is given by

$$p = \frac{6\mu Ur}{c^2} \left\{ \frac{\varepsilon(2+\varepsilon \cos \theta) \sin \theta}{(2+\varepsilon^2)(1+\varepsilon \cos \theta)^2} \right\}. \tag{15.20}$$

If we now divide both sides of this equation by the mean pressure $\bar{p}$ in the bearing, then

$$p/\bar{p} \sim S_0 f(\varepsilon, \theta), \tag{15.21}$$

where the Sommerfeld number[‡] S_0 is given by the ratio $\mu U/\bar{p}L$, and the functional relationship $f(\varepsilon, \theta)$ is given by the chain bracketed term in eqn. (15.20).

Load Capacity

To find the load capacity of the bearing, consider an elementary area dA on the surface of the journal subtending an angle $d\theta$ at O as shown in Fig. 15.8. This area can be written as a vector which is resolved into components $dA \cos \theta$ and $dA \sin \theta$, lying respectively along and perpendicular to the line of centres OO'. The load W can also be resolved into components $W \cos \varphi$ and $W \sin \varphi$ in the same respective directions, so that we can write

$$W \sin \varphi = \int p \, dA \sin \theta = Br \int_0^{2\pi} p \sin \theta \, d\theta = Br \left\{ \left| -p \cos \theta \right|_0^{2\pi} + \int_0^{2\pi} \left(\frac{dp}{d\theta} \right) \cos \theta \, d\theta \right\}$$

$$= 6\mu UBr^2 \int \left(\frac{h-h_0}{h^3} \right) \cos \theta \, d\theta \quad \text{[from eqn. (15.18)].}$$

$$W \sin \varphi = \frac{6\mu UBr^2}{c^2} \left\{ \int_0^{2\pi} \frac{\cos \theta \, d\theta}{(1+\varepsilon \cos \theta)^2} - \frac{h_0}{c} \int_0^{2\pi} \frac{\cos \theta \, d\theta}{(1+\varepsilon \cos \theta)^3} \right\}.$$

[†] See Section 6.7.
[‡] S_0 is defined in Chapter 6 as the ratio of viscous to pressure forces in a bearing.

By following the same procedure used for evaluating pressures in eqn. (15.19), it can be shown that the above equation reduces to

$$W \sin \varphi = \frac{12\pi\mu UB(r/c)^2 \varepsilon}{(2+\varepsilon^2)\sqrt{(1-\varepsilon^2)}} \,. \tag{15.22}$$

To find the attitude angle φ we can write

$$W \cos \varphi = \int p \, d\bar{A} \cos \theta = Br \int_0^{2\pi} p \cos \theta \, d\theta,$$

which eventually reduces to the value zero.[73] Since $W \neq 0$, we conclude that $\cos \varphi = 0$ or $\varphi = \pi/2$. Thus $W \sin \varphi$ becomes W, and eqn. (15.22) gives the load capacity of the bearing directly. We observe that this is a function of the Sommerfeld number and the percentage eccentricity ε. If we substitute $W = 2rB\bar{p}$ in eqn. (15.22), we find that

$$S_0 = \frac{\mu U}{\bar{p}L} = f(\varepsilon), \tag{15.23}$$

where the characteristic length $L = c^2/r$. *Thus the Sommerfeld number in journal bearing applications is a function of the percentage eccentricity only.*

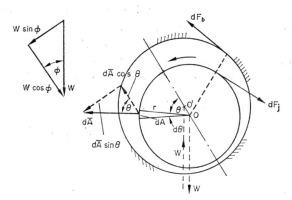

FIG. 15.8. Load-carrying capacity and frictional torque in journal bearing.

For real bearings, the value $\pi/2$ for φ is an unrealistic result which follows from the inclusion of negative pressures in the integration procedure for load capacity. In fact, if we assume that negative pressures cannot exist in the lubricant because of the cavitation phenomenon, we find that $\varphi < \pi/2$. This is found to be the case in a practical bearing.

Friction Torque

During normal operation, the lubricant film in a journal bearing undergoes a continuous shearing process, and this gives rise to frictional forces F_J and F_B which act at the surfaces of journal and bearing respectively. Due to the eccentricity of the rotor, however, these

23*

forces although opposite in direction are not equal. This can be shown by using eqn. (15.6A) at the journal surface ($z = h$) and its counterpart at the bearing surface ($z = 0$). Thus

$$\tau_J = \mu \left(\frac{du}{dz} \right)_{z=h} = \frac{h}{2r} \left(\frac{dp}{d\theta} \right) + \frac{\mu U}{h} \qquad (15.6B)$$

and

$$\tau_B = \mu \left(\frac{du}{dz} \right)_{z=0} = -\frac{h}{2R} \left(\frac{dp}{d\theta} \right) + \frac{\mu U}{h}. \qquad (15.6C)$$

It can be seen at once from these equations that $\tau_J > \tau_B$. Since

$$F_J = Br \int \tau_J \, d\theta \quad \text{and} \quad F_B = BR \int \tau_B \, d\theta,$$

it also follows that $F_J > F_B$. Exact expressions for F_B and F_J can be obtained by substituting for $(dp/d\theta)$ from eqn. (15.18) in eqns. (15.6B) and (15.6C), and they can be shown to be functions of S_0 and ε.[73]

The reason for the difference between the frictional forces F_J and F_B can be seen clearly by considering the lubricant in the bearing as a free body constrained by boundaries and acted upon both by pressure and shear stresses. For the direction of journal rotation shown in Fig. 15.8 the bearing exerts a clockwise torque $M_B = RF_B$ and the journal a counterclockwise torque $M_J = rF_J$ on the body of lubricant. At the same time variable pressure forces created by the geometry of the converging wedge act in a direction normal to the curved boundaries of the lubricant. The new moment caused by these pressure forces is most simply expressed by the loading torque $We \sin \varphi$, since the load W is applied at the journal centre O and reacted at the bearing centre O'. The pressure or load torque is clockwise, and we can therefore equate the clockwise and counterclockwise moments acting on the film thus:

$$RF_B + We \sin \varphi = rF_J. \qquad (15.24)$$

Since $R \doteq r$, we can divide both sides by the product Wr, and write eqn. (15.24) in terms of frictional coefficients

$$f_B + (e/r) \sin \varphi = f_J.$$

We more usually express this equation in terms of generalized frictional coefficients having the form (rf/c). Putting $\varphi = \pi/2$,

$$(r/c) f_B + \varepsilon = (r/c) f_J, \qquad (15.25)$$

so that the difference between the generalized coefficients at journal and bearing is identical with the percentage eccentricity ε.

Michell Bearing

The analysis of journal bearings has so far neglected end leakage effects as in the initial treatment of slider bearings. For $R < B/4$, the error involved is tolerable, but otherwise we must consider the transverse pressure gradient $\partial p/\partial y$ in addition to the circumferential component $\partial p/\partial \theta$. Under such conditions the analysis is relatively complex, and a detailed solution to the general problem may be found elsewhere.[73]

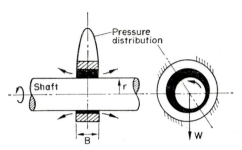

Fɪɢ. 15.9. Diagrammatic sketch of Michell or short bearing.

Michell obtained a solution for the very short bearing $(R \gg B)$ by neglecting the circumferential pressure gradient $\partial p/\partial \theta$ in comparison with $\partial p/\partial y$. Figure 15.9 shows a sketch of the bearing. The Reynolds equation (6.4) for one-dimensional transverse flow takes the following form:

$$\frac{d}{dy}\left(h^3 \frac{dp}{dy}\right) = 6\mu U \frac{dh}{dx}$$

or

$$\frac{d^2p}{dy^2} = \frac{6\mu U}{h^3}\left(\frac{dh}{dx}\right), \tag{15.26}$$

since $h = h(x)$ only. By integrating this equation twice and using the boundary conditions $p = 0$ at $y = \pm\frac{1}{2}B$,

$$p = \frac{3\mu U}{rc^2}\left(\frac{B^2}{4} - y^2\right)\frac{\varepsilon \sin \theta}{(1 + \varepsilon \cos \theta)^3}. \tag{15.27}$$

We observe from this equation that the pressure has a parabolic profile in the transverse or y-direction, as shown in Fig. 15.9, whereas the variation of p in the circumferential or θ-direction is similar to that for a very long bearing. The load capacity and generalized friction factor for the Michell bearing may also be obtained as before. Most practical journal bearings having comparable R and B dimensions fall within the limits prescribed by the very long and very short bearings.

Temperature Rise

We will now obtain an expression for the temperature rise in a journal bearing, since lubricant viscosity and general bearing performance depend critically on this variable. An energy or heat balance for the system[†] may be written as follows:

$$\begin{bmatrix} \text{Frictional} \\ \text{energy} \\ \text{dissipation} \end{bmatrix} = \begin{bmatrix} \text{Rate of heat} \\ \text{removal by} \\ \text{circulating oil} \end{bmatrix} + \begin{bmatrix} \text{Rate of heat} \\ \text{transfer from} \\ \text{oil to bearing} \end{bmatrix}$$

$$FU = \varrho C_p Q\, \Delta T_1 + \bar{h}A\, \Delta T_2.$$

† Under steady-state conditions.

Here F is the mean frictional force in the bearing, ϱ and C_p the density and specific heat at constant pressure of the lubricant, Q the volume flow rate of oil, ΔT_1 the mean temperature rise of the lubricant in passing through the bearing, $\bar{h}$ the convective heat-transfer coefficient at the oil bearing interface, A the area of the latter, and ΔT_2 the mean temperature difference between the oil in the bearing and the bearing surface. Writing $\Delta T = \Delta T_1 = \Delta T_2$ and putting $A = 2\pi RB$ and $F = fW$,

$$fWU = (\varrho C_p Q + 2\pi RB\bar{h})\,\Delta T.$$

Dividing both sides of this equation by the projected area $2RB$ of the bearing and putting $\bar{p} = W/2RB$ and $q = Q/2RB$, the following result appears:

$$\frac{f}{(q/U)} = \Gamma = \frac{\Delta T}{\bar{p}}\left(\varrho C_p + \frac{\pi\bar{h}}{q}\right),\tag{15.28}$$

where the symbol Γ is a temperature number. If we momentarily neglect the second term in the bracket in eqn. (15.28), we can obtain ΔT as follows:

$$\Delta T \doteq \bar{p}\Gamma/\varrho C_p.\tag{15.29}$$

For most oils, the product ϱC_p has the approximate value 1800 kJ/m³ K or 150 lbf/in² R. The temperature number Γ is a function of the Sommerfeld number S_0, and for bearings of finite width it also depends on the ratio R/B of the given bearing.

It follows at once from eqn. (15.29) that an iterative procedure must be followed in a practical journal bearing design problem. We have seen previously that most design param-

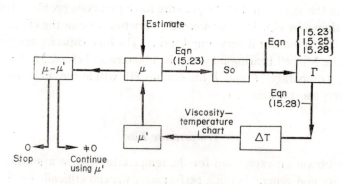

FIG. 15.10. Iterative sequence in journal bearing design.

eters (such as mean pressure, load capacity, friction torque, etc.) are functions of the Sommerfeld number S_0, the latter being proportional to lubricant viscosity μ. Since μ in turn depends critically on the temperature rise ΔT within the bearing and the latter is initially unknown, we must start with a reasonable estimate of μ (or ΔT) for the particular lubricant selected. This estimate is used to complete the loop $\mu \rightarrow S_0 \rightarrow \Gamma \rightarrow \Delta T \rightarrow \mu'$

according to Fig. (15.10), using viscosity–temperature charts where appropriate. If the computed viscosity μ' is not different from the initial guess μ, then the values of S_0, Γ, and ΔT computed while traversing the loop are correct. More probably, $\mu' \neq \mu$, and a second loop must be traversed using μ' as the new estimate. It may be necessary to execute several loops before the difference $(\mu^{i+1} - \mu^i)$ is sufficiently small to assume that the previous loop gave substantially the correct values. During the second and subsequent loops (if any), the second term in brackets in eqn. (15.28) may be included, although the error involved by its omission is still slight.

One of the critical parameters in designing a bearing for a particular application is the dimensionless minimum film thickness (h_{min}/c). Putting $\theta = \pi$ in eqn. (15.17A) it follows that

$$h_{min}/c = 1 - \varepsilon = f(S_0). \tag{15.17B}$$

For a given length-to-breadth ratio in the bearing, we again see that S_0 is the determining factor in establishing the magnitude of (h_{min}/c). As a general guide, this ratio must fall between prescribed limits as follows:

$$\frac{2\delta}{c} < \frac{h_{min}}{c} < 0.35, \tag{15.30}$$

where δ is the peak-to-trough roughness on the surface of the journal or bearing, ranging from 20 μ in rough grinding to about 0.5 μ in extra-fine honing operations. Since the radial clearance c varies from approximately $0.0005\,r$ to $0.003\,r$ (where r is the journal radius), the minimum limit in eqn. (15.30) can vary from $40/3r$ to $1/r$ with r measured in millimetres. We require a minimum limit for (h_{min}/c) to ensure that mixed or boundary lubrication does not exist in the bearing under heavy loading conditions, since part solid contact may well initiate seizure and rapid failure. On the other hand, a maximum limit is imposed to ensure that under light loading instability and oil whip[†] are avoided.

The behaviour of actual journal bearings differs from the simplified theory presented in this chapter because of the effects of elastic deformation of the surfaces, surface roughness, thermal expansion of journal and bearing, turbulence in the lubricant, and unbalance and misalignment of the journal during assembly. Besides these effects, the lubricant viscosity in a real bearing is never quite constant, and variations occur because of pressure, temperature, and rate of shear. When the angle of contact between bearing and lubricant is 360° (as in the case of Figs. 15.6 and 15.8), we speak of a full journal bearing. Partial journal bearings by contrast have an angle of 180° or less, with 120° as the usual value. The advantages of the partial bearing are structural simplicity, convenience in applying the lubricant, and reduced frictional losses. It is necessary, however, that the direction of the load does not change materially during operation. As in the early experiments of Beauchamp Tower (see Chapter 6), the partial bearing is most frequently used in railroad-car applications.

---

† Oil whip is defined as a resonant vibration of a journal in a fluid-film bearing at low eccentricity ratios, and it usually occurs when the speed of rotation of the journal is about twice the first critical speed of the system.

Floating-ring bearings are a modified form of journal bearing designed to reduce frictional shear losses. Here, a thin ring floats freely between journal and bearing, and rotates at a speed which is considerably less than that of the journal. Thus the relative speed between mating surfaces is appreciably reduced. Since power losses vary as the square of the speed but vary only linearly with area, the ring design always results in some reduction in drag.

15.3 Foil Bearings

With the establishment of elastohydrodynamic lubrication as a fundamental branch of lubrication theory[†], the flexibility or elasticity of the bearing or constraining surface acquired significance as a design variable. Thus developments in bearing design took the form of elastomeric and rubberized liners inside the bearing surface to permit a closer conformity between mating surfaces in the minimum film-thickness region. The extreme case of bearing surface flexibility is the foil bearing, shown schematically in Fig. 15.11.[74] This consists

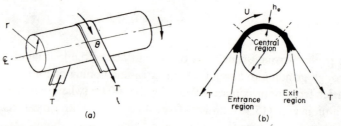

FIG. 15.11. The foil bearing showing (a) general view, and (b) infinitely wide foil.

of a stationary, flexible band or foil wrapped partially around a rotating shaft as shown in Fig. 15.11(a) with an interfacial lubricant between the foil and the shaft. In some cases the shaft is held stationary and the foil moves past it with velocity U, as we see in Fig. 15.11(b), whereas in other circumstances both the shaft and the foil rotate.

Let the tension per unit width of foil be T on each side of the central contact region and let r denote the radius of the journal. The curvature of the foil with respect to the journal may be expressed by the relationship

$$\frac{1}{r'} = \frac{d^2h}{ds^2} \bigg/ \left[1+\left(\frac{dh}{ds}\right)^2\right]^{3/2}, \tag{15.31}$$

where the coordinate s is measured in the circumferential or θ-direction. The relative curvature $1/r'$ in eqn. (15.31) can also be expressed as

$$1/r' = 1/r - 1/R, \tag{15.32}$$

[†] See Chapter 8.

where R is the radius of curvature of the foil. We can usually neglect the term $(dh/ds)^2$ in the denominator of eqn. (15.31), so that from the last two equations

$$\frac{1}{R} = \frac{1}{r} - \frac{d^2h}{ds^2}.$$

(15.33)

Now, the Reynolds equation for incompressible, one-dimensional flow in the s-direction becomes

$$\frac{d}{ds}\left(h^3 \frac{dp}{ds}\right) = 6\mu U \frac{dh}{ds}$$

(15.34)

with the usual notation. Furthermore, the equilibrium of an elemental length of foil $R\, d\theta$ within the angle of wrap shows clearly that the pressure force $pR\, d\theta$ exerted on the element by the shaft is exactly matched by the tension component $T\, d\theta$. Thus

$$p = T/R.$$

(15.35)

By substituting for R from eqn. (15.33) into eqn. (15.35), and then for p into eqn. (15.34), we obtain the foil-bearing equation

$$\frac{d}{ds}\left(h^3 \frac{d^3h}{ds^3}\right) = -\frac{6\mu U}{T}\left(\frac{dh}{ds}\right).$$

(15.36)

It is convenient to write this in non-dimensional form by introducing the following parameters:

$$\varepsilon = \frac{6\mu U}{T}; \quad \xi = \frac{s\, \varepsilon^{1/3}}{h_0}; \quad H = h/h_0,$$

where h_0 is the constant film thickness at the centre of contact [see Fig. 15.11(b)]. With these definitions and after integrating once, eqn. (15.36) becomes

$$\frac{d^3H}{d\xi^3} = \frac{1-H}{H^3}.$$

(15.37A)

This equation is non-linear, and to obtain a solution we assume that the central gap has only very small variations ΔH from the condition $H = 1$, i.e. $\Delta H \ll 1$. Thus, by putting $H = 1 - \Delta H$ in eqn. (15.37A),

$$\frac{d^3}{d\xi^3}(\Delta H) + \Delta H = 0,$$

(15.37B)

for which the solution[75] is given by

$$\Delta H = A\, e^{-\xi} + B\, e^{\xi/2} \cos\frac{\sqrt{(3)}\,\xi}{2} + C\, e^{\xi/2} \sin\frac{\sqrt{(3)}\,\xi}{2}.$$

(15.38)

If we examine this solution carefully, we observe firstly that the constants A, B, and C must be exceedingly small, so that ΔH is small in the central region. The parameter ξ is proportional to distance s measured circumferentially around the bearing from the inlet region in Fig. 15.11. Thus at relatively large values of ξ corresponding to the exit region RS in Fig. 15.12(a), the sinusoidal terms in eqn. (15.38) dominate, and the film thickness exhibits a characteristic "ripple" as shown. At relatively small ξ-values corresponding to the inlet region PQ in Fig. 15.12(a), the first term in eqn. (15.38) dominates, so that the entrance region is exponential in form. The hydrodynamic pressure expressed in the dimensionless form pr/T varies in an approximately inverse manner to film thickness, and exhibits a con-

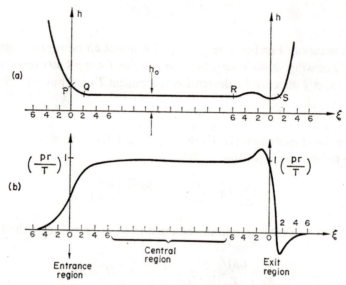

FIG. 15.12. Gap and pressure in perfectly flexible foil bearing.

stant value over the central region followed by a characteristic peak towards the exit, as seen in Fig. 15.12(b). It can be shown that for design purposes[75] the magnitude of the gap h_0 is given by

$$h_0 = 0.643r \left(\frac{6\mu U}{T} \right)^{2/3}. \tag{15.39}$$

This simple formula is important because it applies not only to the infinitely wide, perfectly flexible foil bearing considered above, but also it will remain essentially the same when the effects of foil stiffness, finite width, and lubricant compressibility are taken into account.

Foil bearings are used in all cases where moving webs are supported, such as in the paper manufacturing and textile industries and in steel mills. They also find wide application in information storage devices such as tape recorders, computer reels, and flexible strip memories.

15.4 Rolling Contact Bearings

All bearing types dealt with so far in this chapter are of the fluid-film type, since a hydro-dynamic lubricant film separates the moving surfaces and prevents solid contact. The friction is therefore due to the shear resistance of the lubricant, and has quite a low value. Rolling contact bearings by contrast permit solid contact to exist between the moving surfaces, but the actual area of contact is reduced to near zero dimensions by the use of ball-bearings or rollers. Moreover, these elements accommodate relative motion between surfaces by the

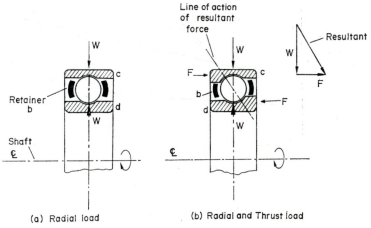

(a) Radial load   (b) Radial and Thrust load

FIG. 15.13. Radial and thrust ball-bearing races.

action of rolling rather than sliding, so that the only frictional forces remaining between the surfaces are due to rolling resistance. Rolling contact bearings are frequently referred to as "anti-friction" bearings, since the rolling resistance of the balls or rollers is usually extremely low. Figure 15.13 shows two common types of ball-bearing races designed to take (a) radial, and (b) combined thrust and radial loading. In both cases, the inner race d is attached perhaps to a rotating member and the outer race c is stationary. During shaft rotation, the ball-bearings rotate between the races, each performing one orbit about the shaft centre line for every two revolutions of the latter. A retaining ring or cage b preserves a uniform spacing between the balls during rotation and rolling. The ball seating within the races is symmetrically about the centre plane of the balls for the case of radial loading, and unsymmetrical for thrust loading to accommodate the transverse load F. Some degree of radial loading W is required in each bearing to ensure that the rolling mode is preserved.

It is interesting to compare the relative friction of journal bearing and rolling contact bearings as a function of journal or shaft speed. For starting conditions and at moderate speeds, the friction of a ball or roller bearing is lower than that of an equivalent journal bearing, as shown in Fig. 15.14. At relatively high speeds, however, the rolling resistance of the contact bearings rises steeply with speed, and the friction exceeds that of a well-designed journal bearing. The curves have inverse characteristics in the sense that the undesirable speed range for the one corresponds to creeping or standstill conditions, and for the other to very rapid shaft rotation.

Ball-bearings theoretically have kinematic point contact, and they are therefore limited to moderate radial loading. If the recommended radial load is exceeded in a particular application, flat-spotting occurs, and the relative friction curves in Fig. 15.14 are no longer applicable. The life of such a bearing is drastically shortened by the combined and cumulative effects of higher friction forces, heating, and wear. Roller bearings are frequently used for

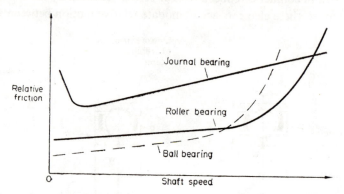

Fig. 15.14. Comparison of friction vs. speed characteristics of journal and rolling contact bearings.

large radial loading because they theoretically exhibit kinematic line contact. However, an inherent disadvantage of the roller bearing is the variation of pressure along the band of contact due to deflections of shafts and mountings. Lubricants are used in rolling contact bearings primarily to prevent rust, corrosion, and the ingestion of foreign matter (such as dust or water), and only secondarily to reduce friction and wear between rubbing parts of the bearing.

15.5 Miscellaneous

Porous Bearings

Porous metal bearings are often used where plain metal bearings are impractical because of inaccessibility for lubrication. They are made of powdered metals which are first compressed and then sintered at high temperature in a reducing atmosphere. The sintering process causes the powdered metal bearings to fuse into a strong porous structure which is then impregnated with lubricant. The void spacing in the bearing usually varies from 16 to 36% of the total bearing volume. Non-gumming lubricants which are strongly resistant to oxidation are used, and maximum bearing temperatures are limited to 66°C. The pores act as minute reservoirs for the lubricant which exudes to the sliding interface through capillary action.

Let us suppose that V_0 is the speed at which the lubricant flows out of the porous surface and q_z the rate of flow per unit area. Then

$$q_z = V_0 = -\frac{\partial p}{\partial z}\bigg|_{z=0} \frac{\Phi}{\mu}, \qquad (15.40)$$

where $\partial p/\partial z$ is the pressure gradient normal to the bearing surface and Φ is a property called permeability[73] which varies with porosity and the size of pore.† Furthermore, from the requirements of continuity, we have for the porous matrix:

$$\nabla q = \frac{\partial q_x}{\partial x} + \frac{\partial q_y}{\partial y} + \frac{\partial q_z}{\partial z} = -\frac{\Phi}{\mu} \nabla^2 p = 0,$$

so that (since $\Phi/\mu \neq 0$)

$$\nabla^2 p = 0. \qquad (15.41)$$

We now make use of eqn. (15.40) to modify the general Reynolds equation (6.4). Thus for incompressible flow and replacing the squeeze contribution by the porosity effect, we have

$$\frac{\partial}{\partial x}\left(h^3 \frac{\partial p}{\partial x}\right) + \frac{\partial}{\partial y}\left(h^3 \frac{\partial p}{\partial y}\right) = 6\mu U \frac{dh}{dx} + 12\Phi \frac{\partial p}{\partial z}\bigg|_{z=0}. \qquad (15.42)$$

Here the stretch term has been neglected. Both eqns. (15.41) and (15.42) can be used to obtain the pressure distribution, load capacity, and generalized friction factor in a porous bearing.[73] As we might anticipate, the load capacity is reduced and the friction greatly increased because of the porous nature of the bearing surface.

In general, a thinner hydrodynamic film is formed in the case of a porous bearing compared with a solid bearing because the "bleeding" effect of the pores tends to drain the film. At the same time, the pores provide lubricating action under conditions where dry contact would otherwise occur. Porous metal bearings have been impregnated with plastic materials as well as lubricants, and teflon provides a consistently low coefficient of friction of about 0.05 when used as the filler material at light loads and speeds. Typical applications of porous bearings include the guide bearings on small generator and distributor shafts, water-pump bearings, and light bushings.

Gas Bearings

We have considered the lubricant used in bearings so far to be an incompressible liquid, but in certain cases gases are used. There are four fundamental differences between gas-lubricated and liquid-lubricated bearings:

---

† Let us suppose that the porosity effect in the vertical or z-direction in a bearing can be represented by a number of capillary tubes of uniform diameter. By comparing eqn. (15.40) and the Hagen–Poiseuille equation for flow through capillary tubes, we find that $\Phi = Nd^2/32$, where N is the number of pores per unit area, each of diameter d. In the general case, the porosity effect is random both in the direction and size of the pores.

(a) *Viscosity Effect*

Since $\mu_{air} = \mu_{oil}/1000$ at average bearing temperatures, the speed of a gas-lubricated bearing must be increased many times compared with a liquid bearing to provide the same Sommerfeld number and load support. Furthermore, the viscosity of a gas increases with temperature in contrast with liquids,[†] so that there is a built-in safety feature against the probability of bearing failure as a result of a sudden rise in operating temperature above the steady-state value.

(b) *Dimensional Accuracy*

The films in gas bearings are appreciably thinner than the films in incompressible lubrication, and the minimum film thickness may have the same order of magnitude as the surface roughness of the journal and bearing. The surface finish in gas bearings must therefore be appreciably smoother, and there can be no "running-in" period (i.e. smoothening of surface peaks, as with oil-bearings). It follows that gas and air bearings are not suited to frequent stopping and starting.

(c) *Compressibility Effects*

Because of density variations in a gas, the Reynolds equation must now take the form of eqn. (6.1). This implies that the wedge term must contain two components as shown in eqn. (6.2) thus:

$$\text{Wedge term} = \underbrace{6h(U_1 - U_2)\frac{\partial \varrho}{\partial x}}_{\substack{\text{Compressibility} \\ \text{term}}} + \underbrace{6\varrho(U_1 - U_2)\frac{\partial h}{\partial x}}_{\substack{\text{Incompressible} \\ \text{wedge term}}} \qquad (6.2)$$

Now, $p = \varrho RT$, or $\varrho = p/R(273+t)$ for an ideal gas, where R is the particular gas constant, T the temperature in K, and t the temperature in °C. *Thus changes of t within the bearing have only a slight effect on* ϱ, *and we conclude that the* $\partial \varrho/\partial x$ *term is due almost entirely to compressibility effects.*

We further note that $\partial \varrho/\partial x$ is normally negative and $\partial h/\partial x$ positive. Thus the incompressible wedge term and the compressibility term in eqn. (6.2) are of opposite sign. It follows that a *gas-lubricated bearing of the slider type must have a larger angle of inclination or wedge angle α than a liquid bearing to compensate for the effects of compressibility.*

(d) *Cavitation and Slip*

No cavitation is possible with gas bearings, and this greatly simplifies the boundary conditions in a journal bearing. There is a closer agreement between theory and experiment, since there is a 360° rather than a 180° journal. When the bearing film thickness becomes comparable in magnitude to the molecular mean free path λ of the gas,[‡] continuum con-

[†] See eqn. (10.39).
[‡] See Section 10.9.

ditions no longer apply, and slip takes place at the boundaries. This requires modifications to the basic Reynolds equation.

In general, the danger of instability and whirl is more pronounced with gas bearings than with liquid lubrication, and the Sommerfeld number must be kept within prescribed limits to avoid this possibility. Compressibility also appears to reduce the maximum pressure in a bearing and to force the position at which it occurs closer to the point of minimum film thickness. The most usual gas used is, of course, air, not only because of its ready availability and low friction but also because it is relatively free from contamination and may be used in high-temperature applications. The absence of contamination is an important consideration if the possibility of dripping, soiling, or spoilage by a liquid lubricant must be eliminated. This applies to food-processing machinery, thread and textile machines, and chemical processing.

CHAPTER 16

MISCELLANEOUS APPLICATIONS

IN THIS concluding chapter we deal with a number of examples in which tribological principles play a fundamental role as before; yet the diverse nature of the applications does not permit ready classification. These applications include flexible seals, layered damping in beams, friction damping, the lubrication of human and animal joints, and a miscellaneous section dealing with the wear of coins, skin disease detection, and friction in typing and printing. We deal with each item separately in the following sections:

16.1 Flexible Seals

Consider the dynamic performance of a flexible lip seal on a lubricated, machined shaft rotating at high speed, as shown in Fig. 16.1. Let the contact width of the lip on the shaft be ΔL and R the shaft radius. The pressure distribution p between the seal and the shaft has

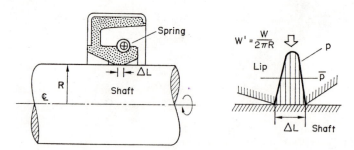

FIG. 16.1. Cross-section of oil seal and pressure distribution.

the form shown on the right-hand side of Fig. 16.1, and the load W' per unit width is given as follows:

$$W' = \int_0^{\Delta L} p \, \mathrm{d}L = \frac{W}{2\pi R} \, . \tag{16.1}$$

Now, in defining a load per unit width in problems of this nature, it is more meaningful to specify that the width selected lies in a direction perpendicular to the relative velocity vector between the moving surfaces. We therefore define

$$W'' = W/\Delta L, \tag{16.2}$$

where W in both equations is the total radial load exerted by the seal on the shaft.

The surface of the shaft exhibits a roughness pattern which is the result of some finishing technique[†] and therefore two-dimensional. It is convenient to represent this roughness effect as a sinusoid with amplitude h_{max} and wavelength λ, as shown in Fig. 16.2(a). The flexible seal drapes readily about the peaks in the shaft profile during sliding action, but fails to make contact in the troughs as indicated. Furthermore, the lubricant on the surface of the shaft will generally not be sufficient to fill the voids between lip and shaft, so that the film breaks up in this region with the bulk adhering to the troughs of the shaft profile. This is clearly indicated in Fig. 16.2(a). During sliding action with velocity V, there is a positive pressure wedge effect which generates hydrodynamic support on the leading edges of the asperities[‡]

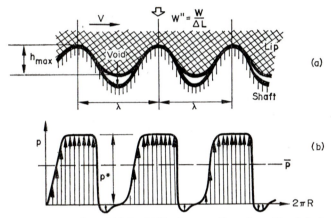

FIG. 16.2. Schematic representation of (a) liquid film between lip and machined shaft, and (b) corresponding pressure distribution.

followed by a constant pressure or foil-bearing region at the summit of each asperity, and finally a very rapid fall in pressure in the diverging film section. Figure 16.2(b) depicts the resulting pressure distribution which is identical with that shown in Fig. 8.14. The pressure and film thickness p^* and h^* over each asperity peak may be obtained following the macro-elastohydrodynamic theory outlined in Section 8.8. The mean pressure $\bar{p}$ is obviously less than p^* and extends over the whole contact area $2\pi R\, \Delta L$.

We observe that the pressure distributions in Figs. 16.1 and 16.2 are at right angles to each other, and therefore no inconsistency arises. The surface wavelength λ has approximately the same order of magnitude as ΔL, so that $R \gg \Delta L$ and a two-dimensional analysis of the problem is permissible. The frictional performance of the lip seal in Fig. 16.1 can

[†] See Table 2.6.
[‡] See Fig. 8.14.

24 M: PAT: 2

be shown[(76)] to be governed by an equation of the form

$$f = \varphi \left(\frac{\mu V}{W''} \right)^{1/3},$$ (16.3)

where the function φ is defined as follows:

$$\varphi = 5 \left(\frac{h_{\max}}{\lambda} \right)^{2/3} \left(\frac{R}{h_{\max}} \right)^{1/3} \exp \left[k \frac{\lambda}{h_{\max}} \left(\frac{\bar{p}}{E'} \right) \right].$$ (16.4)

We observe that the function φ depends in complex fashion on the roughness of the sealing surface and on the viscoelastic properties of the seal material. This function also contains the ratio $(\bar{p}/E')$ which is responsible for the hysteresis component of sliding friction, as seen from eqn. (4.39). The bracketed term in eqn. (16.3) is the well-known Sommerfeld number in lubrication theory.

A significant conclusion on the sealing behaviour of lip seals can be obtained by assigning to φ a critical value φ_c and then plotting f in eqn. (16.3) versus the Sommerfeld number $(\mu V/W'')$. The result appears in Fig. 16.3 as a straight line with positive slope. The value φ_c defines a threshold of dynamic seal performance which marks the boundary between condi-

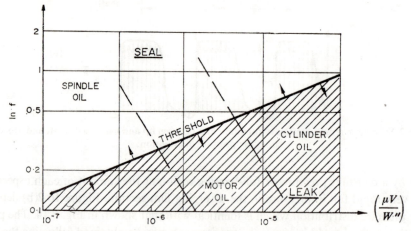

FIG. 16.3. The elastohydrodynamic threshold for dynamic lip-seal performance.

tions of leaking and sealing. For $\varphi > \varphi_c$ the sealing condition is maintained, whereas for $\varphi < \varphi_c$ leakage occurs. Three types of lubricant were used in establishing this property (spindle oil, motor oil, and cylinder oil), and the many experimental data points obtained have not been shown in Fig. 16.3 for reasons of clarity. From eqn. (16.4) it is clear that the value φ_c corresponds to a critical value of pressure $\bar{p} = \bar{p}_c$, above which sealing is effective and below which leakage takes place. We can therefore counteract the elastohydrodynamic separating effect (see Chapter 8) and preserve the sealing condition by increasing the radia load on the seal.

16.2 Layered Damping in Beams

It is unfortunate that the desirable properties of high strength and stiffness in construction members such as beams, girders, and cantilevers seem incompatible with high material damping. The latter is essential to minimize noise propagation effects and to dampen out unwanted vibrations. One common configuration in structures which is designed to reconcile these differences is the sandwich laminate consisting of layers of damping material between the structural members. Figure 16.4 illustrates the principle in the case of a bi-leaf cantilever. In the absence of a damping layer between the leaves, Coulomb damping may occur because of slippage when the cantilever deflects, as shown in Fig. 16.4(a). However,

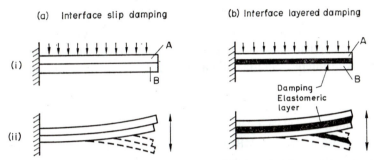

FIG. 16.4. Comparison of (a) interface slip damping, and (b) interface layered damping in a bi-leaf cantilever.

the strength and rigidity of the bi-leaf construction are evident only when the separate members are pressed tightly together, and this reduces the probability of interfacial slippage.

By providing a thin elastomeric layer between the cantilever leaves, beam deflection is tolerated without interfacial slip and shear deformation of the layer occurs, as indicated in Fig. 16.4(b). Thus energy is dissipated internally within the viscoelastic layer. In stress ranges generally encountered, a typical viscoelastic adhesive in shear dissipates as much as 300 times the energy per unit volume dissipated in an average structural material. Some ad-

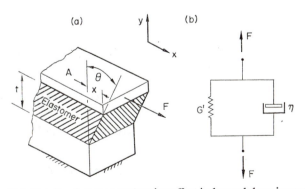

FIG. 16.5. Details of elastomer shearing effect in layered damping system.

hesives can withstand low-frequency cyclic shear strains larger than unity for millions of cycles without fatigue failure. In general, viscoelastic materials must be strained to much higher levels than the structural materials with which they are in contact in order to maximize their damping potential.

Consider an elastomeric material of thickness t which forms an energy-dissipative layer between two parallel rigid members, as shown in Fig. 16.5(a). Relative motion between the members (perhaps as a result of bending, as in Fig. 16.4) causes shear deformation of the layer which creates a resisting force F in the direction indicated. If we assume for simplicity a Voigt model representation of viscoelastic behaviour within the elastomer as shown in Fig. 16.5(b), then we can write[†] for F:

$$F = \underbrace{G'A\theta}_{\text{Spring force}} + \underbrace{\eta A\dot{\theta}}_{\text{Damping force}}, \tag{16.5}$$

where G' is the spring modulus, η the dashpot viscosity in the Voigt model, A the area of shear, and θ, $\dot{\theta}$ the angular strain and rate of strain within the elastomeric material. Only the damping term is of interest here in estimating energy dissipation, and putting $\theta = x/t$ in eqn. (16.5),

$$F_{\text{damping}} = \eta A \dot{x}/t = c\dot{x}, \tag{16.6}$$

where c is the damping coefficient defined in Chapter 6.

We now make use of eqn. (4.24B) which relates frequency of deformation ω, dashpot viscosity η, and the tangent modulus $\tan \delta$ thus:

$$\eta = \frac{G' \tan \delta}{\omega}. \tag{4.22B}$$

Substitution of eqn. (4.24B) into eqn. (16.6) gives the following:

$$c = \frac{G'A \tan \delta}{\omega t}. \tag{16.7}$$

This equation shows clearly that a thin elastomeric layer spread over a wide shear area provides a maximum damping capacity.

If we assume a sinusoidal variation of angular strain θ or linear deflection x with time, according to the relationship

$$\theta = \theta_0 \sin \omega t \quad \text{or} \quad x = x_0 \sin \omega t, \tag{16.8}$$

we can compute the total energy E_d dissipated within the viscoelastic layer per cycle of vibration thus:

$$\left. \begin{aligned} E_d &= \oint c\dot{x} \, dx = \pi \omega c x_0^2 \qquad \text{[from eqn. (16.8)]}, \\ &= \frac{\pi G'A \tan \delta}{t} \cdot x_0^2 \qquad \text{[from eqn. (16.6)]}, \\ &= \frac{\pi \omega \eta A}{t} x_0^2 \qquad \text{[from eqn. (16.6)]}. \end{aligned} \right\} \tag{16.9}$$

[†] See eqn. (4.20B).

We conclude from eqn. (16.9) that a damping layer is most effective in dissipating energy if the geometric ratio (A/t) and the loss[†] modulus G'' are kept as large as possible, and particularly if the amplitude of shearing deformation x_0 is a maximum.

The use of viscoelastic materials to increase structural damping recognizes two distinct categories referred to as constrained-layer- and unconstrained-layer-damping respectively. Figure 16.5(a) is an example of the constrained-layer type, where the constraining effect is provided by the upper rigid member (this is typically a layer of metallic foil). If this member is removed, the result is the unconstrained type of layer which has been used successfully in automobiles and on railroad cars. For a given shear area and layer thickness, the constrained layer-damping design produces greater shear deformation and energy dissipation. In some cases, the thickness t of the damping layer can be reduced still further by providing rigid spacers between the upper and lower members shown in Fig. 16.5.

16.3 Friction Damping

The use of elastomeric layers dissipates energy through the mechanism of internal friction, with zero slip at the solid boundaries between the elastomer and its constraining rigid members. Another form of friction damping permits external or Coulomb friction within a dynamic system to reduce unwanted fluctuations or resonances. Figure 16.6 illustrates the technique of damping rotational speed fluctuations in a spinning shaft with the use of a friction sleeve. This sleeve is attached to the hub or shaft at one end A, and extends concentrically with the shaft until contact is again made at section A–A. In the design shown, a

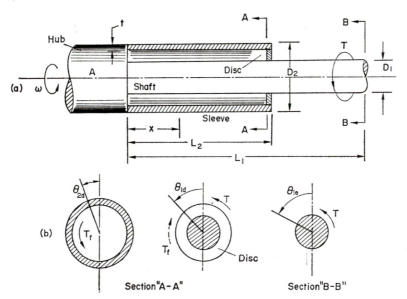

FIG. 16.6. Schematic diagram of friction sleeve on rotating shaft.

[†] $G'' = G' \tan \delta$, from Chapter 4.

disc is rigidly attached to the shaft (by welding or shrink-fitting), and there is usually a snug fit between the disc and the sleeve. The nature of this fit is very crucial to the operation of the system. Thus if the fit is too tight, the shaft and sleeve will move together as a unit and there will be no damping. On the other hand, an excessively loose fit permits a clearance between the disc and the sleeve, so that again there is no damping. With the proper fit design, the damping action will be most effective and the energy dissipated a maximum. Sometimes a thin elastomeric layer is placed between disc and sleeve to avoid fretting corrosion.

To understand the operation of the friction damper,[77] consider the left-hand end of the shaft A fixed momentarily, and let the applied torque T at the right-hand end produce an angular twist of amount θ_{1e} at section B–B as shown in Fig. 16.6(b). Now, if the friction sleeve were not present, the torque T would result in a lesser angle of twist at section A–A, since we are approaching the fixed end A, where the angle of twist is zero. With the friction sleeve in operation, a frictional torque T_f is developed between disc and sleeve, and a net torque $(T-T_f)$ at section A–A produces an angle of twist of amount θ_{1d}. From the fundamental relationship

$$\theta = \frac{Tx}{JG},$$

where G is the shear modulus of the material of the shaft or sleeve, and J its polar moment of inertia, we can write for the shaft at section A–A,

$$\theta_{1d} = \frac{(T-T_f)L_2}{J_1 G_1} = \frac{32(T-T_f)L_2}{\pi D_1^4 G_1}, \tag{16.10}$$

where L_2 is the length of shaft between the "fixed" end A and the disc, and D_1 is the shaft diameter. Between the disc position and the right-hand end of the shaft, there is an additional angle of twist $\Delta\theta$ due to the applied torque T. Thus

$$\Delta\theta = \frac{T(L_1-L_2)}{J_1 G_1} = \frac{32T(L_1-L_2)}{\pi D_1^4 G_1}, \tag{16.11}$$

so that the *total* angle of twist θ_{1e} of the shaft at its right-hand end is obtained by adding eqns. (16.10) and (16.11):

$$\theta_{1e} = \theta_{1d}+\Delta\theta = \frac{32(TL_1-T_f L_2)}{\pi D_1^4 G_1}. \tag{16.12}$$

Similarly, the angular displacement θ_{2d} of the sleeve at A–A due to the friction torque T_f is given by

$$\theta_{2d} = \frac{T_f L_2}{J_2 G_2} = \frac{4T_f L_2}{\pi D_2^3 t G_2}, \tag{16.13}$$

where D_2 and t are the diameter and thickness of the sleeve and G_2 the shear modulus of the sleeve material.

The strain energies U_1 and U_2 acquired by the shaft and sleeve respectively as a result of applying the end torque T are given by

$$U_1 = \tfrac{1}{2}\{(T-T_f)\theta_{1d}+T\varDelta\theta\},\tag{16.14}$$
$$U_2 = \tfrac{1}{2}T_f\theta_{2d},\tag{16.15}$$

and the total strain energy U of the system is

$$U = U_1+U_2 = \underbrace{\tfrac{1}{2}T\theta_{1e}}_{\substack{\text{Maximum strain energy}\\ \text{of shaft without}\\ \text{sleeve friction}}} - \underbrace{\tfrac{1}{2}T_f(\theta_{1d}-\theta_{2d}).}_{\substack{\text{Frictional energy}\\ E_a \text{ from slippage}\\ \text{between sleeve}\\ \text{and disc}}}\tag{16.16}$$

The ratio E_d/U is the ratio of frictional energy expended to the total strain energy of the system, and it can be expressed as follows:

$$\frac{E_d}{U} = 1\Big/\left[\left(\frac{T}{T_f}\right)\left(\frac{\theta_{1e}}{\theta_{1d}-\theta_{2d}}\right)-1\right].\tag{16.17}$$

By substituting for θ_{1e}, θ_{1d} and θ_{2d} from eqns. (16.12), (16.10), and (16.13) and putting $G_1 = G_2 = G$,

$$\frac{E_d}{U} = \frac{1-\left(\dfrac{T_f}{T}\right)\left(1+\dfrac{D_1}{8C_1^3 t}\right)}{a\Big/\left(\dfrac{T_f}{T}\right)+\left(\dfrac{T_f}{T}\right)\left(1+\dfrac{D_1}{8C_1^3 t}\right)-2},\tag{16.18}$$

where $C_1 = D_2/D_1$ and $a = L_1/L_2$. If we differentiate this expression with respect to (T_f/T) and set the result equal to zero, we obtain the optimum value of (T_f/T) which gives maximum damping. Thus

$$\left(\frac{T_f}{T}\right)_{\text{opt}} = a\left[1-\sqrt{\left(1-\frac{1}{ar}\right)}\right],\tag{16.19}$$

where $r = 1+(D_1/8C_1^3 t) = 1+m$. Since the term $1/ar \ll 1$ in this equation, we can express the square root in the form of a series, so that

$$\left(\frac{T_f}{T}\right)_{\text{opt}} \doteq \frac{1}{2(1+m)},\tag{16.20}$$

which is independent of the ratio a. We also note that $m = J_1/J_2$, being the ratio of the polar moment of inertia of the shaft to that of the friction sleeve. Figure 16.7 shows a plot of eqn. (16.20), and we observe that the optimum (T_f/T) ratio decreases progressively according as the shaft becomes stiffer with respect to the sleeve.

Variations of the principle outlined in the torsional friction damper have been used in engine flywheel and other applications. Figure 16.8 shows how the friction damping technique can be used advantageously to dampen torsional vibrations in an engine flywheel system. Both flywheels are pressed against friction discs by means of loading springs and

adjustable nuts. The flywheels themselves are free to rotate on bushings relative to the hub
of the system, the latter rigidly keyed to the shaft. When relatively large changes in rotational
speed of the shaft occur, the flywheels, because of their large polar moment of inertia, tend
to maintain uniform angular velocity. Thus relative slip between hub and flywheels causes
frictional energy dissipation at the surfaces of the friction discs, and this reduces or dampens
the original speed variation of the shaft. The slipping effect at the disc interfaces is not unlike
the action of a friction clutch during engagement. For this application, the torsional stiffness
ratio *m* is very small, and the optimum frictional torque ratio from Fig. 16.7 relatively large.

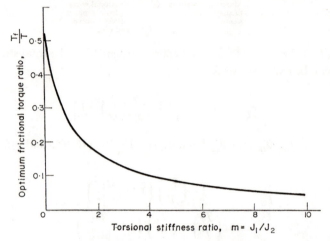

FIG. 16.7. Optimum frictional torque ratio as function of torsional stiffness ratio.

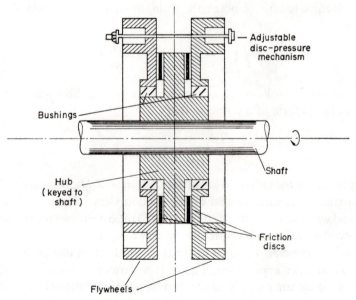

FIG. 16.8. Friction damping in engine flywheel system.

16.4 Lubrication of Human and Animal Joints

One of the most rapidly growing applications of the principles of tribology outlined in Part I is the general field of bio-systems, in particular the lubricating mechanism prevalent in living and artificial human and animal joints. The active study of lubrication in engineering components has a history extending over about 85 years, and scientific work on human joint lubrication has been in progress during the past 40 years or so.[77] It is only within the last 14 years, however, that a relatively complete understanding of the various lubricating mechanisms in joints has emerged.

Figure 16.9(a) shows a diagrammatic representation of a human joint which crudely illustrates common characteristics of the knee, hip, or spine. The load-transmitting structural members are the bones, the ends of which are sufficiently blunt or globular to provide some form of bearing area. In some cases, the opposing bones are best represented by spher-

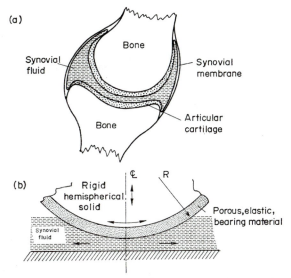

FIG. 16.9. Schematic diagram showing (a) representation of human joint, and (b) model of equivalent bearing. [78]

oidal or ellipsoidal surfaces (hip), and in other cases cylindrical shapes (knee) may be adequate. The bone surfaces are covered with a layer of relatively soft and porous articular cartilage within the joint, and this represents the bearing material. The cartilage on the upper and lower bones is separated by synovial fluid which is enclosed by the synovial membrane and provides the necessary lubricating action between the members.

Indeed, nature appears to have far outclassed the engineer with her natural joints. Thus normally healthy human joints indicate frictional coefficients in the range 0.001–0.03, which is remarkably low even for hydrodynamically lubricated journal bearings or precision rolling-contact bearings (see previous chapter). The earliest studies of the lubricating mechanism in joints suggested a hydrodynamic action, but it soon became apparent that this is inadequate in explaining the extremely low coefficients actually recorded in human joints, particularly since the relative sliding velocity between the bone surfaces is never greater than a few centimetres per second. During the past 14 years, theories of boundary, weeping, and elastohydrodynamic lubrication have been proposed, and it is now apparent that all of these mechanisms play a fundamental role in the lubricating action. Another form of fluid-film lubrication is due to the presence of a squeeze film between the opposing cartilage, which strongly resists relative approach of the bone surfaces, particularly when the film thickness becomes small (about 0.25 μ). Perhaps the overall lubricating mechanism in human and animal joints can be generally described as a squeeze-film action with superimposed boundary, weeping, and elastohydrodynamic effects. The latter in combination with a limited hydrodynamic action provide a replenishment mechanism for the squeeze film.

Properties of Human Joints

Cartilage

Articular cartilage is the smooth gristle which lines the articulating surfaces of the bones in animal joints. Its functions are to absorb the wear which results from joint movement over a period of many years, to minimize joint friction by contributing to a lubricating mechanism, and to transmit some of the highest loads developed in the animal body. The thickness of articular cartilage varies from one joint to another, and often from one position to another on a single joint surface. In the larger joints of young men and women, it may be 4–7 mm in depth, and in smaller joints the average thickness is 1–2 mm.[79]

The structure of cartilage consists of cells distributed throughout a three-dimensional network of collagen fibrils embedded in a ground substance called chondroitin sulphate[80] and filled with a liquid. The liquid component is dispersed throughout the solid framework and associated with it by varying degrees of molecular attraction. Some of the liquid molecules are firmly bound to the fibrous structure, but the majority are simply held in the interfibrillar spaces. These liquid molecules are transported through the cellular matrix by a combination of bulk flow due to pressure gradients and diffusion as a result of local differences in chemical concentration. Two significant physicochemical processes, which occur in cartilage as a consequence of liquid flow in and out of the tissues, are consolidation and swelling. Consolidation is defined as the process which entails a decrease in liquid content when an external compressive load is applied to the cartilage, and swelling implies an increase in liquid content. The rate at which such liquid exchange takes place is of particular interest because it determines the time variation of cartilage thickness. Perhaps the most important feature of articular cartilage is its porosity, and it has been estimated that the average pore size is about 60 Å. These pores play an active part in the mechanism of weeping lubrication at the cartilage surface.

Measurement with a stylus tracing instrument of the surface roughness of cartilage demonstrates that the texture is much more coarse than that in engineering bearings. Figure 16.10 shows a comparison of surface profiles in cartilage taken from (a) young healthy,

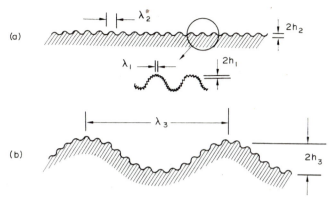

FIG. 16.10. Typical surface roughness measurements of articular cartilage in (a) young healthy, and (b) aged subjects.

Type of roughness	i	Double amplitude $2h_i$ (μ)	Wavelength λ_i (μ)
Micro-roughness	1	0.2 → 0.4	0.5
Macro-roughness	2	2 → 4	30–50
Waviness	3	10 → 20	1000

and (b) aged subjects. It appears that the former exhibits "ripples" or a distinct macro-texture, with a micro-roughness superimposed at the tip of each macro-asperity as indicated. Aged subjects have, in addition, an underlying waviness pattern as sketched in Fig. 16.10(b).

Synovial Fluid

Synovial fluid is a clear, yellowish, and tacky substance found in the cavities of freely movable joints and interacting with the cartilage to provide lubricating action. From an engineering viewpoint, we may regard the fluid as a structure in which mucin forms the walls of a honeycomb network with a watery component in between. In a healthy state, therefore, synovial fluid appears to form a sponge-like structure. Its chemical composition indicates that it is a dialysate of blood plasma with the addition of hyaluronic acid[†] and a small cellular component. The most significant property of synovial fluid is its viscosity, and this appears to be due largely to the hyaluronic acid component. As a boundary lubricant, hyaluronic acid can be expected to influence the frictional characteristics of opposing articular cartilage at separations having the same approximate dimension as its chain-length (i.e. 5000–10000 Å). Synovial fluid is classified as a non-Newtonian liquid having the property of shear-thinning,[‡] i.e. its viscosity decreases almost linearly with shear rate. If the film thickness between cartilage surfaces falls below about 1 μ, the fluid molecules can be expected to profoundly influence sliding behaviour and to behave as a boundary

[†] Hyaluronic acid is a long-chain polymer with high molecular weight.
[‡] Shear-thinning fluids are frequently called "thixotropic".

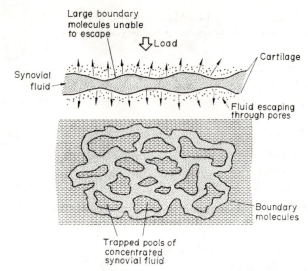

FIG. 16.11. Pictorial representation of the lubrication of cartilage with synovial fluid. [81]

lubricant. It also appears quite likely that under pressure a gel of concentrated synovial fluid having a much higher viscosity[†] than the bulk liquid, forms on the cartilage surface. This is partly due to the seepage of low-viscosity fluid through the sponge-like structure of cartilage during normal loading. The gel-like synovial fluid becomes trapped in the depressions on the cartilage surface, forming reservoirs to maintain boundary lubrication action when necessary. Figure 16.11 shows a pictorial representation of "weeping" action due to the escape of synovial fluid through pores in the cartilage, and the trapping of the gel in isolated pools.

Mechanism of Lubrication

The mechanism of lubrication in animal joints is relatively complex as we have indicated previously. It appears that for loaded joints the contribution of hydrodynamic action to lubrication is negligible. However, under lightly loaded conditions (such as in the swing phase during walking), it is likely that the two cartilage surfaces are separated by hydrodynamic pressure generation. This effect can be considered as a replenishment mechanism for the main squeeze-film action which predominates under loaded conditions.

The squeeze-film effect is given by eqn. (6.11) for parallel approaching plane surfaces, and it is certain that the general form of this equation is applicable to the loaded joint condition. We can see at once both from this equation and Fig. 6.6 that the time of approach becomes exceedingly large when the remaining squeeze-film thickness h attains sufficiently small values. When applied to human joint conditions, it has been found surprisingly that the squeeze-film times observed in experiments are much larger than predicted by the theoretical expression in eqn. (6.11). Thus, putting a typical value of $t = 40s$ and assuming that

---

[†] About 20 poise.

the minimum film thickness corresponds to the diameter of a hyaluronic acid molecule (about 0.5 μ), we find that the average viscosity of the film lubricant is 20 poise. This value is much larger than the bulk viscosity of synovial fluid (*ca.* 0.01 poise), and the result suggests that a very thick substance or gel has formed on the cartilage surface during the squeezing process. This conclusion is consistent with the view that the pores in the cartilage permit the small molecular substances to escape, thus leaving behind the larger gel-like molecules, as described earlier and depicted in Fig. 16.11.

During walking, the duration of loading is usually less than a second before the joint is swung under light load, and it is therefore certain that very little diminution of the squeeze film occurs before replenishment through hydrodynamic action becomes effective. During periods of long standing, the squeeze-film effect generates a thickened gel substance to provide a boundary lubricating effect, so that reasonably low coefficients of friction (about 0.15) are still preserved. Figure 16.12 compares the squeeze-film lubricating effect of dis-

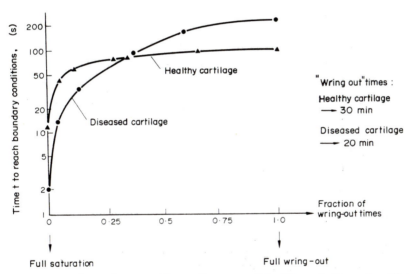

FIG. 16.12. Comparison of squeeze-film performance of healthy and diseased cartilage.[82]

eased and healthy cartilage, expressed as a plot of time *t* to reach boundary film conditions versus the fraction of "wrung-out time". We define wrung-out time in this context as the time taken for the lubricant within the cartilage to degenerate from an initial fully saturated condition to a final state of virtual dryness under the action of squeezing load. The graphs show the times taken for friction values to return to the boundary values after fluid has been introduced between cartilage specimens and glass. The experimental points in Fig. 16.12 were obtained by lifting the cartilage from the glass plate at any stage during the wrung-out process, and then observing the time taken to reach boundary conditions again when the squeezing load is restored. When this is done at all stages of wrung-out, it is observed that it takes considerably longer for boundary conditions to be restored at full wrung out than at the initial condition of full saturation. This is an indication of the articular gelling phenomenon, which greatly increases the lubricant viscosity and approach time in the wrung-out condition. We note that in distinguishing between the performance of healthy and diseased cartilage in Fig. 16.12, the increase in time *t* in going from full saturation to the full wring-out condition is considerably larger for the diseased cartilage. This may be partly due to the differing abilities of healthy and diseased cartilage in permitting the filtration of the less viscous synovial fluid through the porous structure and the resultant effect on the gelling phenomenon.

After long periods of standing, the rims around each pool of trapped fluid in Fig. 16.11 will be in a state of boundary lubrication, and the size of these rims will increase as squeezing progresses. At the same time, elastohydrodynamic effects will be present because the surface irregularities on the surface of the cartilage (see Fig. 16.10) will be compressed to give a surprisingly large area of contact, and this occurs at an earlier stage in the squeezing process than that at which the boundary lubrication effect occurs. The relative order of magnitude of elastohydrodynamic, boundary[†] and hydrodynamic film thicknesses are compared with the surface roughness dimension of articular cartilage in Fig. 16.13. Here, we observe the comparative dimensions of the elastohydrodynamic and boundary lubrication mechanisms and the very small separation due to hydrodynamic pressure wedge formation.

One of the possible applications of these studies on human joint lubrication is finding a means for eliminating some of the crippling diseases which affect the joints of elderly people. Osteoarthrosis is one of these diseases, affecting both the hip and the knee. Although the

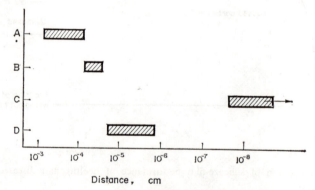

FIG. 16.13. Relative size dimensions in human joint lubrication. A, elastohydrodynamic film thickness; B, length of hyaluronic acid molecule; C, hydrodynamic film thickness; D, surface roughness of articular cartilage.

causes are not yet fully known, the effects are much in evidence. Pain and stiffness are associated with the wearing of cartilage and the breakdown in lubrication. The introduction of an artificial lubricant into the affected joints is one obvious means of preventing further deterioration. If the viscosity of the artificial lubricant is sufficiently high, it will separate the bone surfaces by virtue of its bulk, and this may reduce pain and wear. However, the

---

† Film thicknesses for boundary lubrication in joints are roughly the same as the chain length of hyaluronic acid.

high viscosity will necessitate high shear forces, so that strong muscles are required. It is doubtful whether the muscles of severe osteoarthrosic joints would be able to overcome these high shear forces and permit free joint movement. Another possible application of human joint studies is the design of prosthetic devices and artificial limbs, but further discussion of this particular aspect is beyond the immediate scope of this book.

16.5 Miscellaneous

We include in this final section brief references to a number of remaining applications in tribology. Interest in space exploration during the sixties created a demand for research into the behaviour of materials under extreme environmental conditions, in particular low temperature and pressure. Thus the tribological performance of metals and elastomers under near-vacuum and cryogenic conditions was investigated, and such phenomena as "out-gassing" and seizure acquired a practical significance outside the laboratory. Another significant area of scientific endeavour which received increasing attention during this period can be broadly described as bio-technology, perhaps also stimulated to some extent by the space effort. With scientific and technological research rapidly outpacing medical research, it became inevitable that man would comprehensively apply mathematical thinking and logic to an understanding of how his own elaborate system performed under normal and adverse operating conditions. The lubrication of human and animal joints dealt with in the previous section is one example of technology fulfilling a basic need in the medical field by first providing a basic understanding of the principles involved and then using this information to suggest perhaps improvements in the design of artificial limbs or to restore normal function to arthritic joints, etc. Other equally important examples include the flow of blood in veins and capillary tubes, kidney function and the excretion of waste body fluids, and the study and design of artificial heart valves. These applications are fundamentally concerned with a frictional mechanism, and it is anticipated that work in these areas will reduce the incidence of blood clotting, thrombosis, aneurisms, kidney ailments, and heart failure.

The detection of skin disease by a systematic method of measurement[83] is another less widely known technique in the general field of bio-technology or bio-tribology, still in the development stage. The coefficient of sliding friction on diseased skin differs from that on normal healthy skin, and this has permitted the development of a portable, hand-held, friction instrument to serve as a diagnostic aid in the detection of skin disorders. The instrument has the appearance of a space-age gun and weighs only 850 g. It is equipped at one end with a small ring which rotates continuously over the skin and records the coefficient of friction on a dial mounted in the butt of the gun. The rotating annular ring may be made of steel, nylon, or teflon, and is loaded against the skin by a constant-force adjustable spring and driven by a miniature precision electric motor. The meter is claimed to give extremely repeatable results, and it can be conveniently held in any orientation on the skin. It may be possible to monitor the frictional response of skin subjected to a variety of external injuries such as burns, blisters, and abrasions which often lead to serious infections.

We have dealt with flexible seals in some detail in Section 16.1, and these are of particular interest because of the elastohydrodynamic interaction which offsets sealing ability in the case of lubricated shafts. In general terms, seals may be classified either as static (or stationary) and dynamic. Static seals include pressure rings, gaskets, pipe joints, etc., whereas dynamic seals include piston rings, gland packings, O-rings, flexible seals, etc. As we might anticipate, seals are vitally important mechanical components of any engineering product or system, and their failure to perform as designed may readily cause failure of the entire system. This is particularly true in the development of the Wankel rotary engine, and the striking contrast between its early and present-day performance can be attributed almost entirely to progressive improvements in seal design. In a normal internal combustion engine, the top piston rings function principally as gas seals, while the lower rings scrape excess oil from the cylinder walls; here their different purpose requires different materials and clearances for optimum performance. Other seal types include labyrinth, packing, moulded, and mechanical face seals. In all cases, the primary objective is to eliminate leakage of lubricant from one part of a mechanical system to another, with the secondary objective of excluding dirt and dust deposits.

We conclude with a brief reference to some unusual applications in tribology. The frictional interaction between a typewriter character and a carbon ribbon[84] shows that adhesion and ploughing terms may be included in the overall coefficient of friction. The magnitude of the latter appears to vary with the geometry of the particular character and with the number of sheets of carbon and paper. At light loads a sharp edge perpendicular to the direction of sliding gives very high coefficients ($f > 0.5$), whereas at heavy loads a flat broad edge perpendicular to the direction of sliding gives high coefficients. Another unusual application is the wear of metal by paper, as in the case of paper handling machines, large printing presses, etc. Our initial guess in the case of paper sliding on a metal surface might be that the paper might wear and not the metal, but the opposite is found to be the case. When there is a high relative velocity between paper and metal (about several metres per second) and with large quantities of paper processed, the hard impurities in the paper act as an abrasive which scours the metal surface. This is obviously an important design criterion. Our last unusual example pertains to the wear of coins in circulation.[85] The results indicate that the thickness wear rate approaches a constant value with decreasing coin size for a given material. These values are 0.55% for cupro-nickel, 0.43% for nickel–brass, and about 0.7% for brass. As we might expect, the area exposed to wear increases more rapidly with thickness worn on the "head" side compared with the "tails" side for all coins tested.

APPENDIX TABLES

Shape of flat plate	Schematic	Typical length dimension, L_T	Constant, K
Circular		D	$\dfrac{3\pi}{64}$
Circular with concentric hole		$f(D_1, D_2)^a$	$\dfrac{3\pi}{64}$
Elliptical		$\left[\dfrac{a^3 b^3}{a^2+b^2}\right]^{1/4}$	$\dfrac{3\pi}{2}$
Rectangular (no sideflow)		L	$\dfrac{1}{2}$
Rectangular		$(L^3 B^3)^{1/4}$	Ref. 2

$^a f(D_1, D_2) = D_2\{1 - 2(D_1/D_2) + 2(D_1/D_2)^3 - (D_1/D_2)^4\}^{1/4}.$

TABLE B. APPROACH TIMES FOR ROTATIONALLY-SYMMETRIC, NON-PLANAR SQUEEZE FILMS

Shape of approach surface	Equation for approach time	Schematic and notation
Cylindrical (half-journal bearing)	$t = \dfrac{24\mu b R^3}{\Delta^2 W}\left[\dfrac{H_2}{\sqrt{(1-H_2^2)}}\tan^{-1}\right.$ $\times \sqrt{\left(\dfrac{1+H_2}{1-H_2}\right)} - \dfrac{H_1}{\sqrt{(1-H_1^2)}}\tan^{-1}$ $\left.\times \sqrt{\left(\dfrac{1+H_1}{1-H_1}\right)}\right]$	
Cylindrical (complete journal bearing)	$t = \dfrac{12\pi\mu b R^3}{\Delta^2 W}\left[\dfrac{H_2}{\sqrt{(1-H_2^2)}} - \dfrac{H_1}{\sqrt{(1-H_1^2)}}\right]$	
Spherical (ball-and-socket)	$t = \dfrac{3\pi\mu R^4}{\Delta^2 W}\left[\dfrac{H_2-H_1}{H_1 H_2} + \left(\dfrac{1+H_1^2}{H_1^2}\right)\right.$ $\left. \log(1-H_1) - \left(\dfrac{1+H_2^2}{H_2^2}\right)\log(1-H_2)\right]$	
Spherical (on plane surface)	$t = \dfrac{6\pi\mu R^2}{W}\ln\left(\dfrac{h_1}{h_2}\right)$	
Conical	$t = \dfrac{3\pi\mu R^4}{4W\sin^4\theta}\left[\dfrac{1}{h_2^2} - \dfrac{1}{h_1^2}\right]$	
Truncated cone	$t = \dfrac{3\pi\mu}{4W\sin^4\theta}\left[R_2^4 - R_1^4 - \dfrac{(R_2^2-R_1^2)^2}{\log(R_2/R_1)}\right]$ $\times \left[\dfrac{1}{h_2^2} - \dfrac{1}{h_1^2}\right]$	

General notition: $\Delta = R^1 - R$.
$H = a/\Delta$.
1, 2 denote initial and final states.
b = width perpendicular to paper.
μ = lubricant viscosity.

TABLE C. SPRING AND DAMPING CONSTANTS FOR NON-PLANAR SQUEEZE FILMS

Shape of approach surface	Spring constant, k	Damping constant, c
Cylindrical (half-journal bearing)	$\dfrac{4}{\pi}\left(\dfrac{BRb}{\varDelta}\right)$	$\dfrac{24\mu bR^3}{\varDelta^3}\,\alpha(H)$
Cylindrical (complete journal bearing)	$\dfrac{4}{\pi}\left(\dfrac{BRb}{\varDelta}\right)$	$\dfrac{12\pi\mu b}{(\varDelta/R)^3\,[1-H^2]^{3/2}}$
Spherical (ball-and-socket)	$\dfrac{\pi}{2}\left(\dfrac{BR^2}{\varDelta}\right)$	$\dfrac{3\pi\mu R^4}{\varDelta^3}\,\beta(H)$
Spherical (on plane surface)	$\dfrac{\pi BR^2}{(h+\frac{1}{3}R)}$	$\dfrac{6\pi\mu R^2}{h}$
Conical	$\dfrac{\pi BR^2}{h}$	$\dfrac{3\pi\mu R^4}{2h^3\sin^4\alpha}$
Truncated cone	$\dfrac{\pi B}{h}(R_2^2-R_1^2)$	$\dfrac{3\pi\mu}{2h^3\sin^4\alpha}\,\gamma(R_1, R_2)$

Notation:

$$\alpha(H) = \frac{1}{H^2}\left\{\frac{2-H+H^2}{1-H}+\frac{2}{H^3}\log(1-H)\right\}$$

$$\beta(H) = \left[\frac{1}{1-H^2}\right]\left\{\left[\sqrt{(1-H^2)}+\frac{H^2}{\sqrt{(1-H^2)}}\right]\tan^{-1}\sqrt{\left(\frac{1+H}{1-H}\right)}+\frac{H}{2}\right\}$$

$$\gamma(R_1, R_2) = R_2^4-R_1^4-(R_2^2-R_1^2)^2/\log(R_2/R_1)$$

See Table B for further details of notation used here.

25*

TABLE D. TYPICAL FRICTIONAL COEFFICIENTS

Description of frictional process	Coefficient of friction, f
Shearing of lubricant film	0.0003
Rolling-contact bearings	0.0025
Hydrodynamic journal bearings	0.005
Free rolling resistance of tyres	0.01
Friction on wet ice	0.02
Boundary lubrication Rolling of sphere on rubber Friction of PTFE	0.02–0.10
Friction of glare ice	0.06
Rail–wheel adhesion under damp or dewy conditions	0.09–0.15
Rolling of cylinder on rubber Friction of ski on snow	0.10–0.30
Friction on rough ice	0.12
Rail–wheel adhesion under wet conditions Friction of packed snow	0.20
Friction of like, lubricated metals	0.20–0.40
Rail–wheel adhesion under dry conditions	0.25–0.30
Tyres on wet roads	0.40
Friction of wood Friction of diamond-on-metal in air	0.40–0.60
Friction of clean dry metals in air	0.50–1.50
Tyres on dry roads	0.70
Friction in metal cutting Friction of sandpaper surfaces	1.0
Abrasive wear, or Wear by roll formation	>1.0
Friction of diamond-on-metal in vacuum	1.0–3.0
Friction of clean dry metals in vacuum	>10

REFERENCES

1. BOWDEN, F. P. and TABOR, D., *The Friction and Lubrication of Solids*, Clarendon Press, Oxford, 1950 and 1964.
2. MOORE, D. F., *The Friction and Lubrication of Elastomers*, Pergamon Press, Oxford, 1972.
3. DRAUGLIS, E., *Mech. Engng* **88** (1966) 52.
4. SELWOOD, A., *Wear* **5** (1962) 148.
5. SABEY, B. E. and LUPTON, G. N., Road Research Laboratory, Ministry of Transport Report No. LR 57 (1967).
6. SABEY, B. E., Road Research Laboratory, Ministry of Transport, Report No. LR 131 (1968).
7. HORNE, W. B. and DREHER, R. C., NASA TN D-2056 (1963).
8. MOORE, D. F., *Highway Research Record*, No. 131 (1966) 181.
9. MOORE, D. F., Report No. RC. 88, National Institute for Physical Planning and Construction Research, Dublin, 1972.
10. PEKLENIK, J., C.I.R.P. *Annalen* B. XII, **3** (1965) 173.
11. ABBOTT, E. J. and FIRESTONE, F. A., *Mech. Engng* **55** (1933) 569.
12. GREENWOOD, J. A. and WILLIAMSON, J. B. P., *Proc. Roy. Soc. Lond.*, A, **295** (1966) 300.
13. TIMOSHENKO, S. and GOODIER, J. N., *Theory of Elasticity*, McGraw-Hill, New York, 1951, pp. 372–6.
14. GREENWOOD, J. A., Paper No. 65-Lub-10, ASLE-ASME Lubrication Conference, San Francisco, October 1965.
15. COURTNEY-PRATT, J. S. and EISNER, E., *Proc. Roy. Soc. Lond.* A, **238** (1957) 529.
16. KRAGELSKII, I. V., *Friction and Wear*, Butterworths, Washington, 1965.
17. BULGIN, D., HUBBARD, G. D. and WALTERS, M. H., *Proceedings of the 4th Rubber Technology Conference*, London, May 1962, p. 173.
18. LUDEMA, K. C. and TABOR, D., *Wear* **9** (1966) 329.
19. MOORE, D. F., *Wear* **21** (1972) 179.
20. KUMMER, H. W., *Unified Theory of Rubber and Tyre Friction*, Eng. Research Bulletin B-94, Pennsylvania State University, July 1966.
21. HEGMON, R. R., *Rubber Chem. Technol.* **42** (1969) 1122.
22. WILLIAMS, M. L., LANDEL, R. F., and FERRY, J. D., *J. Am. Chem. Soc.* **77** (1955) 3107.
23. TABOR, D., *Engineering* **186** (1958) 838.
24. STEFAN, J., Sber. bayer. Akad. Wiss., Math-Naturwiss. Kl. **69** (2) (1874) 713.
25. TOWER, B., *Proc. Inst. Mech. Eng. Lond.* **36** (1885) 58.
26. REYNOLDS, O., *Trans. Roy. Soc.* **177** (Part 1) (1886) 157.
27. MOORE, D. F., *J. Fluid Mech.* **20** (1964) 321.
28. MOORE, D. F., *J. Roy. Aero. Soc.* **69** (1965) 337.
29. LICHT, L., FULLER, D. D., and STERNLIGHT, B., *Trans. ASME* **80** (1958) 411.
30. FLOBERG, L., *Trans. Chalmers Univ. Technology (Sweden)*, No. 216 (1959).
31. HAMILTON, D. B., WALOWIT, J. A., and ALLEN, C. M., Paper No. 65-Lub-11, *ASME-ASLE Lubric. Conf., San Francisco, October 1965*.
32. HOTHER-LUSHINGTON, S. and JOHNSON, D. C., *J. Mech. Engng. Sci.* **5** (1963) 175.
33. HUNSAKER, J. C. and RIGHTMIRE, B. G., *Engineering Applications of Fluid Mechanics*, McGraw-Hill, New York, 1947, ch. 13.
34. FULLER, D. D., *Theory and Practice of Lubrication for Engineers*, John Wiley, New York, 1956, ch. 11.
35. GODFREY, D., Boundary lubrication, from *Interdisciplinary Approach to Friction and Wear* (ed. P. M. Ku), National Aeronautics & Space Administration Special Report No. Sp-181 (1968), pp. 335–84.
36. ROWE, G. W., *Proceedings of the Conference on Lubrication and Wear*, Institution of Mechanical Engineering (1957), p. 333.
37. CAMBELL, M. E., LOSER, J. B., and SNEEGAS, E., *Solid Lubricants*, NASA SP 5059, May 1966.
38. CAMPBELL, M. E., *Mech. Engng* **85** (2) (1968) 28.

39. Dowson, D., Paper R1, Elastohydrodynamic Lubrication, *Proc. Inst. Mech. Engrs*, vol. 180, part 3B, 1965–6, p. 7.
40. Dowson, D. and Higginson, G. R., *Elastohydrodynamic Lubrication*, Pergamon Press, Oxford, 1966.
41. Herrebrugh, K., Paper No. 69-Lub-13, *ASLE-ASME Joint Lubrication Conference, Houston, 1969*.
42. Roberts, A. D. and Tabor, D., *Wear* **11** (1968) 163.
43. Hardy, W. and Bircumshaw, I., *Proc. Roy. Soc. Lond.* A, **108** (1925) 12.
44. Blok, H., *Inverse Problems in Hydrodynamic Lubrication*, Internal Report, Technische Hogeschool, Delft, April 1964.
45. Moore, D. F., *Int. J. Mech. Sci.* **9** (1967) 797.
46. Barwell, F. T., *Lubrication of Bearings*, Butterworths, London, 1956.
47. Rabinowicz, E., *J. Appl. Phys.* **32** (8) (1961) 1940.
48. Kerridge, M. and Lancaster, J. K., *Proc. Roy. Soc. Lond.* A, **236** (1956) 250.
49. James, D. I., *Abrasion of Rubber*, McLaren, London, 1967.
50. Schallamach, A., *Wear* **1** (1957/8) 384.
51. Schallamach, A., *Proc. Phys. Soc.* B, **67** (1954) 883.
52. Lipson, C., *Wear Considerations in Design*, Prentice-Hall, 1967, ch. 4.
53. Schallamach, A., *J. Polymer Sci.* **9** (1952) 385.
54. Mann, J. Y., *Fatigue of Materials*, Melbourne University Press, 1967.
55. Grosch, K. A. and Schallamach, A., *Trans. Inst. Rubber Ind.* **41** (T80) (1965); also *Rubber Chem. Technol.* **39** (1966) 287.
56. Grosch, K. A., *Proc. Phys. Soc.* A, **274** (1963) 21.
57. Schallamach, A., *Rubber Chem. Technol.* (1968) 209.
58. Brennan, J. N., The determination of internal friction, Ph.D. dissertation, Pennsylvania State University, August 1952.
59. Ungar, E. E. and Hatch, D. K., *Product Engineering*, Special Report on High-Damping Materials, 1961.
60. Prandtl, L., *Z. angew. Math. Mech.* **8** (1928) 85.
61. Boothroyd, G., *Fundamentals of Metal Machining*, Edward Arnold, London, 1965.
62. *Welding Handbook*, Part 3, Special processes and cutting, Macmillan, 1965.
63. Tylecote, R. F., *The Solid Phase Welding of Metals*, Edward Arnold, London, 1968.
64. Spurgeon, W. M. and Spencer, A. R., *Bendix Technical Journal, Advanced Automotive Technology* **2** (3) (Autumn 1969) 59.
65. Taylor, C. F., *The Internal Combustion Engine in Theory and Practice*, Vol. 1, MIT Press, MIT, Cambridge, Boston, 19, ch. 9.
66. Roberts, A. D., *Engng. Materials & Design* (1969) 55.
67. Schulze, K. M. and Beckmann, L., *Friction Properties of Pavements at Different Speeds*, ASTM Special Technical Publication No. 326, p. 44, June 1962.
68. Moore, D. F., Highway Research Board Special Report 101, Publication No. 1460, Washington, DC, 1969, p. 39.
69. Moore, D. F., *12th International FISITA Congress, Barcelona, Spain, May 1968*, Paper 2–02.
70. Fichaux, H. and Moore, D. F., *Wear* **11** (1968) 51.
71. Koffman, J., *Proc. J. Instn. Loco. Engrs.*, Paper No. 479 (1948), 593.
72. Bekker, M. G., *Theory of Land Locomotion*, University of Michigan Press, Ann Arbor, Michigan, 1962.
73. Pinkus, O. and Sternlicht, B., *Theory of Hydrodynamic Lubrication*, McGraw-Hill, New York, 1961.
74. Blok, H. and Van Rossum, J. J., *Lubric. Engng.* **9** (6) December 1953, 310.
75. Wildmann, M., Paper No. 68-Lub S-43, *ASME Lubrication Symposium, Las Vegas, June 1968*.
76. Hirano, F. and Ishiwata, H., *Proc. Instn. Mech. Engrs*, Paper No. 15, **180** (1965/6) 187.
77. Zimmerman, B., *Product Engng*, No. 22 (1965) 70.
78. Dowson, D., Paper No. 12, *Proc. Instn. Mech. Engrs* **181** (3J) (1967) 45.
79. Barnett, C. H., Davies, D. V., and MacConaill, M. A., *Synovial Joints, Their Structure and Mechanics*, Longmans, London, 1961.
80. Edwards, J., Paper No. 6, *Proc. Instn. Mech. Engrs*. **181** (3J) (1967) 16.
81. Walker, P. S. *et al.*, *Annals Rheumatic Diseases* **27** (6) November 1968, 512.
82. Walker, P. S. *et al.*, *Rheol. Acta* **8** (2) (1969) 234.
83. Harborow, P., Design Unit, Dept. Mech. Eng., University of Newcastle-upon-Tyne, Private Correspondence, October 1972.
84. Bayer, R. G. and Sirico, J. L., *Wear* **11** (1968) 78.
85. Cope, R. G., *Wear* **13** (1969) 217.
86. Geyer, W. and Moore, D. F., Private correspondence, 1973.

INDEX